Refractory Metals in Molten Salts

Their Chemistry, Electrochemistry and Technology

NATO ASI Series

Advanced Science Institute Series

A Series presenting the results of activities sponsored by the NATO Science Committee, which aims at the dissemination of advanced scientific and technological knowledge, with a view to strengthening links between scientific communities.

The Series is published by an international board of publishers in conjunction with the NATO Scientific Affairs Division

A Life Sciences **B Physics**	Plenum Publishing Corporation London and New York
C Mathematical and Physical Sciences **D Behavioural and Social Sciences** **E Applied Sciences**	Kluwer Academic Publishers Dordrecht, Boston and London
F Computer and Systems Sciences **G Ecological Sciences** **H Cell Biology** **I Global Environment Change**	Springer-Verlag Berlin, Heidelberg, New York, London, Paris and Tokyo

PARTNERSHIP SUB-SERIES

1. Disarmament Technologies	Kluwer Academic Publishers
2. Environment	Springer-Verlag / Kluwer Academic Publishers
3. High Technology	Kluwer Academic Publishers
4. Science and Technology Policy	Kluwer Academic Publishers
5. Computer Networking	Kluwer Academic Publishers

The Partnership Sub-Series incorporates activities undertaken in collaboration with NATO's Cooperation Partners, the countries of the CIS and Central and Eastern Europe, in Priority Areas of concern to those countries.

NATO-PCO-DATA BASE

The electronic index to the NATO ASI Series provides full bibliographical references (with keywords and/or abstracts) to about 50,000 contributions from international scientists published in all sections of the NATO ASI Series. Access to the NATO-PCO-DATA BASE is possible via a CD-ROM "NATO Science and Technology Disk" with user-friendly retrieval software in English, French, and German (©WTV GmbH and DATAWARE Technologies, Inc. 1989). The CD-ROM contains the AGARD Aerospace Database.

The CD-ROM can be ordered through any member of the Board of Publishers or through NATO-PCO, Overijse, Belgium.

3. High Technology – Vol. 53

Refractory Metals in Molten Salts

Their Chemistry, Electrochemistry and Technology

edited by

David H. Kerridge
Department of Engineering Materials,
University of Southampton,
Southampton, United Kingdom

and

Evgeny G. Polyakov
Institute of Chemistry,
Kola Science Centre,
Russian Academy of Sciences,
Apatity, Russia

Springer-Science+Business Media, B.V.

Proceedings of the NATO Advanced Research Workshop on
Refractory Metals in Molten Salts
Apatity, Russia
August 12-17, 1997

A C.I.P. Catalogue record for this book is available from the Library of Congress.

Printed on acid-free paper

ISBN 978-90-481-5054-0 ISBN 978-94-015-9135-5 (eBook)
DOI 10.1007/978-94-015-9135-5

Originally published by Kluwer Academic Publishers in 1998.
Softcover reprint of the hardcover 1st edition 1998

TABLE OF CONTENTS

Preface

This **NATO Advanced Research Workshop** was devoted to a specialized topic in molten salt chemistry and was held in an exotic location (as far as Westerners were concerned) well within the Arctic Circle. It nevertheless facilitated a fruitful week, both of science and of human contacts.

The 42 oral presentations and posters from nine countries enabled the 59 participants to learn a great deal about many areas of recent research in the molten salt chemistry of refractory metals, while making new contacts as well as renewing old friendships. The time of informal contact led to the beginning of a number of new research cooperations with interchange of personnel. Thus the twin aims of advancing science and improving East-West understanding were both amply fulfilled. Indeed a warm and happy family atmosphere was very tangible doth during the scientific sessions and the social events, which participants, accompanying persons and local staff all enjoyed.

This opportunity of living for a short time within the Arctic Circle was a novelty for most Westerners, who generally appreciated the very warm weather (the hottest for 20 years according to some residents), as well as the beautiful surroundings of very green birch/pine forest, rushing rivers, vast lakes and rounded mountains, frequently illuminated by wonderful sunsets. The evening barbeque beside Lake Imandra (100 km long) and the coach tour beside the beautiful White Sea dotted with islands in the Kandalaksha Recreational Area (National Park), to sample Pomor culture, dancing and fresh salmon soup, were high spots of the social programme. Scientifically the Institute of Chemistry was visited for a summary of its history, together with tours of the molten salt laboratories and plating workshop. Never to be forgotten was the drive up to the huge opencast apatite mine at the top of Khibini mountain, when we were enveloped in a snowstorm.

In summary the ARW provided a very positive and worthwhile experience, both scientifically and culturally, which will be long remembered by those who took part. The present volume of Proceedings, containing the most significant of the contributions to the ARW, will disseminate the scientific information more widely and thus enhance both the benefit, and the circle of those who have benefitted, from this meeting.

The Co-Directors and all the participants wish to thank most warmly the Science Committee of NATO, together with the Assistant Secretary-General and Programme Director (Scientific and Environmental Affairs), who made the ARW possible, as well as the following organisations who supported the meeting financially.

Russian Ministry for Science and Technology
Russian Foundation for Basic Research
Chepetsky Mechanical Plant
Apatity Town Administration
Danfoss A/S, Denmark
Durand Ltd., Guernsey

David H. Kerridge
Evgeny G. Polyakov Co-Directors, ARW

LOGO OF THE APATITY ARW

NATO ADVANCED RESEARCH WORKSHOP

'Refractory Metals in Molten Salts'

АПАТИТЫ

August 12th–17th 1997

APATITY · RUSSIA

NATO ADVANCED RESEARCH WORKSHOP

The ear of wheat, copied from the large concrete sculpture marking arrival in the town, symbolises the importance of Apatity (founded 1966) as the home of the miners producing apatite ore from the top of the Khibini mountain.

This ore, about 200.10^6 tonne/annum at maximum output, is approximately 10% apatite, which is taken south by train and converted into superphosphate with sulphuric acid. This is an essential fertiliser for the wheatfields of central Russia.

ORGANISING COMMITTEE

Co-Directors

David H Kerridge, Dept of Engineering Materials,
University of Southampton,
Southampton, U.K.

Evgeny G Polyakov, Institute of Chemistry,
Kola Science Centre,
Russian Academy of Sciences,
Apatity, Russia.

Committee Members

Niels J Bjerrum, Dept of Chemistry,
Technical University of Denmark,
Lyngby, Denmark.

Werner Freyland, Dept of Physical Chemistry and Electrochemistry,
University of Karlsruhe,
Karlsruhe, Germany.

Frederic Lantelme, Electrochemistry Laboratory,
Pierre and Marie Curie University
Paris, France.

Terje Ostvold, Institute of Inorganic Chemistry,
Norwegian University of Science and Technology,
Trondheim, Norway.

Thermodynamics of niobium chlorides and mixtures with alkali chlorides

M. GAUNE-ESCARD
IUSTI, CNRS UMR 6590, Technopôle Chateau-Gombert
5 rue Enrico Fermi, 13453 Marseille cedex 13, FRANCE
mge@iusti.univ-mrs.fr

INTRODUCTION

Refractory metals and particularly niobium and tantalum are obtained by molten salt electrolysis. Several authors /1-6/ investigated the electrochemical reduction of niobium ions in chloride or chloride + fluoride melts. However the current efficiency during electrolysis is strongly dependent on the "average" valence of the bath. Different reduction steps are proposed with niobium valences ranging from V to I . This is supported by the existence of many stoichiometric compounds. However, very few if not nothing is known on thermodynamics.

BIBLIOGRAPHY SURVEY

Bibliography survey was conducted on the thermodynamics of the pure niobium halides and their mixtures with alkali chlorides.

Few data exist on niobium halides due to the great reactivity of these compounds which makes it extremely difficult to obtain oxychloride-free halides. Also experimental investigations are not easy since they have high vapor pressure and low melting points compared to the alkali halides melts.The data relative to the pure niobium halides reported on thermodynamic tables /7,8/ were mostly obtained by estimation .

Phase diagrams were reported for the mixtures formed by niobium halides and alkali halides They are often questionable for the same reasons and disagreement exists between the results of different authors on the same system.

D.H. Kerridge and E.G. Polyakov (eds.), Refractory Metals in Molten Salts, 1-7.

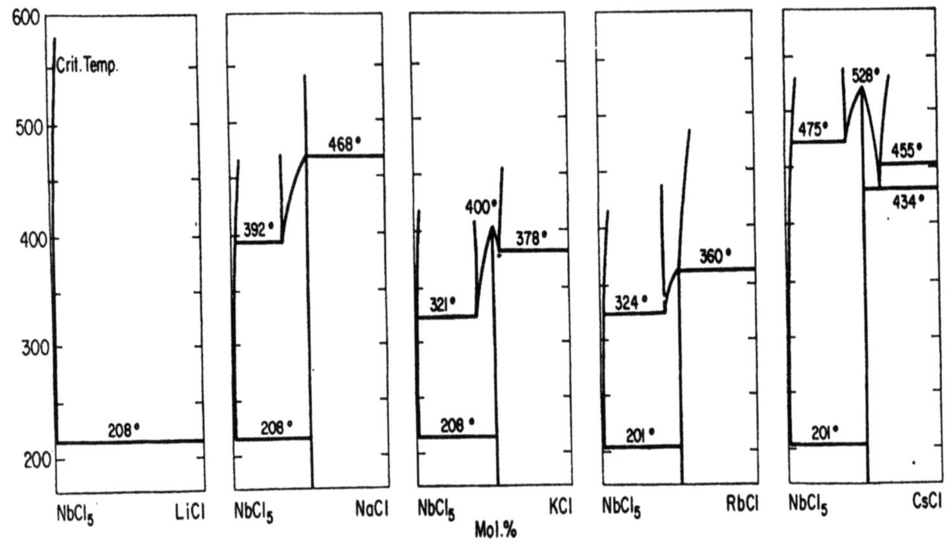

Fig.1 :	**Fig.2** :	**Fig 3** :	**Fig.4** :	**Fig.5**:
LiCl-$NbCl_5$	NaCl-$NbCl_5$	KCl-$NbCl_5$	RbCl-$NbCl_5$	CsCl-$NbCl_5$

Mixtures formed by $NbCl_5$ niobium pentachloride and the MCl alkali halides (Fig. 1-5 from / 9/) are not very stable and exhibit liquid miscibility gap in the low $NbCl_5$ composition range. $MNbCl_6$ stoichiometric compounds form, except with LiCl, but congruent melting is only observed for $CsNbCl_6$, at a temperature lower than the melting points of components.

Mixtures of the MCl alkali halides with $NbCl_4$ niobium tetrachloride and $NbCl_3$ niobium trichloride show increased stability and the phase diagrams exhibit congruently melting M_2NbCl_6 /10/ and M_2NbCl_5 stoichiometric compounds /11/ (Fig. 6-9).

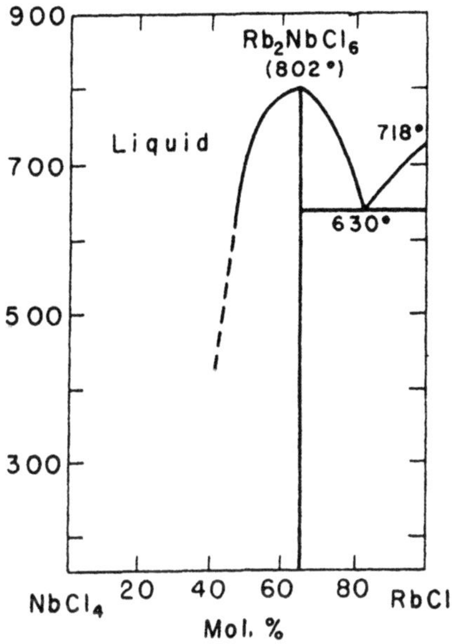

Fig. 6 :
RbCl-$NbCl_4$

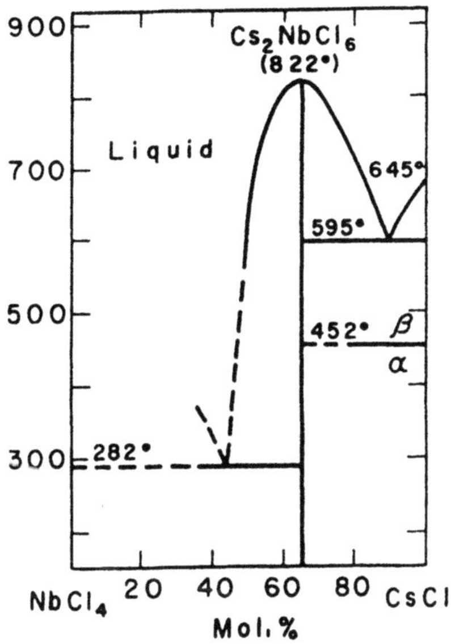

Fig.7 :
CsCl-$NbCl_4$

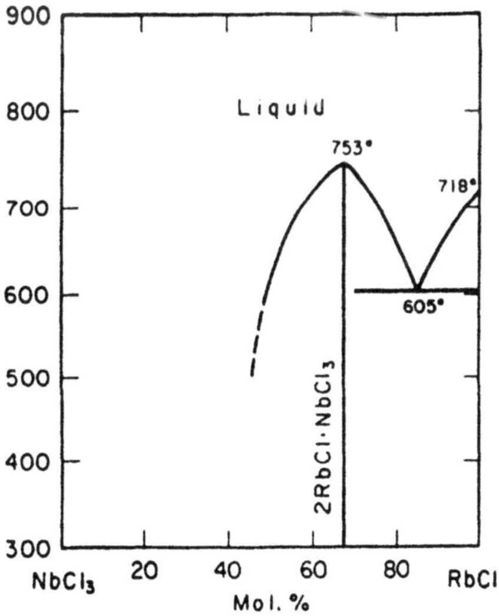

Fig. 8 :
RbCl-$NbCl_3$

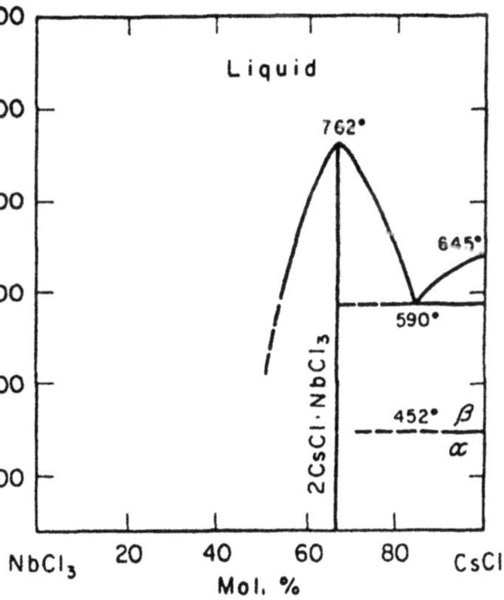

Fig.9 :
CsCl-$NbCl_3$

EXPERIMENTAL

Experimental determinations were conducted on several niobium /V/ compounds. Two calorimetric techniques were used (drop calorimetry and differential scanning calorimetry DSC) to measure the temperatures and enthalpies of phase transitions (solid->solid and solid->liquid). Solution calorimetric determinations were also initiated.

$NbCl_5$

The values measured for temperature and enthalpy of fusion, 478 K and 30 kJ/mol respectively, compare reasonably well with those, 478-483 K and 28.9 - 33.9 kJ/mol respectively, obtained by estimation /8, 12-13/.

The heat capacities of solid and liquid $NbCl_5$ were also determined :

$$Cp, s\ (443\ K) = 209\ J/mol.K$$
$$Cp, l\ (483\ K) = 335\ J/mol.K$$

They differ noticeably from those estimated previously, 148 and 262 J/mol.K respectively /7/.

$KNbCl_6$

A calorimetric method was chosen to observe the formation of the $KNbCl_6$ compound and "in situ" synthesis was carried out from the two constituants KCl and $NbCl_5$.

The experimental thermograms obtained, however, indicate that $KNbCl_6$ formation is probably more complex than a simple dissolution of solid KCl into $NbCl_5$: *on heating* , an endothermic peak is observed at 478 K, which corresponds to the fusion of $NbCl_5$, then the thermal effect becomes highly exothermic and ends near 498 K after an endothermic peak.

The incongruent melting compound $2KCl.3NbCl_5$ was reported to form between 413 K and 483 K /12/ : the features observed experimentally could correspond to the formation of this compound according to the reaction:

$$3\ NbCl_5 + 3\ KCl \rightarrow 2KCl.3NbCl_5 + KCl \qquad 478\ K<T<483\ K$$

and then to its decomposition above 483 K, decomposition to which juxtaposes the formation of the $KNbCl_6$ compound (endothermic effect):

$$2KCl.3NbCl_5 + KCl \rightarrow 3\ KNbCl_6 \qquad 483\ K<T<498\ K$$

These effects do not show anymore after futher heating-cooling cycles. A new thermal effect is observed reproducibly instead, near 444 K. This may be interpretated as corresponding to the $\gamma \rightarrow \beta$ transition of $KNbCl_6$ observed near 459 K (heating) and 433 K (cooling) /13/.

The formation enthalpy of the compound $KNbCl_6$ (478<T<498 K):

$$NbCl_5\ (l) + KCl\ (s) \rightarrow KNbCl_6\ (s)$$

was determined as ΔH_{form} = -14 kJ/mol.

All the previous experiments were realized with sealed quartz ampoules, under argon atmosphere.

When the experimental volume is unlimited, a decomposition of $KNbCl_6$ can be noticed: heating under argon flow produces chlorine and condensation of yellow needles of $NbCl_5$, in the cool part of the assembly. The probable reaction is:

$$2KNbCl_6 \rightarrow K_2NbCl_6 + NbCl_5 + (1/2)\ Cl_2$$

It is very slow.
Also attack of platinum by molten $KNbCl_6$ can be noted, probably by the reaction with chlorine.

$RbNbCl_6$

This compound was investigated by Huber /9/ and, more recently, by Sadoway /13, 14/. The incongruent melting, reported at 633 K, was not confirmed by the later who, furthermore, indicated the $\alpha \rightarrow \beta$ transition . The calorimetric results obtained in the present work confirm this point. They are given in Table 1 and include the melting enthalpy value still not measured so far.

Table 1

Ttransf (K)	**ΔHtransf (kJ/mol)**	**Tfus (K)**	**ΔHfus (kJ/mol)**	**Ref.**
-	-	633	-	/9/
578	20.2	653	-	/13,14/
581	19.7	667	9.4	this work

$CsNbCl_6$

The $CsNbCl_6$ compound was reported to undergo a $\alpha \rightarrow \beta$ transition in the solid state /13-15/. We have determined by DSC the temperature and enthalpy of this phase transition and also those corresponding to melting (Table 2).

Table 2

Ttransf	**ΔHtransf**	**Tfus**	**ΔHfus**	**Ref.**
(K)	**(kJ/mol)**	**(K)**	**(kJ/mol)**	
563	17.1	808	15.7	/13,14/
573	-	801	-	/15/
568	21	796	21.5	this work

CONCLUSION

The thermodynamic data determined in the present work, together with the enthalpy measurements in progress, will make it possible to get more insight into the stability of niobium species and correlatively to obtain new routes leading to the optimization of electrochemical processes.

REFERENCES

1) G.W. Mellors, S. Senderoff
J. Electrochem, Soc. 112, 266, 1965
2) T. Yoko, R.A. Bailey
Proc. 1st Int. Symposium on molten salt chemistry and technology, 20-22 April, 1983, Kyoto,
Japon, p. 111
3) M. Chemla, V. Grinevitch
Bull. Soc. Chim., 853, 1973
4) V.I. Konstantinov, E.G. Polyakov, P.T. Stangrit
Electrochimica Acta 23, 713, 1978
5) P. Taxil, J. Mahenc
J. Appl. Electrochem. 17, 261, 1987
6) Z.Y. Qiao, P. Taxil
J. Appl. Electrochem. 15, 259, 1987
7) "JANAF" Thermochemical tables".
Third Edition, US Department of Commerce, NBS.1985

8) Barin I., Knacke O.
"Thermochemical properties of inorganic substances". Springer Verlag, 1973.
9) K. Huber, E. Jost, E. Neuenschwander, M. Studer and B. Roth.
Helv. Chim. Acta, 41, 2414, 1958.
10) V.V. Safonov, B.G. Korshunov, Z. Shevtsova.
Zh. Neorg. Khim., 7, 8, 1979, 1962.
11) V.V. Safonov, B.G. Korshunov, T.N. Zimina.
Zh. Neorg. Khim., 11, 4, 906, 1966.
12) A.P. Palkin, N.D. Chikanov.
Russ. J. Inorg. Chem. 7, 705, 1962.
13) D.R. Sadoway, S.N. Flengas.
Can. J. Chem., 56, 2013, 1978.
14) E.E. Shawki, S.N. Flengas, D.R. Sadoway.
Can. J. Chem., 67, 1193, 1989.
15) I. S. Morozov.
Primenenie Khlora V Metallurgii Redkikh i Tsvetnykh Metallov, Nauka, Moscow, 1966, p.222

INTERRELATION BETWEEN THE COMPOSITION AND STRUCTURE OF METAPHOSPHATE POLYHEDRA OF MOLYBDENUM AND TUNGSTEN IN THE CASE OF ELECTRODEPOSITION OF METALS AND THEIR ALLOYS

N.I.BURYAK,V.A.BANDUR,S.V.VOLKOV
V.I.Vernadskii Institute of General and Inorganic Chemistry of the National Academy of Sciences of Ukraine
Pr.Palladina 32-34, 252680 Kiev-142, Ukraine

Absract

By electron and ESR spectroscopy, the reduction of MoO_3 and WO_3 to Mo(V) and W(V) in $NaPO_3$ and $NaPO_3$ - NaCl melts was observed. It was found that the products form tetragonally distorted $[M^VO_6]$ polyhedra of D_{4h} symmetry. The composition - structure relationships revealed for the Mo(V) and W(V) polyhedra allowed us to perform electrolytic deposition of the metals in the individual state or as alloys by electrolysis of the metals.

Keywords: EAS; ESR; Mo(V), W(V), polyhedra; $NaPO_3$ melts; Electrodeposition of molybdenum-tungsten alloy

The relationships between the composition and structure of the metaphosphate melts, especially containing such difficulty fusible metals as molybdenum and tungsten, are of essential interest for direct electrolytic preparation of the individual metals and their alloys. Certain d-metal complexes arising in metaphosphate melts exhibit unique properties of melted catalysts for partial oxidation of methane [1-3] and electrolytes for direct isolation of metal alloys [4,5]. Of particular, promise is the electrolysis of technological products of their conversion, i.e., solutions of MoO_3 and WO_3 in sodium metaphosphate melts.

D.H. Kerridge and E.G. Polyakov (eds.), Refractory Metals in Molten Salts, 9-15.

Using electron absorption spectroscopy (EAS) and ESR we studied the structure of oxide-salt fusions and quenched $NaPO_3$-MoO_3 melts, the first of which contains up to 50wt.% of dissolved MoO_3 at 720^{0}C and the second , 14.9wt.% of WO_3 at the same temperature. EAS study of the melts was performed on a "Specord UV-VIS" spectrophotometer modified for high-temperature measurements [6]. The melts were homogenized by heating to 900-950^{0}C. Degree of oxidation of molybdenum and tungsten ions in $NaPO_3$ was determined by electron and ESR spectra (spectrometer RE-1306) of samples rapidly quenched from 900 to 20^{0}C.

Coordination compounds of sexavalent molybdenum and tungsten possessing $4d^0$- and $5d^0$-electron configuration exhibit no d-d transitions and can be characterized by charge-transfer bands in the near UV region [7]. Nevertheless, when MoO_3 or WO_3 is added to a $NaPO_3$ melt, a distinct d-d spectrum appears. The nature of the absorption bands and their location suggest formation of molybdenum (V) and tungsten (V) coordination polyhedra with a detectable evolution of oxygen (Fig.1). The presence of Mo(V) and W(V) ions in the melt is also confirmed by ESR spectra. Due to experimental difficulty of ESR measurements in melts at high temperatures(>650^{0}C), rapidly quenched samples were used for this purpose as previously. We have found that the ESR spectra for liquid and instantly quenched solid melts are identical [8]. The ESR spectra obtained confirm the reduction of Mo(VI) and W(VI) in the $NaPO_3$ melt just to Mo(V) and W(V) as a signal that only one unpaired electron is

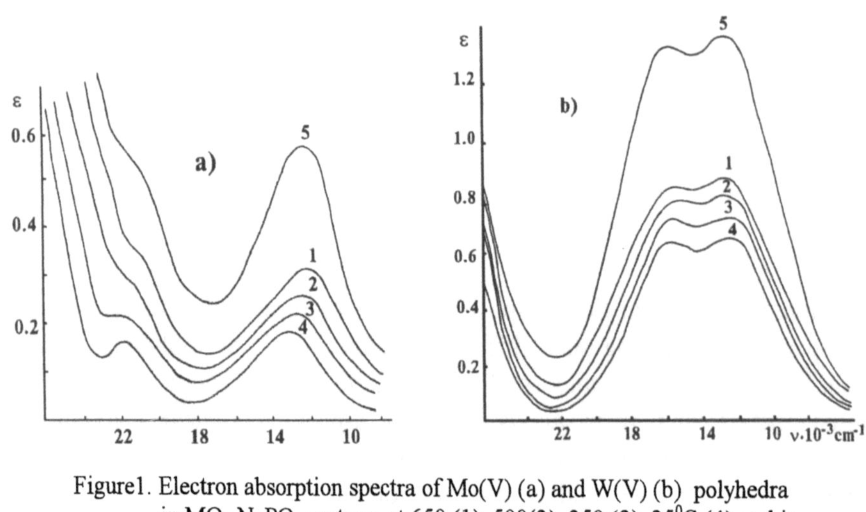

Figure1. Electron absorption spectra of Mo(V) (a) and W(V) (b) polyhedra in MO_3-$NaPO_3$ systems at 650 (1), 500(2), 250 (3), 25^{0}C (4) and in MO_3-$NaPO_3$-NaCl mixtures at 650^{0}C (5).

unambiguously detected ($4d^1$- and $5d^1$-electron configuration). This is connected with specific properties of the metaphosphate-melt medium since the oxygen atoms in phosphate chains $(NaPO_3)_m$ bear an excessive negative charge and are electron donors so that the melt can act as a reductant stabilizing transition metals in lower oxidation states [9]. Lande factors determined for Mo(V) and W(V) polyhedra in the quenched $NaPO_3$-MO_3 melts are characterized by $g_{\parallel}$ and $g_{\perp}$ values of to 1.874, 1.915 and 1.656, 1.730, respectively. Electron states of Mo(V) and W(V) having $4d^1$- and $5d^1$-electron structures are degenerate in fields of cubic symmetry due to the configurational instability of the ground and excited states (Jahn-Teller effect). Because of this, Mo(V) and W(V) polyhedra can give theoretically two to four absorption bands, depending on the degree of disturbance of cubic symmetry.

Electron spectra of Mo(V) and W(V) in liquid and corresponding quenched solid melts contain two bands related to Mo(V)[12000(13000), 21000(21700) cm^{-1}] and W(V) polyhedra [14800(15200), 12000(11700) cm^{-1}](Fig.1). According to [7,10,11], the electron spectra of Mo(V) and W(V) in $NaPO_3$ and their temperature dependence (a decrease in band intensity at lower temperatures) are typical of Mo(V) and W(V) quasioctahedral coordination polyhedra. Coordination of Mo(V) and W(V) ions in molten and glassy $NaPO_3$ should be determined by its structure presented mainly by $(NaPO_3)_m$ polyphosphate chains of different length exhibiting a higher tendency to coordination to d-metal ions as compared to cyclic metaphosphates with the same number of phosphorus atoms. For this reason the quasioctahedral coordination of Mo(V) and W(V) ions can be accomplished by $[PO_3]_n$ groups belonging to one chain or to some neighboring chains, thus providing coordination number 6 of the metals. The character of the electron spectra and ESR data indicate that the most acceptable coordination for Mo(V) and W(V) ions in $NaPO_3$ may be a distorted octahedron presumably of D_{4h} symmetry. A doublet nature of absorption bands observed for W(V) in $NaPO_3$ is connected with an essential distortion of the octahedral structure of $[W^VO_6]$ polyhedra rather than with spin-orbital interactions, which are considerable for W(V). This suggestion is supported by the dependence of absorption band splitting on temperature. Indeed, for W(V)-phosphate polyhedra studied, the splitting is 2800 cm^{-1} at 650^{0}C and 3500 cm^{-1} at 20^{0}C, that is, with increase in temperature the bands approach each other. The spin-orbital interaction, however, would result in an increase of the $\delta \sim \sqrt{T}$ splitting.

Then, in accordance with the energy-level diagram for Mo(V) and W(V) ions of D_{4h} symmetry (Fig.3), the following assignment of the absorption bands experimentally observed for liquid and quenched solid melts can be proposed: 12000,13000 cm^{-1}(for Mo(V) polyhedra), 12000,11700 cm^{-1}(for W(V) polyhedra) relate to the $B_{2g} \rightarrow E_g$ transition; 21000,21700 and 14800, 15200 cm^{-1} to the $B_{2g} \rightarrow B_{1g}$ transition. The relative position of the ground $^2T_{2g}$ level orbital doublet E_g and singlet B_{2g} was determined on the basis of spectroscopic splitting factors in ESR spectra. If this spin doublet refers to the ground state, then $g_{\parallel} = g_{\perp} = g_e$, then for the singlet we have $g_{\parallel} \neq g_{\perp} < g_e$ [12]. From ESR spectra obtained for sodium metaphosphate-based quenched systems containing Mo(V) and W(V) the following values were determined: $g_{\parallel} = 1.874$, $g_{\perp} = 1.915$, $g_{\parallel} = 1.656$, $g_{\perp} = 1.730$, respectively. Thus, the B_{2g} singlet lies below the E_g doublet in the tetragonally shortened pyramid of the [MoO_6] chromophores. The temperature changes in band intensities of Mo(V) and W(V) polyhedra are associated with Mo(V)- and W(V)-induced reversible polymerization ↔ depolymerization processes occurring in $NaPO_3$ and hence with redistribution of electron density in the chain -M -(O-P-O...), where M= Mo(V) or W(V) [13].

It was of interest to clarify the behavior of MoO_3 and WO_3 dissolved in the most depolymerized $NaPO_3$ melt containing 30mol.% of NaCl [14]. Absorption bands of Mo(V) and W(V) in the $NaPO_3$-NaCl melt are identical to those observed in $NaPO_3$ despite a high concentration of NaCl added(up to30 mol.%).

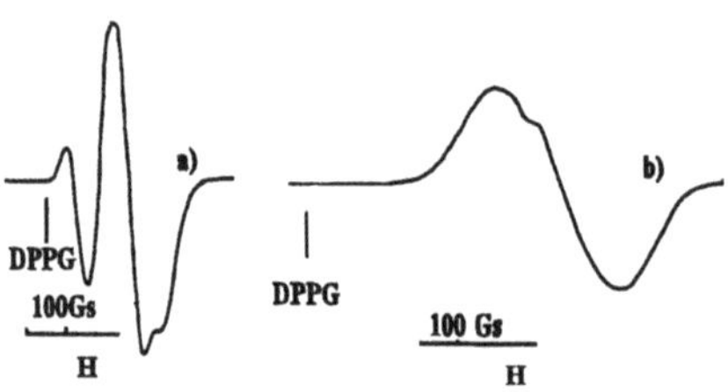

Figure2. ESR spectra of Mo(V) (a) and W(V) (b) in quenched $NaPO_3$ and $NaPO_3$-NaCl melts

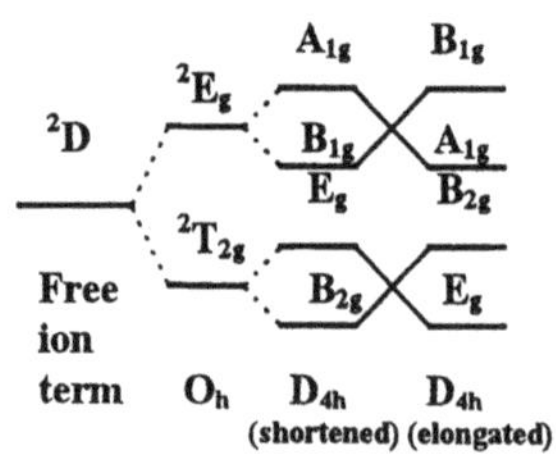

Figure.3. Energy level diagram for ions of d^1-electron configuration in octahedral field (O_h) with tetragonal distortion of D_{4h} symmetry [7]

However, the molar extinction coefficients somewhat differ in magnitude (Fig.1).This is due to fact that in all cases Mo(V) and W(V) ions coordinate solely with PO_3^- fragments of the melts. The inference is supported also by identify of ESR spectra for Mo(V) and W(V) in $NaPO_3$ and $NaPO_3$-NaCl system distorted polyhedra [Mo^VO_6] and [W^VO_6] presumably of D_{4h} symmetry are formed. The increase in molar extinction coefficients of Mo(V) and W(V) polyhedra in going from $NaPO_3$ to $NaPO_3$-NaCl melts can be associated with changes in covalency of the M-O-P bond. Such a behavior can be accounted for by structural changes in the $NaPO_3$ melt due to formation of more simple depolymerized phosphate groups and even anions such as PO_3Cl^-, when NaCl is added [14]. It is just differences in coordination properties of depolymerized phosphate groups that tells on molar extinction coefficients of Mo(V) and W(V) polyhedra due to formation of a more covalent bond between M(V) and depolymerized phosphate ligands.

As is seen, in the $NaPO_3$-NaCl system similar to Mo(V) and W(V) polyhedra of the same composition and structure are formed that predetermines possible similarity of their electrochemical behavior. Indeed, in the system studied the values of tungsten and molybdenum reduction potentials were proved to be almost identical and more positive as compared to the phosphorus reduction potential (Fig.4). In addition, solubility of molybdenum and tungsten trioxides was found to be fairly high and the binary system metaphosphate-sodium chloride rather low-melting.

A trend to partial chemical reduction of Mo(VI) and W(VI) dissolved in the metaphosphate-based melts to Mo(V) and W(V) should also be noted, which is favorable for electrochemical reduction of the d-metals.

Initially, by current-voltage measurements we determined deposition- potential ranges for molybdenum, tungsten, phosphorus, and tungsten-molybdenum alloys. In Fig.4 are presented current-voltage curves for electrodeposition of molybdenum and tungsten trioxides added separately or in combination to a $NaPO_3$-NaCl(mol.%)melt. The measurements were carried out at 700^0C using nickel or stainless steel plates as a cathode and a glassy-carbon beaker as an anode. The electrolysis was performed under galvanostatic conditions at 1.0-1.8 V voltage applied to the bath.

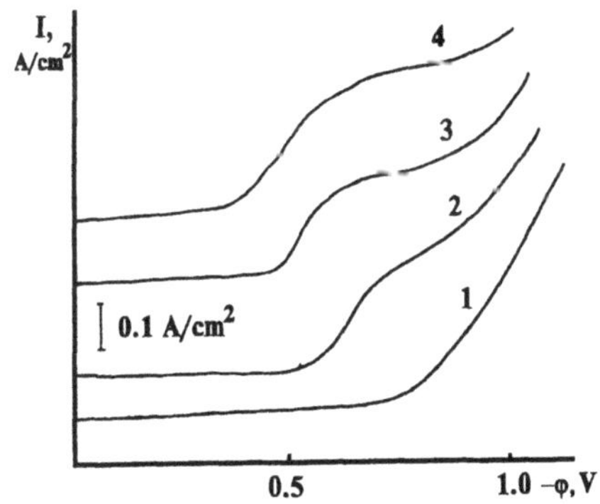

Fig.4. Current-voltage characteristics of molten systems at 700^0C:
1)$NaPO_3$-NaCl(background); 2)$NaPO_3$-NaCl-MoO_3;
3)$NaPO_3$-NaCl-WO_3; 4) $NaPO_3$-NaCl-MoO_3-WO_3.
Cathode - Pt, V = 0.1 V/s.

Electrodeposition of neat molybdenum proceeds at a potential about 100 mV more positive as compared to deposition of neat tungsten. When both MoO_3 and WO_3 are present in the melt, the process of combined electrodeposition is described in the current-voltage curves by one wave (on the potential axes) the limit current of which is equal to a sum of partial limit currents for molybdenum and tungsten deposition. Potentiostatic electrolysis (at the wave potential corresponding to combined electroreduction of MoO_3 and WO_3) gives a deposit which is a molybdenum-tungsten alloy.

References

1. Volkov S.V.,Buryak N.I.,Bandur V.A.(1989) New one-stage methods of methane conversion with melts and low-powered IR-lasers utilization, *XIV Mendeleev Cong.Gen. and Appl. Chem.Absrt.*,Nauka, Moscow,p.225.
2. Volkov S.V.,Buryak N.I.,Bandur V.A.(1991) Careful (partial) oxidation of methane gas on molten metal complexes, *Mat. Int. Symp. on Molten Salt Chem. and Technol.*, Trans.Tech.Publ., Paris, pp.275-278.
3. Volkov S.V.,Bandur V.A.,Buryak N.I.(1991) Methane conversion in molten metaphosphate catalysts, *Mat. Int. Conf. "C_1-C_3" Hydrocarbon Conversions"*,Institute of Catalysis, Novosibirsk, pp.25-26.
4. Tarasova K.P.,Baraboshkin A.N.,Martem`yanova Z.S., and Bychin V.P. (1981) Electrodeposition of molybdenum and molybdenum-tungsten alloys from molybdenum-tungsten melts, *Zashchita Metallov*, **17**, 371-374.
5. Bezarashvili L.R., Volkov S.V., Buryak N.I.,Tsiklauri O.G.,Bandur V.A. (1990) Structure of metaphosphate based melts and electroreduction of molybdenum-tungsten alloys, *IV All-Union Sem."Electroreduction of Polyvalent Metals in Ionic Melts,* Metsniereba, Tbilisi, p.29.
6. Volkov S.V., Buryak N.I.(1971) Methodic of investigation of high-temperature electron absorption spectra of molten salts, *Teoret. i Eksperim. Khimiya*, **7**, 275-278.
7. Volkov S.V. and Yatsimirskii K.B.(1977) *Spectroscopy of Molten Salts*, Naukova Dumka, Kiev.
8. Volkov S.V.,Buryak N.I.,Bandur V.A.(1987) Coordination of Co(II) in phosphate and chloride melts, *Zhurn. Neorgan. Khimii*, **32**, 2634-2638.
9. Tananaev I.V., Voskresenskaya N.K., Dzhurinskii B.F. et al.(1972) Absorbance spectra of glasses and melts and properties exchange of molten systems $Co(PO_3)$-$NaPO_3$ and $Co(PO_3)$-KPO_3, *Izv.AN SSSR, Neorgan. Materialy*, **8**, 1474-1477.
10. Lever A.(1987) *Inorganic Electron Spectroscopy*, Mir, Moscow.
11. Ballhausen C.J.(1989) *Introduction to Ligand Field Theory*, Mir, Moscow.
12. Wertz J.,Bolton J.(1969) *Electron Spin Resonance*, Mir, Moscow.

13. Buryak N.I.,Bandur V.A.,Volkov S.V. et.al.(1989) Electron absorption spectra, ESR and structure of coordination polyhedra of titanium (III) in molten and quenched systems of sodium phosphates and halogens,*Teoret. i Eksperim. Khimiya,* **25**, 610-614.

14. Bezarashvili L.R., Buryak N.I., Bandur V.A., and Volkov S.V.(1990) IR-spectra and behavior of sodium metaphosphate based systems in molten and "quenched" states,*Ukr.Khim.Zhurn.*,**56**, 115-118.

High Performance Coating of Steel with Tantalum Boride Layers

H. Vetters, IWT Bremen, Badgasteinerstr. 3, D- 28359, Germany
J. Christensen, DANFOSS A/S, DK- 6430 Nordborg, Denmark

Abstract:

Steel bars (AISI 316L and AISI 1015), coated with tantalum by the FLINAK process, have been borided under low pressure and the properties evaluated. Dense layers have been obtained with thicknesses ranging from 1.5 to 30 microns. The coating of bulk material (bolts, rods and sheets of sintered tantalum) has been used for boriding experiments. Boriding with less activated dopant pastes in a hydrogen atmosphere showed the formation of thin layers consisting mainly of TaB. Metallographic examination showed dense covering and no flaws. Comparative tests of microhardness and residual stress formation showed the existence of compressive stress fields in the outer rim of the TaB-film. The TaB_2- coating obtained by means of amorphous boron powder treatment lead to the formation of cracks caused by tensile stress in the near surface regions. Depending on the treatment parameters the interface formation varied between a planar and interdendritic shape. Larger parts of complex geometry have been coated and comparative tests made. Even at small notches the stability of these thin boride coatings was verified. Generally it was possible to transfer the laboratory process parameters to an industrial score, so that this data is relevant for constructional details.

1. Boriding of tantalum coated steel bars

1.1 Process description

Steel bars, coated with tantalum by the FLINAK [1)] process, have been borided under low pressure and the properties evaluated. The process (Fig. 1) is conducted by consecutive coating procedures. To define the boride processing data for the diffusion treatment of refractory metals preliminary investigations have been carried out, such i.e.

- Evaluation of the constituents of the boride dope,
- Pre-treatment of the dope,
- Determination of the temperature-time process data.

D.H. Kerridge and E.G. Polyakov (eds.), Refractory Metals in Molten Salts, 17-29.

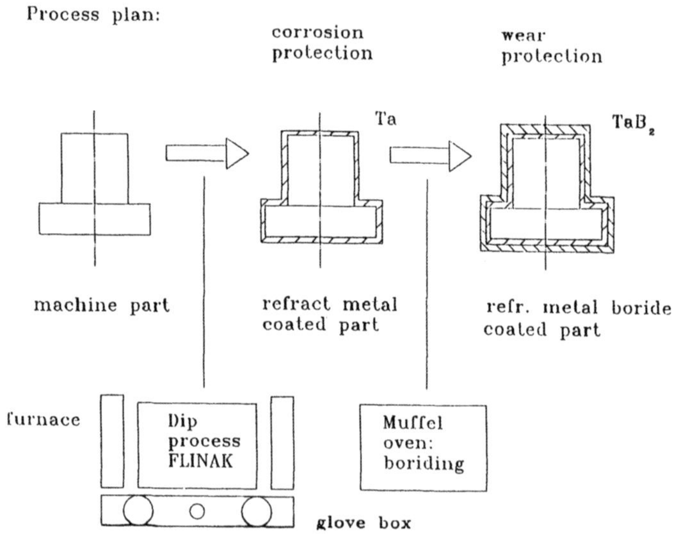

Fig. 1: coating procedures

Powder pack boriding at low pressure in a vacuum chamber was studied by laboratory tests using dopes of:

1.) 70 (mass-%) amorphous boron + 30 (mass-%) Al_2O_3,
2.) a boron enriched paste with boron carbide and silicon carbide [2]. In addition, boriding by glow discharge under hydrogen atmosphere was tested at 850°C [3].

The test material was sintered tantalum (99,7) and tantalum coated steel (AISI 316L, AISI 1015). Two types of reactors have been used:

a) muffle furnace with a minimum pressure of 0.5 bar and reaction temperatures up to 1200°C.
b) reactor chamber with a pressure of 1,33 mPa and a reaction temperature up to 950°C .

[1] "FLINAK": eutectic mixture of LiF, NaF, and KF used as melt solvent , DTH, Denmark
[2] "EKABOR": Elektroschmelzwerk Kempten, Wacker Chemie, Germany
[3] This procedure is an intermediate between gas and plasma processing

1.2 DETERMINATION OF THE BORIDING PARAMETERS:

The coating thickness was determined by means of metallographical examination of cross sectioned parts. The thickness (d) of the boride layer correlated with the exposure time has a dependence:

$$d^2 = k. (t - t_o) \qquad (1)$$

indicating a parabolic law for the increase of thickness. The expression t_o indicates the incubation time, and $(t - t_o)$ the boriding duration. The facor k is determined experimentally. The graphic plot d^2 versus exposure time (fig. 2) shows the increase of the layer thickness with higher treatment temperatures. Values of k for to the treatment temperature are given (Tab. 1).

T[°C]	d^2 [10^{-5}.cm^2]	k [10^{-9}.cm^2.s^{-1}]
950	1,225	1,7
1000	1,444	2,0
1050	1,849	2,6
1100	4,235	5,9
1200	9,025	12,5

Tab. 1: correlation of temperature (T), d^2 (d: coating thickness), and k-factor

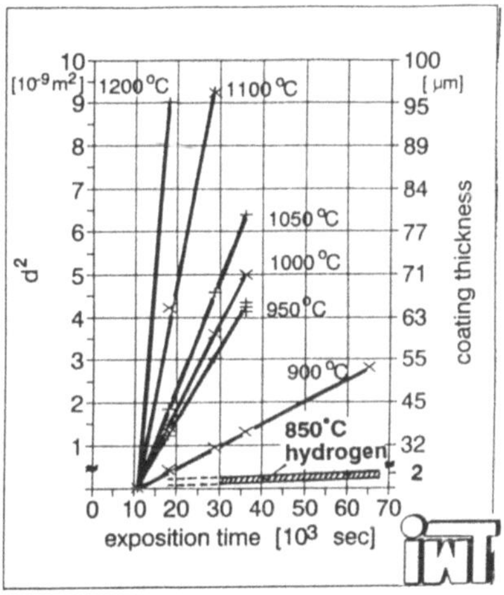

Fig.2: (d^2) versus exposition time

The activation energy (Q) following the Arrhenius-law is determined with:

$$Q = R. \ln (k_2/k_1)/[(T_2)^{-1} - (T_1)^{-1}] \qquad (2)$$

where R represents the gas constant, $T_{1,2}$ the boriding temperature of two treatment ranges determined in Kelvin, and k the corresponding values. The value is extrapolated as:

$$Q = 128 \text{ kJ/g-At} \qquad (3)$$

Thus, the data for boriding (Tab. 2) with amorphous boron dope were determined over various ranges.

Duration:	1 hr < t < 5 hrs
Pressure:	10^{-1} Pa (10^{-3} Torr)
Temperature:	900°C < T < 1200°C

Tab. 2: data for diffusion boriding

Limits are set in choice of the temperature range because the heat treatment of the tantalum coated steel parts must not exceed 950°C. After that the duration of boriding was the only variable to be set.

1.3 CHEMICAL REACTIONS:

Identification of the chemical species present in the reactor atmosphere was carried out using a vacuum receiver connected to a quadrupole mass spectrometer (Fig.3).

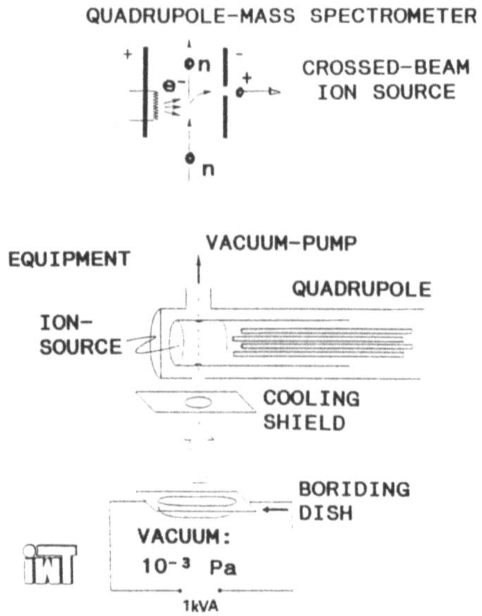

Fig.3: Assembly with quadrupole mass spectrometer

The temperature was controlled by means of a pyrometer and in situ thermocouples. The reactions have been studied comparatively on low carbon steel and tantalum. Reactions with paste boriding at 900°C are observed after 3 minutes of treatment and indicate the formation of a 1,5 µm layer. The reaction equilibria are suggested as:

(a) $26\ KBF_4 \rightarrow 26\ KF + 26\ BF_3$
(b) $26\ BF_3 + B_{13}C_2 \rightarrow 39\ BF_2^{1+} + 39\ F^{1-} + 2\ C$
(c) $39\ BF_2^{1+} + 26\ Me + 39\ F^{1-} \rightarrow 13\ Me_2B + 26\ BF_3 + 19\ 1/2\ F_2$
$B_{13}C_2 + 26\ KBF_4 + 26\ Fe \rightarrow 26\ BF_3 + 13\ Me_2B + 2\ C + 26\ KF + 19{,}5\ F_2$
$B_{13}C_2 + 39\ F_2 \rightarrow 13\ BF_3 + 2\ C$ (4)

where Me symbolises metal, and B_4C is structured as $B_{13}C_2$ and:

(a) $KBF_4 \rightarrow KF + BF_3$
(b) $4\ BF_3 + 3\ SiC + 3/2\ O_2 \rightarrow 3\ Si\ F_4 + 3\ CO + 4B$
(c) $19{,}5\ Si\ F_4 + 2\ B_{13}C_2 + 7\ O_2 \rightarrow 26\ BF_3 + 6\ SiO_2 + 2\ CO + 13{,}5\ Si$
$2\ B_{13}C_2 + 3\ SiC + KBF_4 + 8{,}5\ O_2 \rightarrow 4B + 13{,}5\ Si + 6\ SiO_2 + 5\ CO$ (5)

Pretreatments such as reductive preglowing under hydrogen atmosphere have also been examined. Further investigations have been conducted with the tantalum coated parts and two different types of coatings were found:

a) TaB_2- layer on tantalum coated steel, where an intermediate layer remains, which is obtained by means of powder pack boriding in a low pressure chamber;

b) TaB- film on tantalum, (1,5 microns), which is obtained by vapour boriding under H_2-atmosphere for one hour duration.

2. Postprocessing

2.1 REMOVAL OF DOPANTS, CLEANING PROCEDURES:

The dopant of the layer of type (a) can be removed by means of air blasting after extraction of the filler material, which leads to annoying additional work which increases the costs. Boriding with hydrogen atmosphere (layer, type b) has been tested under comparative conditions. This procedure can be undertaken with a paste consisting of amorphous boron, and B_4C added with activators, such as KBF_4. The adherent dope can be removed directly after heat treatment by air blasting under exhausters, so that the dope can be used again after a defined recycling procedure. This process can be implemented easily to a routine process of surface treatment.

2.2 SURFACE CHARACTERIZATION:

The steel types to be coated: AISI 316L (austenitic quality) and low carbon steel AISI 1015 where designed as cylindrical bolts and dip coated. The samples, as described above have been borided under low pressure and under hydrogen. The samples show a yellowish shine (Fig. 4) and no macrodefects.

Fig. 4 Steel bars, coated with tantalum and borided

2.3 Roughness measurement:

The surface roughness was inspected (Tab. 3) and a slight increase after boriding found.

Roughness values	Rt	Rm	Rz	Ra
	μm	μm	μm	μm
tantalum (ground)	7.4	6.75	6.13	0.83
tantalum (borided)	8.2	7.95	6.83	1.15

Tab 3: roughness values

2.4 Metallographical examination:

As examined by optical microscopy, change in tempering duration at the preselected temperature of 950°C leads to the formation of different layer types (Tab. 4).

Duration	thickness	characteristics	remarks
(hrs)	(μm)		
6	37.5 ± 2.5	closed pores	cracks
3	5.5 ± 2.5	closed pores	no cracks
1	1.5 ± 0.05	no pores	no cracks

Tab 4: characterisation of boride coatings

The cross section of the coating shows homogeneous and dense formation (figs. 5 ,6).

Fig. 5 Ta coated(left) steel-bolt (polished)

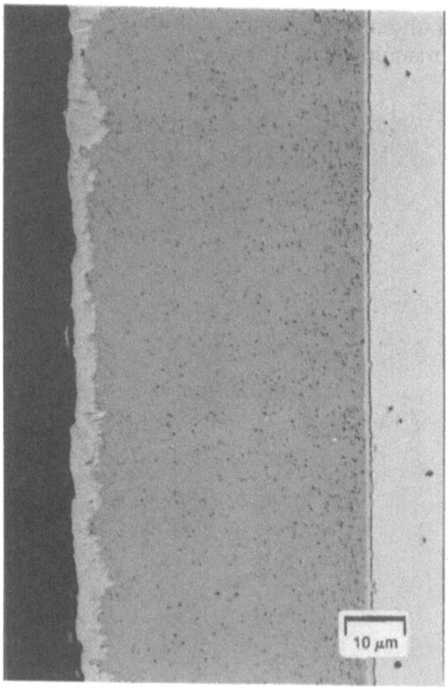

Fig 6 boride layer, Ta-coating, interface, steel

The glow discharge optical spectroscopy (GDOS) of the A 301L sample indicated the coating composition (Figs.7 ,8). The results of the depth profile analysis proved carbon enrichment in the interface of the steel substrate and the tantalum layer.

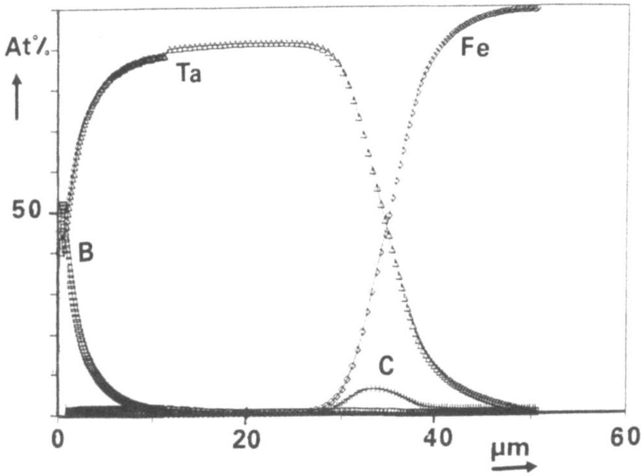

Fig. 7 GDOS analysis of Ta, Fe and C versus depth

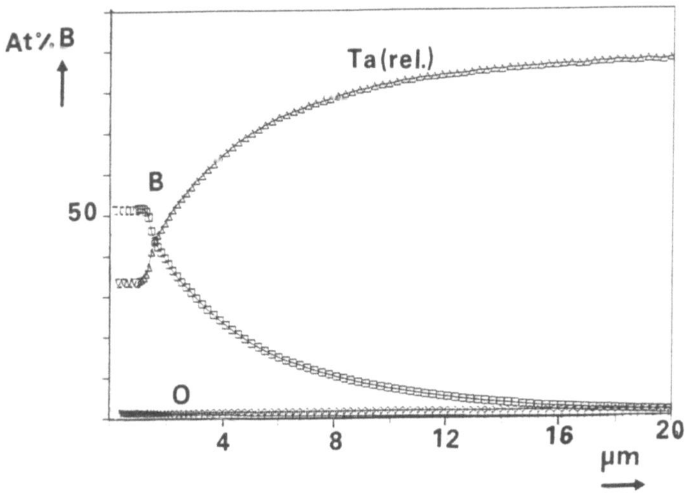

Fig. 8 GDOS analysis of the outer rim

2.5 MICROHARDNESS INDENTATION TEST
The microhardness was examined on the cross sectioned parts by means of Vickers indentation tests (Fig.).

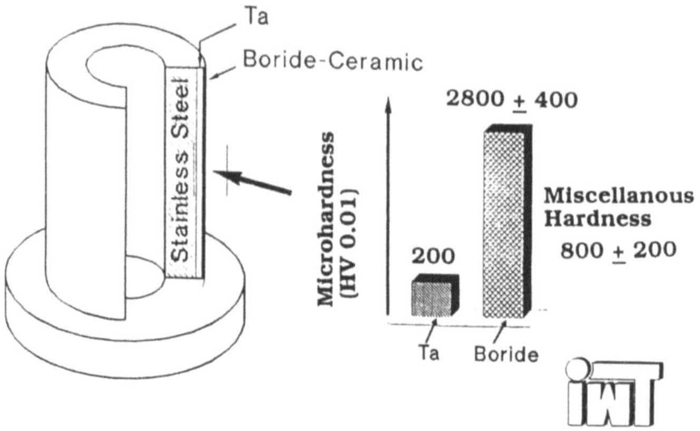

Fig. 9 Hardness values of the Ta coated and borided steel sample

The mean hardness value of the layer can be estimated as about 3000 HV0,05 (error ± 30%). The miscellanous hardness between layer and substrate is about 800 HV1.

2.6 RESIDUAL STRESS MEASUREMENTS:
Experiments to test the residual stress of the borided tantalum layer on steel have been started with an X-ray diffractive goniometer. Thick layers of TaB_2 showed tensile stress crack formation. In the boron depleated TaB films compressive stress has been detected (Fig. 10).

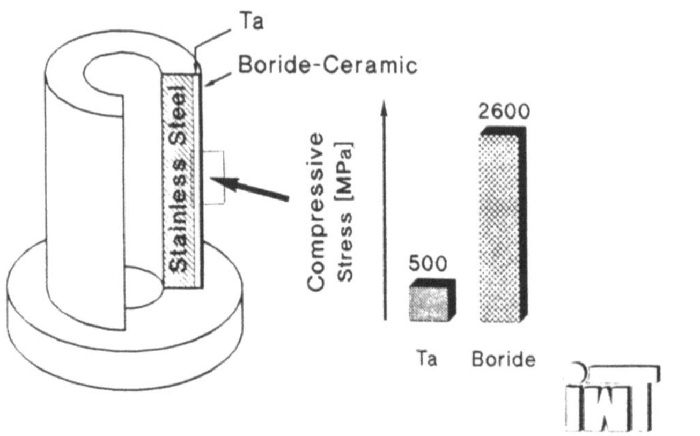

Fig. 10 Residual stress formation in the borided rim

2.7 CORROSION TESTS:
The resistance of borided samples against aggressive chemical reagents have been tested (Tab. 5) and following assumptions have been made.

Reagent	Temp. [°C]	time [hrs]	attack: Ta	attack: TaB_2
			no(--), slight(+), etched(++), dissolved(+++)	
50 ml H_2O+50 ml HF + 50 ml H_2SO_4	22 (RT)	2	+	--
50 ml H_2O+50 ml HF + 50 ml H_2SO_4	60	2	++	--
HF conc.	22	2	+	--
HF conc.	100	3	++	--
HF conc. + HNO_3 (1:1)	22	2	++	--
HF conc. + HNO_3 (1:1)	100	0,1	+++	+++
NaOH (30%)	80	2	+	--

Tab 5: chemical attack test of tantalum and tantalum-boride

The liquid acid was examined by means AAS spectroscopy for boron. If no residuels were detected, the qualification of no attack (--) was determined. Slight attack is a change in colour, but no real etching. The partial dissolution of tantalum in comparison to the boride layer after exposure to boiling HF for three hours is visible in the remaining boride rim around the tantalum (Fig. 11). Attack after 6 min. could be seen by macroscopic inspection.

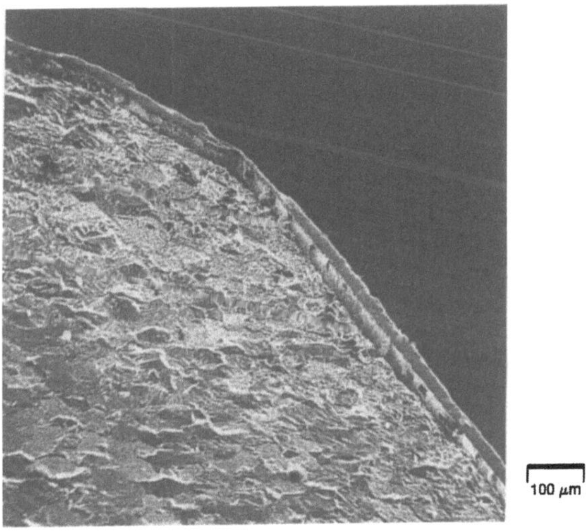

Fig. 11 cross section of the borided Ta- bolt [SEM- micrograph 100 x]

2.8 TEST OF THERMAL STABILITY

Thermal stability in air and vacuum (Tab. 6) were tested on a thermal balance under vacuum and air environments for 3 hrs.

Environment	Temp.	weight loss
	[°C]	[mg]
air, normal pressure	950	3
air, normal pressure	1200	9
vacuum [10 mPa]	1200	--

Tab. 6: weight loss at high temperatures versus environment

2.9 WEAR TEST

Samples of borided tantalum have been tested with the pin on disk method. The disk for testing was borided steel. The test parameters have been: 20°C, 50% humidity, the normal force F_N = 2.5 N, the rotating speed 0.1 m.sec^{-1}, the friction number f: 0.7, the test length s: 1000 m. 15 samples each have been tested.

boriding time	wear volume [10^{-6} mm^3]	
[hrs]	min	max
1	0,2	0,5
3	0,7	0,9
6	0,6	0,9

Tab. 7: wear volume (pin) versus boriding time

The result (Tab. 7) shows, that the thicker coatings have an increase in wear volume (thickness > 10 µm) which corresponds to a deterioration of about 30 %.

3. Discussion:

Boriding of refractories leads to the formation of ceramic like layers [1-12]. The chemical bonding of these borides extends from stoechiometric compositions such as Me_6B to MeB_{12} to non stoechiometric composites[2]. Beside treatment of solid metals [1, 4, 10, 11] electrolytical processes have also been used [7]. Boriding methods from the solid state as revealed in this study [4, 5] give improved environmental protection and also the possibility of recycling the dope.

The surface layers are mainly characterised by the formation of mono- and diboride arrays[2, 5]. As found [8], the boron depleated composites have the tendency to form ceramic films of high density and avoid crack formation by prevailing compressive residual stress. Therefore the surface films show excellent adhesion enhancing the resistance against abrasion[8, 12].

By TEM- analysis of refractory boride layers the formation kinetics have been formulated [13]. The first stage of layer formation occurs by monoboride precipitation at preferred orientation by forming needle shaped orthorhombic crystals. Superlattices enable the formation of more boron enriched complexes where the boron concentration of the most stable diboride is exceeded during the diffusion process[14]. As boriding acts as a thermochemical process controlled by diffusion [1, 9, 11], this study has developed a new procedure between solid and gaseous state treatment [9]. Further developments show that activation by BF_3 at low pressure can be used for plasma processes [15].

Earlier, diborane has been used [16] but because of ist high poisonous nature thismethod failed. However, the BF_3 activated plasma process can be used successfully [17].

The thermodynamics are guided by the free enthalpy (ΔG) wherefrom the chemical reactions at a preselected pressure can be derived. Regarding the GIBBS- HELMHOLTZ eqn.:

$$d\left(\frac{\Delta G}{T}\right) = -\frac{\Delta H}{T^2}dT \qquad (6)$$

the free enthalpy necessary for reaction is:

$$\Delta G_T = \Delta G_T^0 + RT\sum \ln p_i^{\nu_i} \qquad (7)$$

verifying the reaction at low pressure (p). Regarding equns. (4) and (5), the reductive BF_3 reaction with hydrogen can be expreesed as

$BF_3 + 3H \rightarrow B + 3HF$ with $\Delta G_{1173[K]} = -283$ kJ (8)

$BF_3 + 3H + Me \rightarrow MeB + 3HF$ with $\Delta G_{1173[K]} = -304$ kJ (9)

where Me is the reacting metal as calculated for iron [17] and is possibly transponible to the Ta-reaction.

Thus the values for ΔG at low pressure can be estimated by:

$$\Delta G = \Delta G_T^0 + RT\ln\frac{p_{HF}^3}{p_{BF_3}\cdot p_H^3} \qquad (10)$$

concerning the equivalence of the pressures $p_H = p_{HF}$, the boriding condition depends upon the partial pressure of the boron trifluoride as:

$$\Delta G = \Delta G_T^0 + RT\ln\frac{1}{p_{BF_3}} \qquad (11)$$

The GIBBS free energies of the reactions as calculated for Fe are strongly negative, which enhances boron transfer into the surface and, as proved by coating experiments, leads to a reliable layer formation.

The mechanical properties of the boron depleated refractory borides give high stability and strength favouring use in chemical equipment by coating crucibles exposed to liquid salt. The formation of compressive stresses in the near surface regions strongly inhibits crack formation and contributes to wear and corrosion resistance. The hardness is also influenced by residual stresses. The interface between boride and tantalum is locked by the formation of micro dendrites. By means of these structural components the coating adhesion is enhanced in competition with even structured layer interfaces.

Under vacuum conditions the thermal stability of boride layers is very high exceeding temperatures of 2500 K , and remaining extremely hard [6]. Oxydative environments reduce this stability dramatically to temperature limits of 900K where the boride dissolves. A positive aspect of this enhanced reactivity in air is the possibility of recycling substrate material by oxydative tempering at 1000 K.

Acknowledgement:

This research work has been supported by funds of the European Research Programme **BRITE, Contract No. 3359, "HIPERMAT"** to whom the authors want to express their gratitude.

References:

[1] Samsonov, G.V.: Protective coatings from refractory Compounds on Metal and Graphite, Zaschitnya Pokrit. Metallakh, 6 (1972), 5-24 [engl. traNsl.]
[2] Post, B., Glaser, F.W., Moskowitz, D.: Transition Metal Diborides, Act. Met.,2 (1954), 20-25
[3] Golego,N.N et al.: Wear resistance of diffusion boride coatings in vacuum and gaseous media, Prot. Coat. of Met. 5 (1973),257-260
[4] Krzyminski, H.: Borieren refraktärer Metalle I, Haerterei Techn. Mitt. 28 (1973), 2,100-104
[5] Kunst, H.: Borieren refraktärer Metalle II, Konstitution und Eigenschaften der boridschichten Haerterei-Techn. Mitt. 28 (1973), 105-112
[6] Binder, F.: Refraktäre metall. Hartstoffe, Radex Rdsch.: 4 (1975), 531-557
[7] Flinn, D.R.: Electrodeposition of erosion resistant titanium diboride coatings, Rep. Of invest. 8332, Bureau of Mines, U.S.Dep.of Interior,1-29 (1979)
[8] Vetters, H., Mayr, P.: Gefüge und Eigenschaften borierter Refraktärmetallboridschichten, Haerterei-Techn. Mitt. 40 (1985),5,198-210
[9] Lu, Ming-Jiong: Die Bildungs- und Wachstumskinetik monophasiger Boridschichten, Haerterei-Techn. Mitt. 38 (1983), 156-169
[10] Schaaber, O.: Die Erzeugung von Randschichten höchster Härte mit borhaltigen Mitteln, Haerteri-Techn. Mitt. 15 (1960), 1,44-48
[11] Minkevic, A.N.: Durch diffusion erhaltene Boridschichten auf Metallen, Haerterei-techn. Mitt. 17 (1962), 3, 141-147

[12] Durakevich, I.S., Effect of phase composition of boride case on residual stresses, Transl. Metalloved.i.Termicheskaya Obrabotka Metallov 10 (1968), 16-19
[13] Vetters H., et al.. TEM analysis of microcrystalline boride layers, Microstructural Science (ASM internat. Publ. Metals Park Ohio 1988), 16, 499-507
[14] De Graeff, M. et al.: The evolution of metastable borides in a Ti-Al-B-alloy, Acta met. Mater.,40 (1992), 3395-3406
[15] Wierzchon, T., Pokrasen, S., Karpinski, T.: Plasmaborieren-Faktoren, die die Keimbildung der Boridschicht auf Stahl bedingen, Haerterei-techn. Mitt. 38 (1983), 57-62
[16] Casadesus, P.: Mise au point d'un traitement de borurartion des aciers par bombardement ionique avec le diborane, thèse doctorat Ecole des Mines, Nancy, (1977)
[17] Hunger, H.-J.: et al.: Plasmaaktiviertes Gasborieren mit Bortrifluorid, Haertere-techn. Mitt. 51 (1997), 1, 39-45

ELECTROCHEMICAL AND UV-VISIBLE SPECTROSCOPIC STUDIES OF REFRACTORY METALS IN CHLORIDE MELTS

J. BOUTEILLON, J.C. GABRIEL, T. GALVEZ, T.R. GRIFFITHS*, P.KORONAIOS*, J.C. POIGNET, D. VINCENT*, D. VULPESCU

Laboratoire d'Electrochimie et de Physicochimie des Matériaux et Interfaces, UMR 5631 INPG-CNRS, associé à l'UJF, ENSEEG, Domaine Universitaire, BP 75, 38402 Saint Martin d'Hères, France
**School of Chemistry,The University of Leeds, Leeds, LS2 9JT, UK*

1. Introduction

Molten Salt electrolysis is an excellent route for electrowinning high purity refractory metals [1,2], and a few industrial or pilot installations exist.
However the process control is made difficult since knowledge of the exact nature and concentrations of electroactive species is still speculative, and no analytical characterisation method is available for in situ determinations.
In practice, analytical determinations are usually performed once the molten phases have been cooled and solidified.
The application of in situ spectroscopic techniques is now possible by the use of the new generation of spectrophotometers which employ optical fibres to take the light to and from the sample.
The group of Leeds has pioneered this application [3], having had experience investigating the ligand field spectra of species in molten salts.
The main aim is to acquire knowledge of the nature, structure and concentrations of the electroactive species present in the molten electrolytes, at the working temperatures, during electrochemical reduction of refractory metal and other transition metal ions in molten alkali chloride mixtures.
These determinations should be made both in the bulk of the solution and at the interface between the electrode and the solution, using electrochemical and spectroelectrochemical measurements, either separately or together.
I will present here the studies of K_3MoCl_6 and $CrCl_3$ -LiCl-KCl electrolytes where both electrochemical and UV-Visible techniques are used.
First let us remind about molten salt UV-Visible spectroscopy and shortly present the experimental devices.

1.1. MOLTEN SALT UV-VISIBLE SPECTROSCOPY

A good review of the simplified theory of molten salt spectroscopy is given by Gruen [4], while a description of the electronic states of atoms and ions may be found in the book by Atkins [5], and a description of the effects of symmetry on electronic spectra is given by Cotton [6]. A short discussion of the use of spectroscopy as an analytical tool is given by Smith [7], while a review of the techniques of molten salt spectroscopy is

D.H. Kerridge and E.G. Polyakov (eds.), Refractory Metals in Molten Salts, 31-42.

given by Griffiths [8]. Electronic spectroscopy covers the near-IR (1300-700 nm), the visible range (70-400 nm) and the near-UV (400-200 nm). The transitions which are caused in this wavelength range are mainly electronic:
(1) d-d transitions which involve transitions between different arrangements of electrons in the d-orbitals of metal ions, sensitive to the symmetry of the complex ion or the environment of the metal ion, and which can often be used to determine the coordination round a metal ion.
(2) charge-transfer spectra which involve the movement of electrons between the metal and the ligand orbitals.
The Beer-Lambert law, ($I_{out} = I_{in}.\ 10^{-\varepsilon.c.l}$), relates the amount of light , I_{out}, passing through a sample, to path length, l, to concentration of the absorbing species, c, and to molar absorption coefficient, ε.
Molten salt solvents begin to absorb in the UV region, and the wavelength at which a salt begins to absorb strongly depends on the salt. These wavelengths (the wavelength of the absorption edges) are reviewed by Gruen [4]. Alkali metal chlorides are transparent to about 300-350 nm, which is acceptable for many studies. The UV absorptions in molten halides are charge-transfer bands. In these, the proton causes an electron to be transferred from the halide ion to an orbital on the alkali metal ions surrounding it.

1.2.EXPERIMENTAL SET-UP

The spectroscopic study of solutions of K_3MoCl_6 in molten LiCl-KCl eutectic has been undertaken in Leeds at Dr. T. Griffiths 's laboratory.
These measurements have been made by using a Cary 14 H cuvette. Meanwhile the molybdenum electrochemistry was studied in our laboratory, in Saint Martin d'Hères.
The setting up of a new UV-Visible spectrophotometer and a fibre-optic interface was carried out in the laboratory when the study of the $CrCl_3$-LiCl-KCl molten electrolytes was undertaken.
The electrochemical device has been described before [9] and will not be reported here.
A GBC 918 Spectrophotometer (wavelength range: 190-900nm) equipped with two light sources, a 50 W quartz-tungsten-halogen lamp and a 30 W deuterium lamp was used.
A Czerny-Turner monochromator, moving diffraction grating is in front of the sample, so that only monochromatic light passes through the sample, was employed. A photo multiplier tube, model R446, was used as detector.
Hellma 662.002-QX (wavelength range: nominal 220-1500nm) quartz optical fibres (length 2m each, diameter 600μm), with collimating lens at the end of each fibre were employed. The maximum temperature allowed is 150°C due to the metal piece at the end of the optical fibres.
The heating device includes an external furnace and an internal one.
The external furnace is fitted with two water-cooled probe holders, with adjusting screws to allow the probes to be adjusted to face each other. Each probe holder has two sets of three screws. Once collimating probes are properly aligned, it should be possible for them to remain stable for several weeks. Thanks to the water cooling system the probes can be placed about 5cm inside the probe holder tubes up to 550°C.
The outside of the internal furnace, as well as the bottom, is water-cooled, to reduce the amount of heat escaping and heating the spectrophotometer.
Both the spectrophotometer and the furnace are mounted on a heavy marble table, so they should be isolated from general vibrations.

It is necessary to align both the transmitting and the receiving probes. However, once the probes are aligned, they can remain aligned for a long time.
Finally a box has been made to cover the external furnace, and thus to reduce the amount of stray light entering the system.

2. Chemical and electrochemical studies of K_3MoCl_6-LiCl-KCl solutions

Chemical and electrochemical studies of K_3MoCl_6-LiCl-KCl solutions began some years ago [9-11] .
These solutions were found to be generally quite stable, but later low concentrations of molybdenum(III) in this eutectic were found, somewhat surprisingly, to decompose.
In this study we have examined this decomposition reaction. It could be a disproportionation reaction,

$$5\,[Mo(III)] \Longleftrightarrow 3\,[Mo(V)] + 2\,Mo(O) \qquad (1)$$

The lower concentration limit above which decomposition is negligible is around .05M.
To explain the stability at higher concentrations Senderoff [9] and Inman [10] have proposed the existence of an equilibrium between di- and mononuclear Mo(III) complex ions, namely, $Mo_2(III) \Longleftrightarrow 2Mo(III)$ and that only the mononuclear species Mo(III) is the electroactive species.
An equilibrium constant for this proposed reaction has been reported by White et al. [11] as .012 mol dm^{-3}.
The aim of this study is to compare several physico-chemical measurements, employing electrochemical and electronic absorption spectroscopy techniques, and thereby obtain a better understanding and knowledge of the chemistry of the LiCl-KCl-K_3MoCl_6 solutions.
The spectroscopic study has been undertaken in Leeds using a Cary 14 H cuvette.
Meanwhile the molybdenum electrochemistry was studied in our laboratory, in Saint Martin d'Héres.
Let us begin with the electrochemical study.
Assuming the mononuclear-dinuclear equilibrium is taking place, the total molybdenum concentration, C_T, and the electroactive Mo(III) species concentration C_{Mo} can be related by:

$$C_T = C_{Mo}\,[\,1 + 2\,C_{Mo}\,K^{-1}] \qquad (2)$$

Where K is the equilibrium constant.
The total molybdenum concentration was determined volumetrically, whereas the electroactive molybdenum species concentration was determined from the cathodic peak current read on the voltammogram related to the electrochemical reduction of the K_3MoCl_6 solutions.
The curves presented in figure 1 proved that the solutions of K_3MoCl_6 with molybdenum concentration higher than .05 mol/l are rather stable while they decompose quickly when the molybdenum concentration is lower. The initial rise in molybdenum concentration could be attributed to the slowliness of K_3MoCl_6 dissolution. At the beginning there is also a gap between the total molybdenum concentration and the electroactive molybdenum one, but after a long time the difference disappeared and the two concentrations become either equal or proportional.
The above measurements have clearly evidenced the unity of the molybdenum electroactive species, but it is not possible to affirm if it is in equilibrium with one or several electroinactive molybdenum complexes.

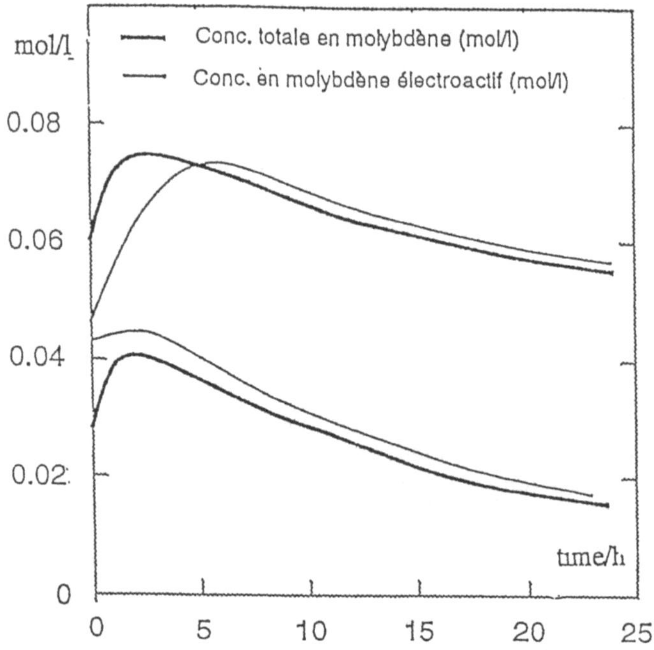

Figure 1. Total (___) and electroactive (–) molybdenum concentration

Indeed the best electrochemical method to solve this indetermination is the chronopotentiometry techniques. The experimental results are reported in figure 2, they are consistent with Sand's law.

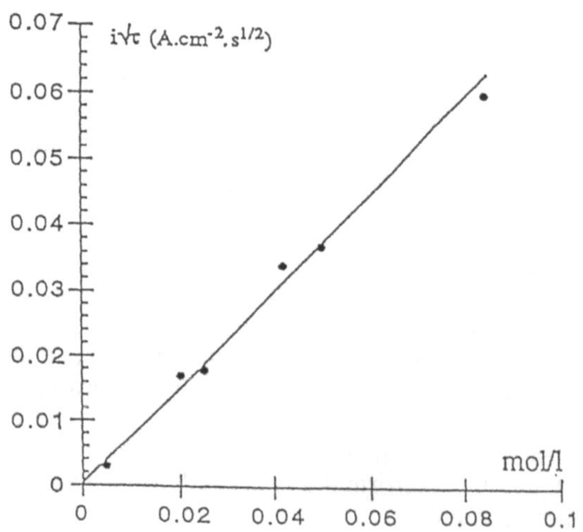

Figure 2. Chronopotentiometric study of K_3MoCl_6 reduction

The conclusion from the electrochemical study was that either the molybdenum electroactive species are reduced directly or the chemical reaction preceding the electronic exchange is too fast to be observed.
Let us pass on to the spectoscopic study of the K_3MoCl_6 solutions.
The measurements have been done using electrolyte prepared in the same way as for the electrochemical study.
Fortunately spectra could be measured between 300 and 800 nm, for solutions having concentrations above and below the concentration limit established above for the disproportionation reaction.
Figure 3 shows typical spectra taken shortly after the K_3MoCl_6 added had dissolved. The spectra obtained at high molybdenum concentration exhibited maxima at 425, 552 and 685 nm. They resembled that reported by Scheffler [12] in chloroaluminate room temperature melts. The spectrum obtained at low molybdenum concentration has similar peak maxima but also a fourth additional peak absorbance potential at 340 nm. The peak position steadily shifts with time to lower wave lengths, until the band maximum can no longer be resolved. This peak has not previously been reported in molybdenum solutions.

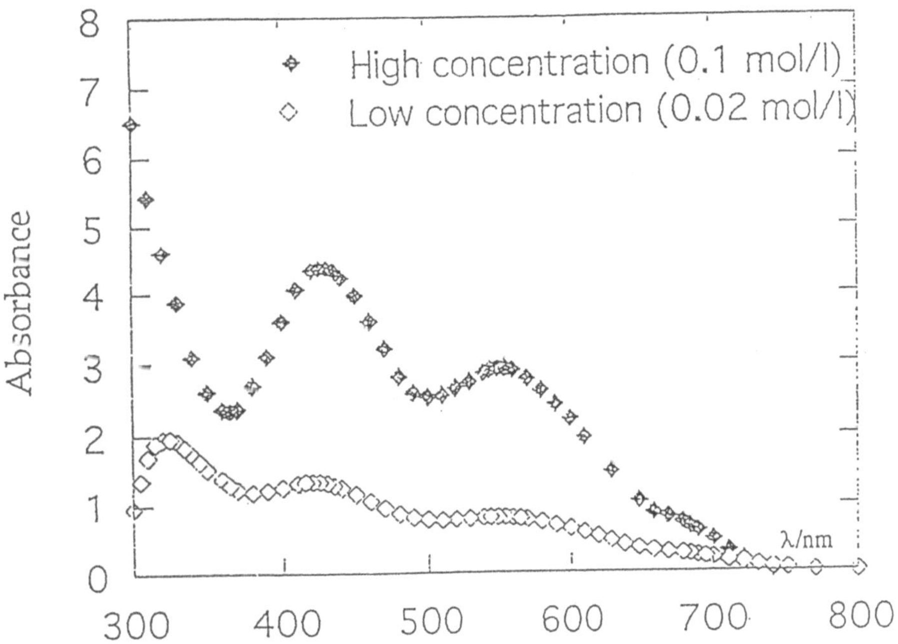

Figure 3. K_3MoCl_6 -LiCl-KCl UV-Visible Spectra

Figure 4 shows the decrease with time of the 550 and 420 nm absorbance peaks. The same trend may clearly be seen as was noticed for the electrochemical studies. The above electrochemical results indicate disproportionation of Mo(III) species solutions at concentrations below 0.05 mol/l. The complementary results got with UV-Visible spectroscopy is the presence of the absorption band initially at 340 nm in low molybdenum solution concentration.

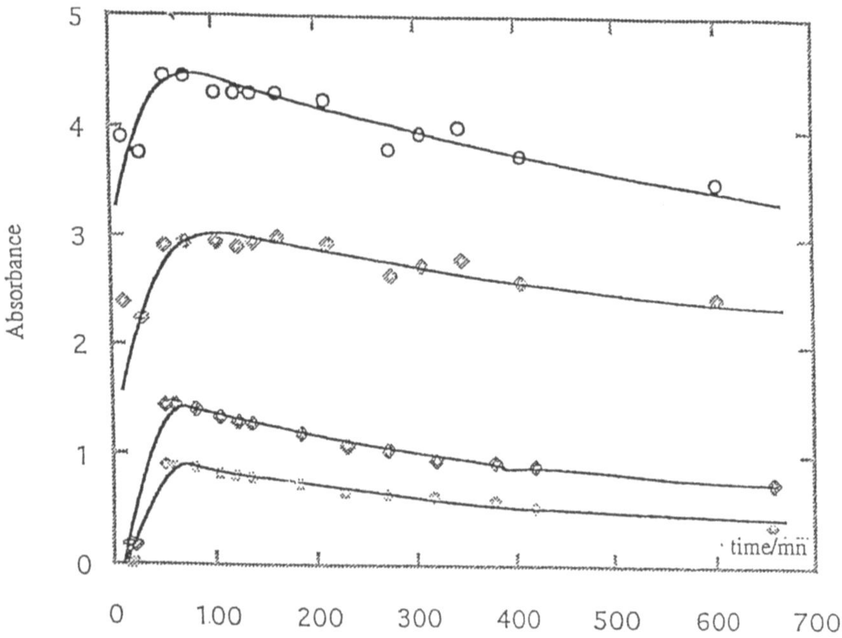

Figure 4. The decrease with time of absorbance peaks.

◇ 0.02 mol/l (550 nm) ◈ 0.1 mol/l (550 nm)

◆ 0.02 mol/l (420 nm) ○ 0.1 mol/l (420 nm)

In comparison with the spectra recorded by Scheffler [12] in chloroaluminate melts, $[MoCl_6]^{3-}$ has been identified as the only stable Mo(III) species. The available results from electrochemical studies point out to Mo(0) as one of the products, and the other is postulated to be Mo(V). The absorption spectra of the $[MoCl_6]^-$ octahedral complex ion have not been fully reported, however the rising absorption observed in the 300 nm region is consistent with the formation of this Mo(V) species.

3. The influence of oxygen on chromium electrocristallisation

The fundamental objective of the research was to analyse the effect of additives onto both the chromium electroactive species, and the chromium electrocristallisation phenomena. The practical objective was to improve the quality of the chromium

deposits obtained from the molten alkali chloride-chromium chloride electrolytes by acting onto the nature of the electroactive species.
The influence of various compounds: fluoride ions, ammonia, oxygen has been studied [11, 13]. According to the authors, the introduction of fluoride ions into the chlorides melts has no effect on the chromium species coordination. Nevertheless a beneficial influence resulted in the compacity of the chromium deposits, due to the fluxing agent properties of fluorides. White [11] reported that oxygen has no effect on the $CrCl_3$ solutions whereas it reacts with $CrCl_2$, according to :

$$12\ CrCl_2 + 3\ O_2 \text{------>} 2\ Cr_2O_3 + 8\ CrCl_3 \quad (3)$$

Benslimane [13] noticed that flowing oxygen through the molten alkali chloride electrolytes containing CrCl3 improves the quality of the chromium deposits, but didn't give any explanation.
In the present work, UV-Visible spectroscopy coupled with electrochemistry were used to analyse the influence of oxygen on the chromium chloride solution properties.
A series of voltammograms obtained 1, 18, 24 and 45 hours after the $CrCl_3$ dissolution in molten LiCl-KCl are reported in figure 5. They evidence the slowliness of the $CrCl_3$ dissolution process.

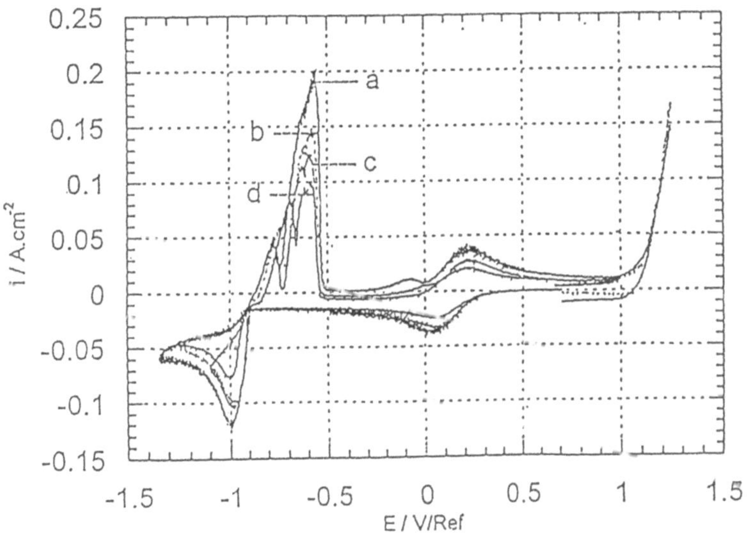

Figure 5. A series of voltammograms obtained in the LiCl-KCl-$CrCl_3$ mixture, Temperature: 450 °C ; sweep rate:500mV/s ; electrode: vitreous carbon ; $CrCl_3$ concentration added: .0105 mol/l
The curves have been obtained: (a) 45 h, (b) 24h, (c) 12h, (d) and just after the addition.

When the $CrCl_3$ was completely dissolved in the solvent, the effect of oxygen on the electrochemical properties of the electrolyte was characterised by recording voltammograms along the time. The curve (a) on figure 6 was recorded 20 minutes after the beginning of an air flow over the electrolyte, whereas curve (b) was obtained after 54 hours. One can observe the decrease of the peak current density and the increase of the crystallisation overvoltage.

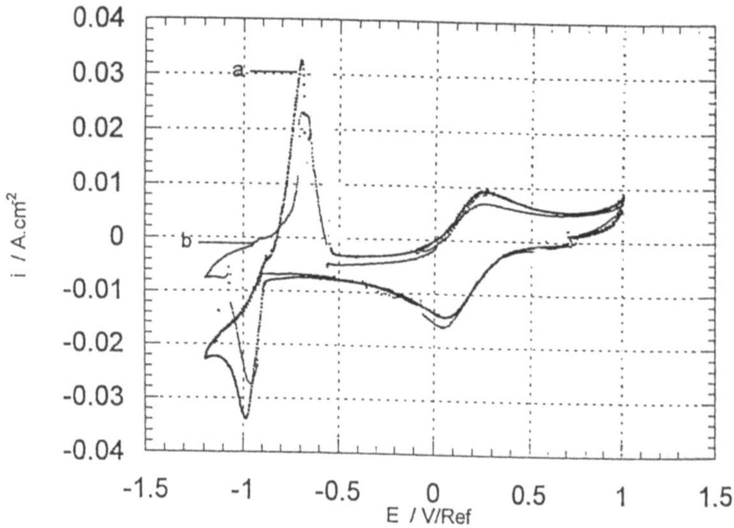

Figure 6. Voltammograms obtained in the LiCl-KCl-$CrCl_3$ mixture under oxygen flow (p_{O2} = .21 atm.)
curves (a) and (b) have been obtained 20 mn and 54 h after oygen exposure beginning

The conclusion was that the decrease of the peak current density could be related to the decrease of the Cr(III) species concentration, due to the reaction between oxygen and the Cr(III) species.
It was also observed that simultaneously, the darkness of the electrolyte increased, which could be related to the formation of a precipitate.

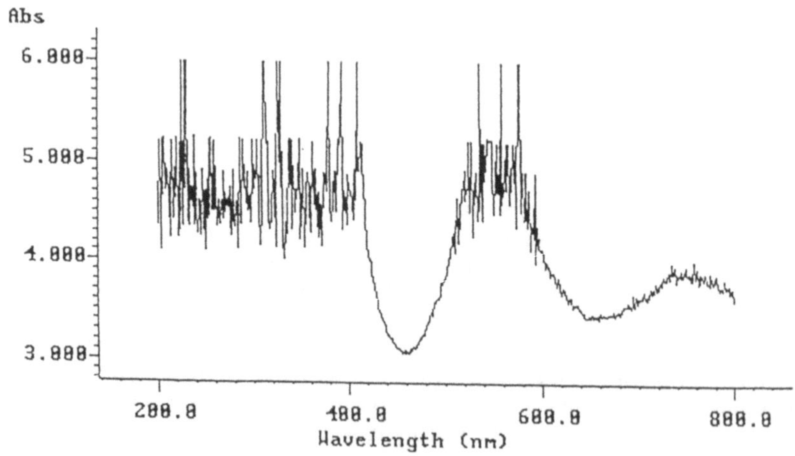

Figure 7a. Spectrum recorded in a LiCl-KCl-$CrCl_3$ mixture, 90 mn after oxygen exposure beginning

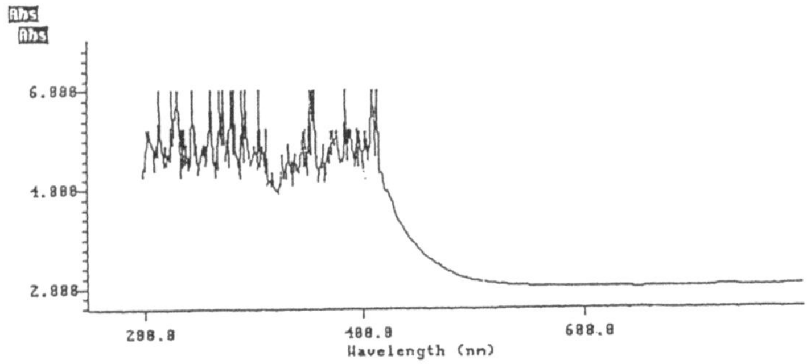

Figure 7b. Spectrum recorded in a LiCl-KCl-$CrCl_3$ mixture, 17 hours after oxygen exposure beginning

Then the reaction between oxygen and the Cr(III) species was followed by recording UV-Visible spectra of the melts.
The procedure was to pick up samples of the molten electrolyte in the electrochemical cell, then to transfer the frozen electrolyte samples in the optical cell, and to introduce the optical cell inside the furnace linked to the spectrophotometer. The UV-Visible spectra could be recorded as soon as the electrolyte was melted again.
Two spectra are reported in figure 7. The spectrum a was recorded for a sample picked up 90 minutes after oxygen exposure beginning. The two absorption bands at 550 and 780 nm visible on this spectrum are characteristic of the stable Cr(III) species.
The spectrum b was obtained 17 hours after the has been transferred in the optical cell. The two absorption waves characteristic of the Cr(III) species have vanished, but no new absorption band appeared.

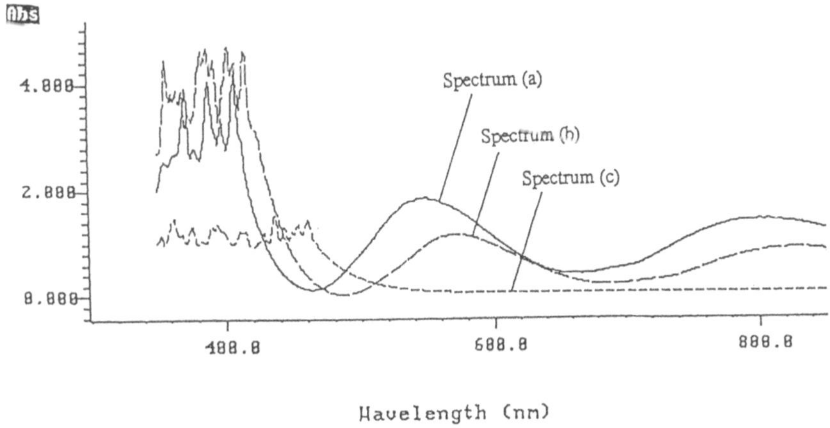

Figure 8. Spectra (a), (b), (c) have been obtained in a LiCl-KCl-$CrCl_3$ molten mixture, immediately, 8h, and 30h after oxygen exposure beginning

Similar results were obtained previously by Koronaios (figure 8). Spectrum (a) and spectrum (b) were obtained before and after exposure of the solution to oxygen. The spectrum reported figure 9a was recorded for a sample picked up 7 hours after oxygen exposure beginning. One can se that the two absorption bands related to Cr(III) species have vanished, but a new absorption band arises at 380 nm. The absorption band at 380 nm appears more clearly on the spectrum figure 9b which was recorded 17 hours after the sample has been transferred.

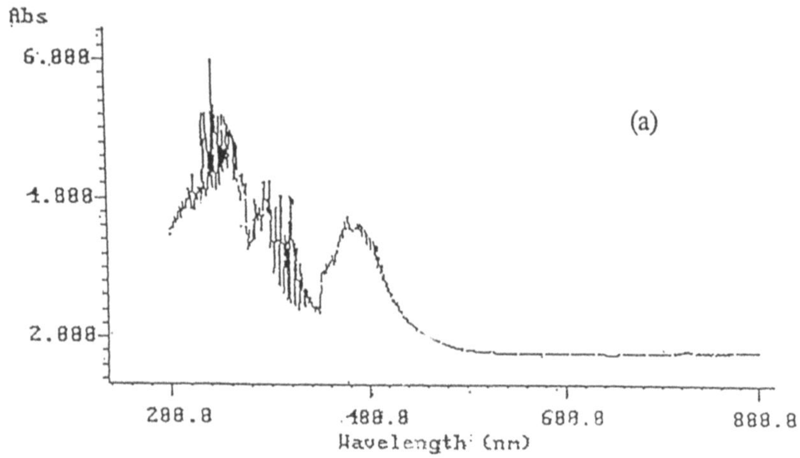

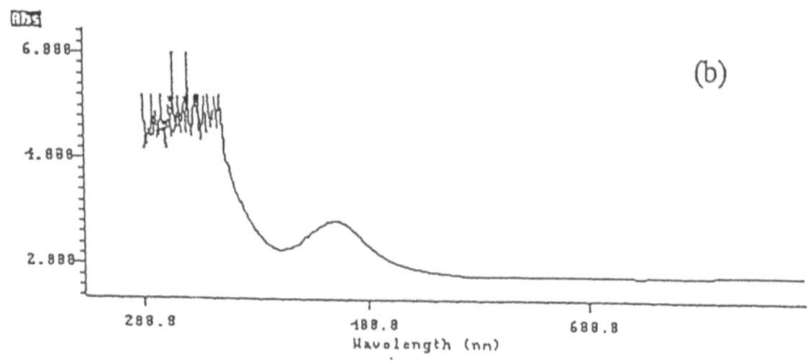

Figure 9. Spectra (a) and (b) have been obtained in a LiCl-KCl-$CrCl_3$ molten mixture, 7h after oxygen exposure beginning and 17h after the sample has been transferred in the UV-Visible device.

At this level the conclusion was that the reaction between oxygen and chromium(III) species resulted in a new chromium soluble species.
Let us go to a new series of experiments where oxygen will be bubbled inside the electrolyte.
The effect of oxygen was followed by recording a series of voltammograms along the time. The voltammogram reported in figure 10 was obtained 5 hours and half after the

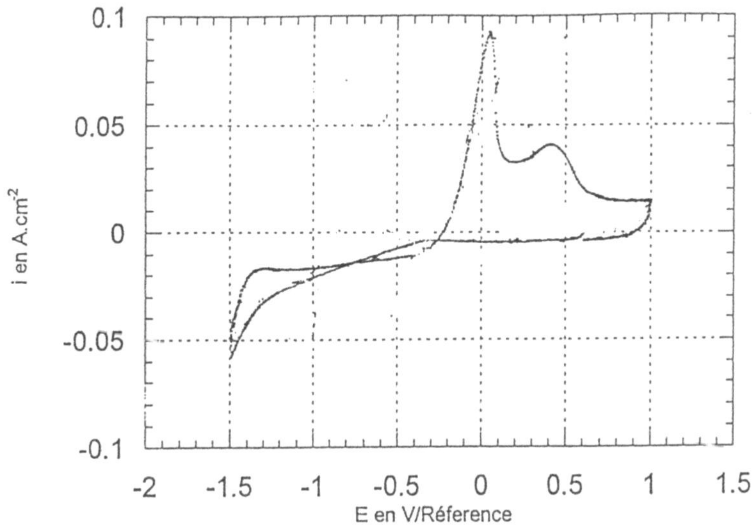

Figure 10. Voltammograms obtained obtained in a LiCl-KCl-$CrCl_3$ molten mixture, 20 mn and 5h30 after the beginning of oxygen bubbling.

beginning of the oxygen bubbling. One can see, in comparison with figure 1 that the signal characteristic of the Cr(III) species reduction has completely disappeared.
At the end of the experiment, the oxygen bubbling has resulted in the formation of dark precipitate in the solution. The precipitate has been collected and characterised as Cr2O3 by X-ray analysis.
The results presented here allowed us to postulate that the Cr(III) species react with oxygen according to the scheme :

$$2\ Cr^{3+} + 1{,}5\ O_2 \text{ ----> } 2\ Cr^{6+} + 3\ O^{2-}$$
$$2\ Cr^{3+} + 3\ O^{2-} \text{ ----> } Cr_2O_3$$

4. Conclusion

The chemical properties of the LiCl-KCl-K_3MoCl_6 solutions have been studied by spectroscopic and electrochemical techniques.
Processes which take place during the dissolution of the molybdenum salts are very slow. It is necessary to wait until the solution has got stable before making reproducible electrochemical measurements.
Composition and therefore stability of the electrolyte as a function of the molybdenum concentration has been studied:
At high concentrations (>5.10^{-2} M), the stable species $[MoCl_6]^{3-}$ has been identified as the only stable Mo(III) species. At low concentration a Mo(V) species resulted from the disproportionation of Mo(III) species, it could be assumed to be the $[MoCl_6]^-$ octahedral complex ion.

The effect of oxygen on the $CrCl_3$-LiCl-KCl molten electrolyte has been characterised either by electrochemistry or UV-Visible spectroscopy .
The electrolyte was studied by voltammetry. A reaction between oxygen and Cr(III) species resulting in Cr_2O_3 precipitation was evidenced, and also the influence of the reaction occurrence on the chromium electrocristallisation.
UV-Visible absorption spectra of the molten electrolyte when oxygen was introduced gave information about the occurrence of a new chromium soluble species. The new absorption band which appears in the spectra was attributed to the Cr(VI) species, and the following reaction mechanism was assumed:

$$2\,Cr^{3+} + 1{,}5\,O_2 \text{ ----> } 2\,Cr^{6+} + 3\,O^{2-}$$

$$2\,Cr^{3+} + 3\,O^{2-} \text{ ----> } Cr_2O_3$$

The interest of coupling the electrochemical and the UV-Visible spectroscopy investigations has been demonstrated. The objective is now to pass on to the study in situ of the electrochemical process occurring near the electrode surface jointly by electrochemistry and UV-Visible spectroscopy.

References

1. Sadoway R.D. (1991) *J. of Metals, July, pp.15-19*
2. Bouteillon J., Gabriel J.C., Poignet J.C., Roman J.M.,(1993) *Proceed. of the International Symposium "Molten Salt Chemistry and Technology,"*The Electrochem. Soc. Inc. Pennington, NJ, pp. 458-471
3. Griffiths T.R., Philipps N.J., (1991), *J. Electrochem. Soc.,* 138, pp. 3575-3580
4. Gruen D.M. (1964) *Fused Salts,* McGraw-Hill, NewYork.
5. Atkins P.W. (1978) *Physical Chemistry,* Oxford University Press, Oxford.
6. Cotton F. A. *Chemical Applications of Group Theory,* Wiley Interscience.
7. Smith G. P. (1978), *Molten Salt Chemistry,* Reidel Publ Co, Dordrecht.
8. Griffiths T. R. (1984), *Molten Salt Techniques,* Plenum Press, New York.
9. Senderoff S, Brenner A. (1954), *J. Electrochem. Soc.,* 101, pp. 16
10. Inman D., Vargas T. (1987), *J. of Applied Electrochem. , 17, pp.* 270-282
11 White S.H., Twardoch U.M. (1987), *J. of Applied Electrochem,* 17, pp. 225-242
12. Scheffler T.B., Hussey C.L., Seddon K.R., Kear C.M. and Armitage P.D., (1983), Inorg. Chem. 22, pp. 2099
13. Benslimane K.(1992), *Thesis,* Université Paris VI, Paris.

HIGH TEMPERATURE ^{93}Nb NMR STUDY OF LIQUID $NbCl_5$ AND ITS MIXTURES WITH VARIOUS ALKALI CHLORIDES

E. SCHELLKES, M. HOLZ AND W. FREYLAND

Institute of Physical Chemistry and Electrochemistry,

University of Karlsruhe, D-76128 Karlsruhe, Germany

Abstract

Liquid $NbCl_5$ and its mixtures with different alkali halides have been investigated for the first time by ^{93}Nb NMR at temperatures up to 900K. In liquid $NbCl_5$ quadrupolar relaxation dominates. Both the chemical shift and the relaxation rate exhibit a characteristic temperature dependence indicative of the $Nb_2Cl_{10} \leftrightarrow NbCl_5$ dimer-monomer equilibrium. In $(CsCl/NaCl)_{eut}$ mixtures a strong increase of the ^{93}Nb relaxation rate is observed on melting. This is not the case in the corresponding LiCl/KCl eutectics where the line width decreases with increasing temperature in the melt. An analogous difference is observed in the resonance shift. Furthermore we report first results of $NbCl_5$ reduction in alkali halide melts either by Nb metal addition or by electrochemical reduction in an *in situ* NMR-electrochemical high temperature experiment.

D.H. Kerridge and E.G. Polyakov (eds.), Refractory Metals in Molten Salts, 43-60.

1. Introduction

Electrowinning and electroplating of refractory metals like niobium and tantalum from molten salt electrolytes is of increasing technological interest. In their first successful experiments in 1965 Senderoff and Mellors have used alkalifluoride melts [1], whereas more recent studies have concentrated on alkalichloride electrolytes, see e.g. [2-5]. In order to optimize the electrochemistry of these processes and to produce coherent coatings a detailed knowledge of the reduction mechanism and complex formation is fundamental. In the study of these problems various electrochemical methods and some spectroscopic techniques, in particular, Raman spectroscopy have been employed so far. NMR measurements on the electrodeposition of refractory metals in molten salts at elevated temperatures to the best of our knowledge have not been reported.

From the NMR point of view, ^{93}Nb is a favourable nucleus due to its high NMR sensitivity although it has a quadrupolar moment of $-0.28 \cdot 10^{-28}$ e $\cdot$ m^2. In this study we report the first ^{93}Nb NMR investigation of liquid $NbCl_5$ and its mixtures with various alkali halides at temperatures up to 900 K. In addition, preliminary results of an *in situ* NMR-electrochemical experiment are briefly described. This experiment represents a novel electrochemical NMR application. The combination of NMR and electrochemistry (DCNMR) has attracted increasing interest in recent years [6,7]. The main objective of this communication is to demonstrate that NMR experiments of these complex melts can be performed with high enough precision to deduce the interesting structural (via chemical shift and relaxation rate) and dynamical (via relaxation rate) information. A first qualitative discussion of the NMR results in comparison with relevant Raman data is given [8-10].

2. Experimental

A new high temperature probehead has been constructed for NMR measurements up to 900 K. The rf-coil made from molybdenum wire of 0.5 mm diameter is fixed on a boron nitride cylinder which is located inside the furnace tube. The sample volume which is detected by the NMR coil is 10 mm in diameter and 15 mm in height. The furnace consists of a molybdenum wire resistance heater noninductively wound on an alumina tubing. A dc power supply has been used to avoid parasitic signals. The whole system is placed inside a water cooled vacuum recipient fitting into the cryomagnet. For recent reviews on high temperature NMR techniques see e.g. [11,12].

Measurements have been performed with a Bruker AM 300 FT-NMR spectrometer with a 7.05 T wide-bore cryomagnet at frequencies of 73.450 MHz for pure $NbCl_5$ and 73.440 to 73.327 MHz for $NbCl_6^-$ samples. The chemical shifts are expressed relative to the signal of $NbCl_6^-$ in CsCl/NaCl(eut) at 425°C (for details see section 3). The premelted salt mixtures have been sealed in quartz ampoules with 10 mm outer diameter and a height of about 20 mm, larger height for the two phase system $NbCl_5$/LiCl/KCl. In the case of reduction experiments cleaned pieces of niobium foil have been added. The NMR-electrochemical cell consists of two glassy carbon electrodes opposite to each other in a distance of about 15 mm connected with molybdenum wires and sealed into a quartz tube.

A special set-up had to be constructed to perform differential thermoanalytic (DTA) experiments on $NbCl_5$-MX-M'X mixtures. Due to the high corrosivity of $NbCl_5$ and its high vapour pressure at elevated temperatures (melting point: 204 °C, boiling point: 254 °C, critical point: 534 °C and 46 atm) graphite containers coated with glassy carbon and sealed by a graphite foil have been used. In order to balance the high vapour pressure of the samples measurements have been performed inside a high pressure vessel with argon as pressurizing medium. Sample quantities of about 100 mg have been measured relative to an alumina reference. A signal resolution of about 0.1 °C has thus been achieved. In both NMR and DTA experiments chemicals of Alfa-Ventron Company with the following purity have been used: $NbCl_5$ of 99.9985 % purity (metals basis) and alkali chloride crystals of highest purity (99.999 %) which have been purified further by recrystallization.

3. Results and Discussion

NIOBIUM PENTACHLORIDE

The ^{93}Nb NMR resonance of pure liquid $NbCl_5$ has been measured as a function of temperature between the melting point and 500 °C. Fig.1 shows the most interesting NMR parameter, the signal half width $\Delta\nu_{1/2} = (\pi T_2)^{-1}$ (upper part) and the corresponding chemical shift (lower part). Included in Fig.1 are typical NMR signals obtained in these experiments demonstrating the high signal-to-noise ratio. Two types of data sets are visible in both figures: open symbols refer to data obtained at a constant cooling rate ($\approx$ 4 K min^{-1}), those with solid symbols have been measured at thermal equilibrium. The reproducibility of measurements is indicated by the different symbols which belong to separate experiments. The relative errors within one run are comparable in magnitude to the size of symbols. The chemical shifts in Fig.1b are expressed relative to the relatively narrow signal of $NbCl_6^-$ in CsCl/NaCl(eut) at 425°C, since all reference systems described in literature like $NbCl_5$ in acetonitril [13] are not suitable here.

The results of Fig. 1 show the following remarkable features. At all temperatures only one resonance line with a width of the order of $\Delta\nu_{1/2} = 10^3$ Hz has been observed, the experimental chemical shift resolution thus being $\Delta\nu/\nu_o \sim 10$ ppm. A strong temperature dependence is observed both for the relaxation rate T_2^{-1} and the chemical shift. The decrease in $\Delta\nu_{1/2}(T)$ with increasing temperature apparently is connected with an increase in the chemical shift δ. The latter behaviour is a well known fact for niobium and vanadium complexes. In the following discussion we focus on the behaviour of the ^{93}Nb line width and thus of the relaxation rate.

In principle, three contributions to the NMR relaxation rate have to be considered: relaxation due to paramagentic impurities, the nuclear dipole-dipole relaxation and quadrupolar relaxation (QF-relaxation). Excluding fast exchange in this system, the first is not of interest here because it would lead to invisible broad lines. To give a theoretical estimate of the relaxation rates we assume that the extreme narrowing condition is valid, namely $\omega_o\tau_c << 1$, where τ_c is the correlation time of the relaxation process. An estimate for the rotational correlation time τ_c is given by [14]

$$\tau_c = 4\pi\, a^3 \eta / 3\, kT \qquad (1)$$

where a = molecular radius, η = viscosity. Taking the highest viscosity for

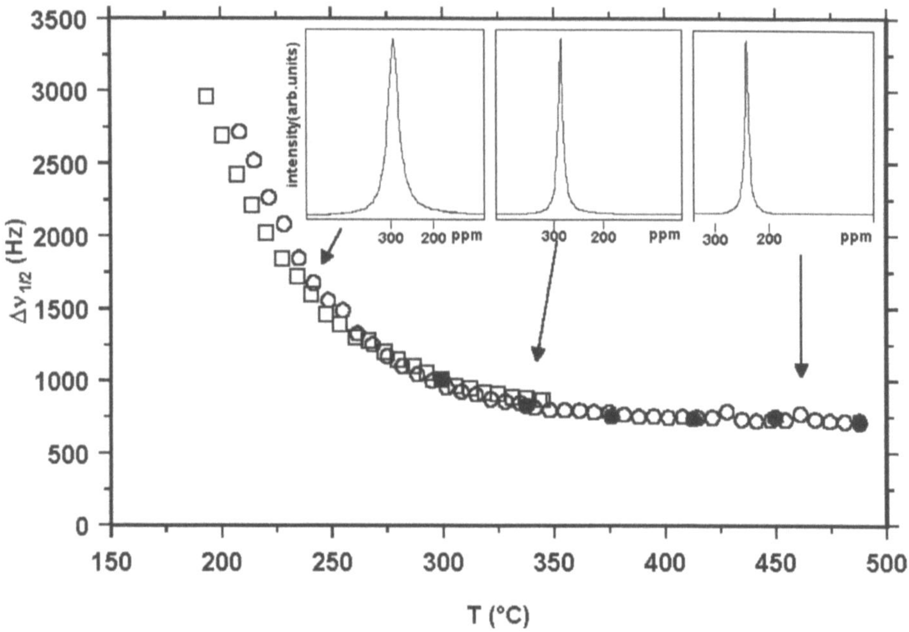

Fig. 1a: Temperature dependence of ^{93}Nb line width in liquid niobium pentachloride. Open symbols are measured at constant cooling rate at different runs, solid symbols at thermal equilibrium. The inset shows typical NMR-signals at three different temperatures.

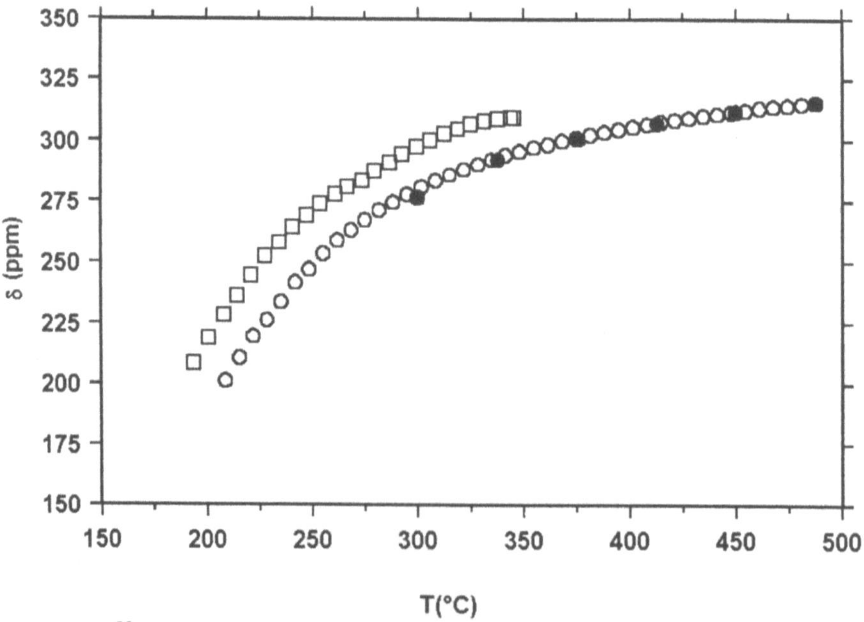

Fig. 1b: ^{93}Nb chemical shifts. All other details correspond to Fig.1a)

$NbCl_5$ [15] near the melting point eq.1 yields $\tau_c \sim 10^{-11}$ s, justifying with $\omega_o \sim 10^8$ Hz the assumption above. The dipole-dipole relaxation rates are orders of magnitude lower than the experimental values of Fig. 1. Therefore we can conclude that quadrupolar relaxation dominates in liquid $NbCl_5$. For the ^{93}Nb nucleus the QF relaxation rate is is given by [16]:

$$\frac{1}{T_2} = \frac{3\pi^2}{10}\frac{2}{27}\left(1 - \frac{\eta_{as}^2}{3}\right)\left(\frac{e^2qQ}{h}\right)^2 \tau_c \qquad (2)$$

where η_{as} = asymmetry parameter, q = eV_{zz} = zz-component of electric field gradient tensor, eQ = nuclear quadrupol moment. Taking the quadrupolar coupling constant from the solid [17] and τ_c above yields relaxation rates in the same order of magnitude as the experimental data, supporting the conclusion that we are dealing here with pure QF relaxation.

In discussing the temperature dependence of the relaxation rates two aspects shall be considered here. According to eq. (1) and (2) $\Delta\nu_{1/2}$ should be proportional to η/T. Both quantities are plotted on a logarithmic scale versus the reciprocal temperature in Fig. 2, whereby the viscosity data of Ref. [15] have been used. Apparently, both quantities exhibit the same temperature dependence up to T = 250 °C. Applying eqs.1 and 2 a quantitative agreement between both data sets requires a quadrupolar coupling constant of (e^2qQ/h) = 45 MHz. For solid $NbCl_5$ (e^2qQ/h) = 78 MHz was measured [17]. An apparent activation energy of E_A= 30 $kJmol^{-1}$ is determined close to the melting point. It is larger than that of a common liquid like H_2O (21 $kJmol^{-1}$), although its viscosity is only 10 % higher than that of water. At higher temperatures the behaviour of $\Delta\nu_{1/2}$ clearly deviates from that of the viscosity. It approaches a weak temperature dependence above 350 °C, where predomi-

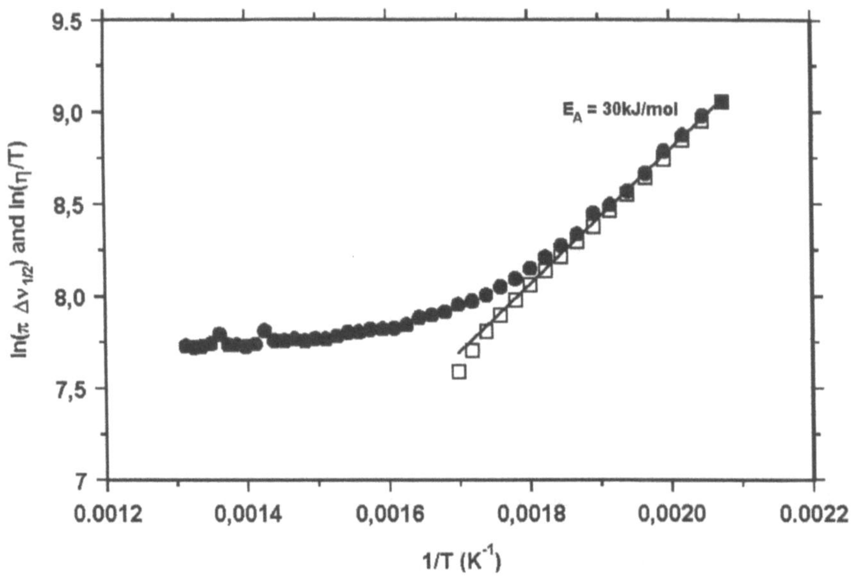

Fig. 2: Logarithmic plot of relaxation rate, $\pi\Delta\nu_{1/2}$, and viscosity-temperature ratio, η/T, versus reciprocal temperature of liquid $NbCl_5$. Solid symbols are data of T_2^{-1} of pure $NbCl_5$ at constant cooling rate, open sysmbols η / T.

nantly monomeric species prevail as determined by Raman spectroscopy [9,10,18]. The second aspect emphasizes the monomer-dimer equilibrium. Solid $NbCl_5$ forms Nb_2Cl_{10} dimeric units [19]. The dimers dissociate in the liquid with increasing temperature but are still present up to 350°C [9,18]. If monomers and dimers have different relaxation rates, $T_2^{-1}(m)$ and $T_2^{-1}(d)$, respectively, the total relaxation rate may be approximated as:

$$\frac{1}{T_2} \approx \frac{1-\alpha}{T_2(d)} + \frac{2\alpha}{T_2(m)} \tag{3}$$

Where α = degree of dissociation. The temperature dependence of α is determined by the equilibrium constant, yet has not a simple exponential T-dependence. In any case both contributions have to be taken into account for a quantitative interpretation of the temperature dependence of the measured relaxation rates. The striking similarity in the temperature behaviour of relaxation rates and chemical shifts suggests that the latter are also influenced by the dissociation equilibrium. Between the limits of dimers near the melting point and monomers at the highest temperature measured a clear shift to higher frequencies is observed. However, the absolute values of the chemical shift cannot be given due to the lack of bulk susceptibility data.

SOLID AND LIQUID $NbCl_5$-MCl-M'Cl-MIXTURES

For the investigation of $NbCl_5$ in alkali halide melts low melting eutectic mixtures have been selected due to the corrosivity of $NbCl_5$ at temperatures above 600 °C. Mixtures of interest are CsCl/NaCl (T_{eut}=490 °C) and LiCl/KCl (T_{eut} = 375 °C). In these two systems studies of the electrochemical reduction mechanism of Ta(V) and Nb(V) have recently been performed in our group [5,20]. The phase diagrams of these ternary systems are not reported in literature, only the binary mixtures have been measured up to now [21,22,23]. Fig.3 shows the phase diagram of the pseudo binary system $NbCl_5$-CsCl/NaCl (eut) up to a concentration of x_{NbCl_5} =0.45 as measured with DTA. These measurements have been limited to temperatures below 600 °C. A characteristic of this system is the solid-solid transition at 295 °C. High and low temperature solid state modifications are also reported for the binary systems $NbCl_5$ in CsCl, NaCl and KCl. $NaNbCl_6$ and $KNbCl_6$ reveal a structure consisting of dimers in the low temperature modification comparable to $NbCl_5$. $CsNbCl_6$ is a stable, congruent melting compound and consists already in the low temperature modification of $NbCl_6^-$ units.

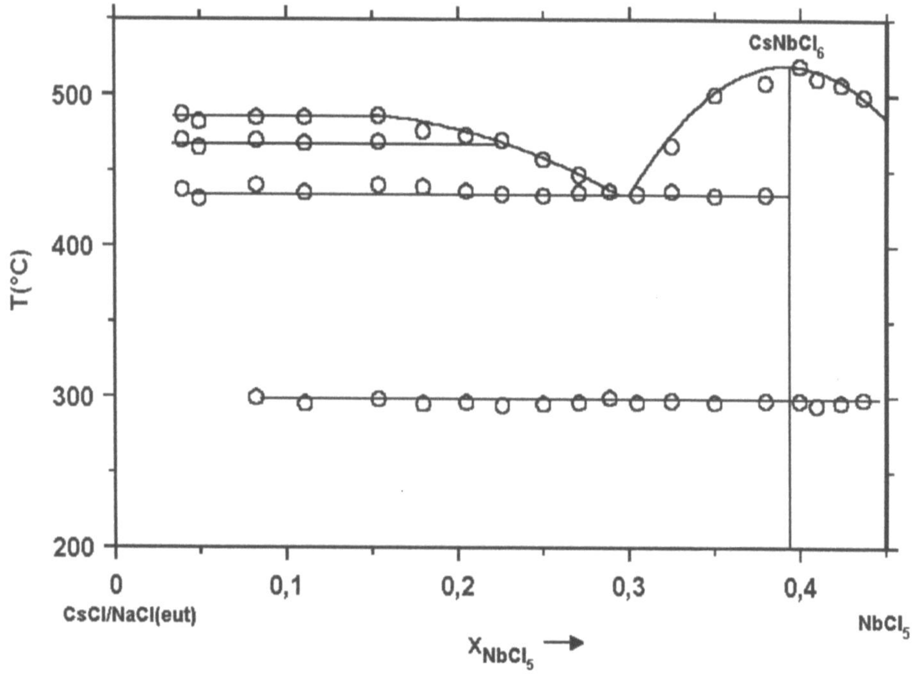

Fig. 3: Detail of the phase diagram of the pseudo binary mixture of $CsCl/NaCl$(eut) and $NbCl_5$ as measured by DTA.

The high temperature structure of all compounds has not been analysed up to now, but for all binary systems $NbCl_6^-$ the unit seems to be the dominating element [24]. They are characterized by a fast ionic conduction with $\sigma = 0.4$ $\Omega^{-1}cm^{-1}$ for $NaNbCl_6$ and an elevated conductivity between $\sigma = 0.0012$ $\Omega^{-1}cm^{-1}$ and 0.006 $\Omega^{-1}cm^{-1}$ for $KNbCl_6$ [21]. At 435 °C the $NbCl_5$-$CsCl/NaCl$(eut) system forms a liquid and is completely molten above 485 °C. The triple point around $x_{NbCl_5} = 0.23$ is not yet clear. At a concentration of $x_{NbCl_5} = 0.395$ the n_{Nb}/n_{Cs} ratio is exactly one and the system forms the congruently melting compound $CsNbCl_6$ which reflects the stability of this complex in the melt.

Fig. 4 shows the temperature dependence of the ^{93}Nb line width of various Cs^{+}-containing $NbCl_5$ - salt mixtures. All data points were taken in thermal equilibrium expect the values for 48% $NbCl_5$ in CsCl which have been measured at constant cooling rate. Error bars representing the reproducibility of several measurements at constant temperature are indicated. All systems including that with oxide addition revealed a single NMR resonance line. Referring to the crystallographic analyses of the solid [24] and to Raman spectra of the melts [25], the Nb^{5+}-ions in these systems form $NbCl_6^-$ complexes. So we expect that the ^{93}Nb line is due to that complex. For the sample with addition of Na_2O a mixture of $NbCl_6^-$ and $NbOCl_5^{2-}$ ions must be considered. Narrow NMR resonance lines are expected for complexes with high symmetry because of the strong sensitivity of the quadrupolar relaxation rate to the electric field gradient at the nucleus (eq.2). A distortion of the symmetry leads to a strong broadening of the line. Due to this apparently the asymmetric $NbOCl_5^{2-}$ ion yields invisible broad lines. We first focus on the high temperature modification of the systems in the range between 280°C and 435°C. Surprisingly all systems reveal narrow signals of remarkable quality which show up the same weak temperature dependence of the line width within experimental errors. Only the system of 48 % $NbCl_5$ in CsCl which consists almost completely of $CsNbCl_6$ differs slightly from the other systems. This compound exhibits in the solid state half widths which are significantly smaller than the line widths found for liquid $NbCl_5$. This NMR signal is a clear indication for a solid state structure of high symmetry with respect to the environment of the Nb nucleus.

The line width of the systems with a concentration of $NbCl_5$ $x < 0.20$ increases strongly on melting at 435°C. At 500 °C the signal becomes too broad to be evaluated and disappears at higher temperatures. With respect to the dominating quadrupolar interactions this huge increase of the relaxation rate must be caused by a significant change of the electric field gradient at the nucleus (eq.2). Since this property is depending on the symmetry of the charges in the environment of the nucleus under observation the geometry must be distorted presumably due to Cs^+-ions still strongly coordinated to the $NbCl_6^-$ complexes in the melt. The line width of the system with 48%$NbCl_5$ exhibits only a slight change since merely a very small amount of the substance is liquid between 435°C and 500°C.

These conclusions drawn from the relaxation data are supported by corresponding changes in the chemical shifts. In the solid negative shifts of $\delta \approx -20$ ppm relative to the reference frequency have been observed. They are of comparable magnitude for all mixtures studied and exhibit a weak temperature dependence. On melting a clear shift to lower frequencies of $\delta \approx -100$ ppm occurs. A significantly different behaviour is observed in the NMR parameters of $NbCl_5$ in LiCl/KCl-eutectics in comparison to the systems presented above. The phase diagram of this system is unknown yet. Apparently, both in the liquid and the solid phase demixing occurs. While the lower density phase is diluted in $NbCl_6^-$-complexes and rich in Li^+-ions, the higher density phase is concentrated in $KNbCl_6$. In the following we focus on

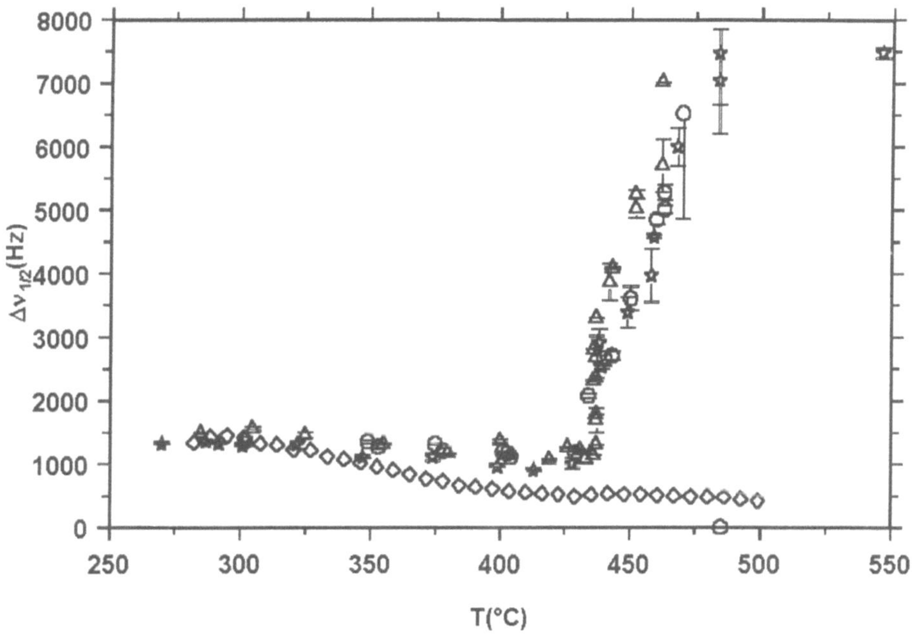

Fig. 4: Half-width of the ^{93}Nb line in solid and liquid mixtures of $NbCl_5$ in various CsCl-containing salts. Circles: 19.1% $NbCl_5$ in CsCl/NaCl (eut) +Na_2O (O/Nb=0.45), triangles: 8.2% $NbCl_5$ in CsCl/NaCl (eut), stars: 8.1%$NbCl_5$ in CsCl, all measured at thermoequilibrium; diamonds: 48%$NbCl_5$ in CsCl measured at constant cooling rate

the higher density phase. Fig. 5 shows the temperature dependence of the line widths of solid and liquid systems in comparison to measurements with added oxide impurities. Shown are results from different temperature runs representing the reproducibility of measurements. A strong temperature dependence of $\Delta\nu_{1/2}$ is observed in the solid, however, no drastic change occurs on melting and the line width is further reduced on heating the melt reaching a value of $\Delta\nu_{1/2} \approx 10^3$ Hz at 600 °C. This behaviour is in contrast to that of the CsCl containing systems described above. Obviously in these melts the $NbCl_6^-$-complex is not strongly distorted by the alkali cations which relates to the instability of the $KNbCl_6$ compounds. As seen in Fig. 5 the

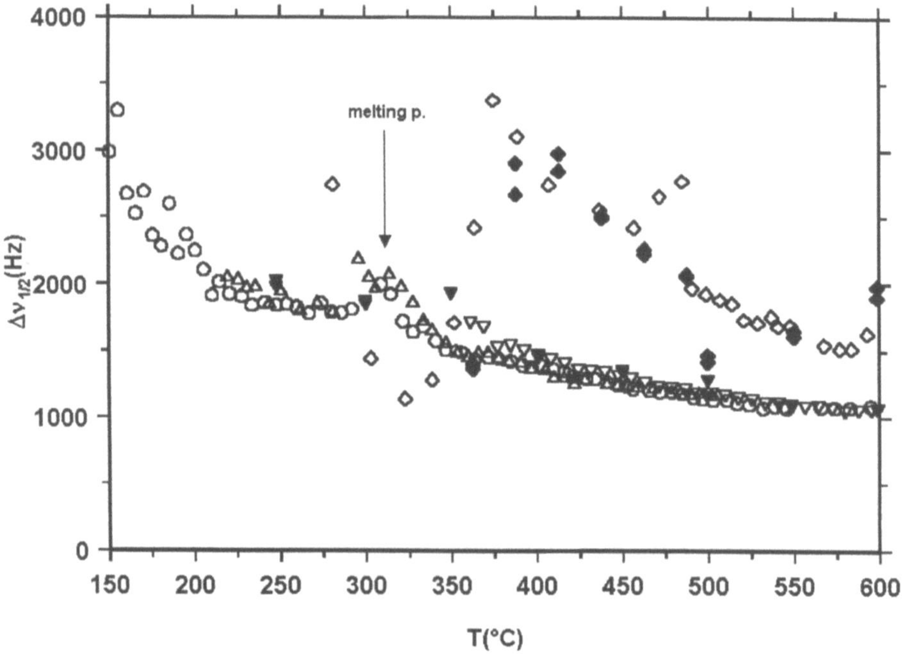

Fig. 5 same as in Fig. 4, but $NbCl_5$-LiCl/KCl(eut) mixtures. Circles and open triangles: various runs of 7%$NbCl_5$ in LiCl/KCl(eut) at constant cooling rate, solid triangles: same at thermal equilibrium, diamonds: 5%$NbCl_5$ in LiCl/KCl(eut)+ Na_2O (O/Nb=0.38), open: measured at constant cooling rate, solid:at thermal equilibrium.

influence of oxide addition is twofold: in the solid the signal is broadened to an extent to become invisible, in the melt the relaxation rates are enhanced but are clearly decreasing with increasing temperature. Without going into details of the characteristics of the corresponding chemical shift data it is interesting to note that the chemical shift shows a monotonous slight increase over the whole temperature range, but no discontinuity on melting. A shift of -75 ppm exists at low temperatures with a temperature dependence towards higher frequencies with increasing T. The shift of the oxide containing melts has a minimum value of about -150 ppm and its temperature coefficient parallels that of the pure $NbCl_5$ LiCl/KCl - eutectic melts.

Nb(V) REDUCTION

Two kind of experiments have been performed to study Nb(V)-reduction *in situ* by NMR: niobium metal addition and electrochemical reduction. Adding a niobium foil to a CsCl/NaCl - or LiCl/KCl - eutectic melt with low concentration of $NbCl_5$, two resonances are observed in general, the first with the NMR parameters as shown in Fig. 4 and Fig. 5, and a second line with lower intensity separated from the first. As a function of time the intensity of the first signal continuously decreases. It is improbable that the second line is due to Nb(IV) since Nb(IV) is paramagnetic leading to a high broadening of the resonance line and making it undetectable. An assignment of the second signal remains an open question. Reducing the temperature to the solid region the remaining Nb(V) yields a signal of weaker intensity but with the same parameters as the corresponding Nb free solid. Thus the presence of Nb(IV) does not influence the Nb(V) resonance in the high temperature solid. An electrochemical reduction of 17,5 % $NbCl_5$ in CsCl/NaCl-eutectic containing Na_2O(O/Nb= 0.5) has been investigated in the electrochemical cell at a constant temperature in the molten state by applying constant potentials. It has been observed that the integrated intensity of the ^{93}Nb line decreases in proportion to the accumulated charge, whereby the half width and chemical shift remain constant. As can be seen in Fig. 6 the observed reduction of the integral NMR intensity of the Nb(V) signal is consistent with a reduction of $(Nb^{5+}Cl_6)^-$ to $(Nb^{4+}Cl_6)^{2-}$. This example demonstrates that the NMR probe can be used to study *in situ* the reduction process quantitatively. Further work in this direction is in progress.

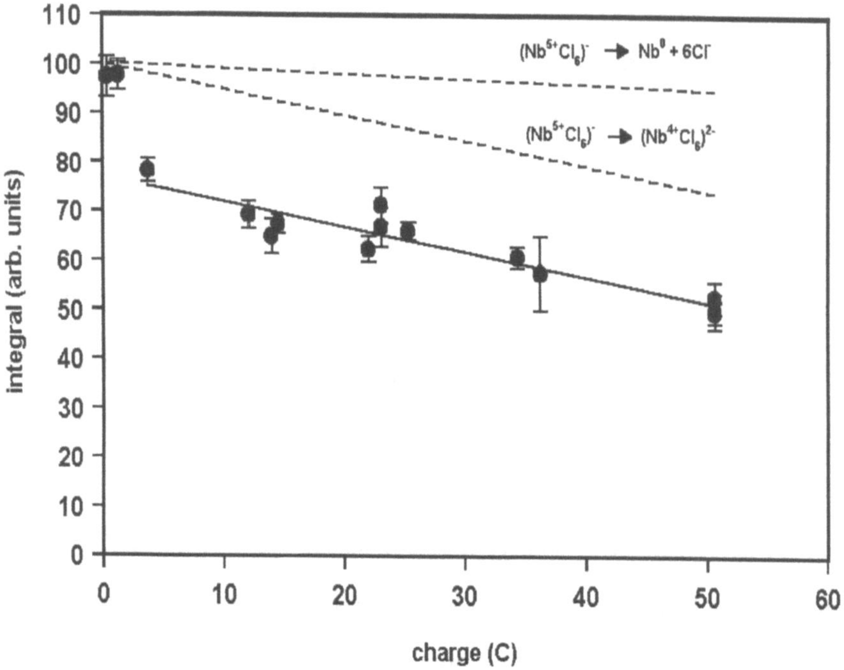

Fig. 6: Integral of the ^{93}Nb line versus accumulated charge as obtained in the NMR-electrochemical experiment at a constant temperature of T= 474°C. Dashed lines: calculated values for different reduction processes.

ACKNOWLEDGEMENT

We thank J. Dubielzig for his help in the DTA experiments and Dr. R. Haselmeier for the help in the NMR experiment. Discussion with Th. Koslowski on the field gradient calculation is appreciated. Financial support of this work by DFG through SFB 195 and partly by an EC-network No CHRX-CT93-0142 is gratefully acknowledged.

4. References

1. Mellors, G.W. and Senderoff, S., *J. Electrochem. Soc.* **112** (1965) 266; ibid. **114** (1967) 556
2. Kuznetsov, S.A., Morachevskii, A.G. and P.T. Stangrit, A.G., *Soviet Electrochem.* **18** (1982) 1357
3. Polyakova, L.P., Kosilo,B.I., Polyakov, E.G. and Smirnov, A.B., *Soviet Electrochem.* **24** (1988) 825
4. Lantelme, F., Barkoun, A. and Chevalet, J., *J. Electrochem. Soc.* **140** (1993) 324
5. Bachtler, M., Freyland, W., Voyiatzis, G.A. and Papatheodorou,G.N., *Ber.Bunsenges. Phys. Chem.* **99** (1995) 21
6. Holz, M., *Chem. Soc.. Reviews,* **23** (1994) 165
7. Wu, J., Day, J.B., Franaszczuk, K., Montez, B., Oldfield, E. ,Wieckowski, A., Vuissoz, P.-A. and Ansermet, J.-P., *J. Chem. Soc., Faraday Trans.,* **93**(6) (1997) 1017
8. Bues, W., Demiray, F. and Oye,H., *Z. Phys. Chem.*, Neue Folge, **84** (1973), 18
9. Krebs,B., Janssen,H., Bjerrum, N.J., Berg, R.W. and Papatheodorou, G.N., *Inorg. Chem.* **23** (1984) 164
10. Rosenkilde, Ch., Voyiatzis, G.A., Jensen, V.R., Ystems, M. and Østvold, T., *Inorg. Chemistry*, **34** (1995), 4360
11. Warren,W.W., Jr., in *Molten Salt Techniques*, Vol. 4; Lovening, D.G., Gale, R.J. Eds. (Plenum Press, New York 1992)
12. Stebbins, J.F., *Chem. Rev.* **91** (1991), 1353

13. Rehder, D., in *Multinuclear NMR*, Mason, J. Ed. (Plenum Press, New York 1987)

14. Carrington, A. and McLachlan, A.D., *Introduction to magnetic resonance* (Harper & Row, New York, 1967)

15.Nisel'son, L.A. and Pustil'nik, A.I., *Izv. Akad. Nauk* 1963, 110 (C.A. 59(1963) 5792)

16.Abragam, A., *The principles of nuclear magnetism* (Clarendon Press, Oxford, 1961)

17.Reddoch, A.H., *J. Chem. Phys.* **35** (1961), 1085

18. Schellkes, E. et al., to be published

19. Zalkin, A. and Sands, D.E., *Acta Cryst.* **11** (1958) 615

20. Stöhr, U., Bandi,P.R., Matthiesen,F. and Freyland, W., *Electrochem. Acta* (1997) in press

21. Palkin, A.P. and Chikanov, D.N., *Russ. J. Inorg. Chem.* **7** (1962) 705

22. Huber, K., Jest, E., Neuenschwander,E., Studer,M., Roth, B., *Helv. Chem.Acta* **41** (1958) 2411

23. Schellkes, E., *thesis*, University Karlsruhe 1998

24. Gaebell,H.-Chr., Meyer, G. and Hoppe,R., *Z. Anorg. Allg. Chem.* **493** (1982) 65

25. Voyiatzis, G.A., Pavlatou, E.A., Papatheodorou, G.N., Bachtler, M. and Freyland, W., *Proc. Int. Symp. Molten Salt Chem. Techn. Electrochem. Soc.* **93-9** (1993) 252

ELECTROCHEMISTRY OF REFRACTORY METALS : Hf, Mo, Cr

ADHOUM, N., ARURAULT, L., BOUTEILLON, J., COTARTA, A., GABRIEL, J. C. and POIGNET, J. C.

Laboratoire d'Electrochimie et de Physicochimie des Matériaux et des Interfaces, UMR 5631 INPG-CNRS, associé à l'UJF, ENSEEG, Domaine Universitaire, BP 75, 38402 .Saint Martin d'Hères France

In this paper, results of studies performed in our laboratory, concerning the electrochemistry of hafnium, molybdenum and chromium in molten alkali chlorides with additions of fluorides will be outlined. The electrochemistry of niobium will not be discussed, because niobium electrowinning is extensively dealt with in this conference.

1. Electrochemistry of hafnium

1.1. INTRODUCTION

A brief survey of the literature shows that multistep reduction of Hf(IV) ions such as Hf(IV) → Hf(III) →Hf(II) → Hf, or Hf(IV) → Hf(II) → Hf have been reported. With several other authors, we have only observed a one step reaction in our experimental conditions.

Solutions of $HfCl_4$ in molten NaCl-KCl were prepared and studied at temperatures from 750°C to 900°C. These solutions were obtained by bubbling gaseous $HfCl_4$ in the melt, or by adding directly the powdered salt. In the latter case, pre-electrolysis was performed, and resulted in the cathodic deposition of the main impurity, i.e. iron. The same result was obtained by immersing a hafnium rod in the melt.

Linear sweep voltammetry was used, with the convolution technique. A silver chloride-silver reference electrode was employed. The working electrode was a tungsten wire (1 mm in diameter). The counter electrode was either a carbon rod or the graphite crucible.

Figure 1 shows a series of voltammograms with variation of the cathodic limit. The single reduction step (A) is associated to the reoxidation peak A'. The cathodic limit is due to Na^+ ions reduction, which on the oxidation side result in the re-oxidation of a Hf-Na amalgam (peak B'). We can now step on further with the study of the kinetics of the reaction Hf / $HfCl_6^{2-}$.

Figure 2 shows voltammograms with varying sweep rate. For sweep rates inferior to 2 Vs^{-1}, the peak current is proportional to the square root of the sweep rate, but it is no longer the case at higher sweep rates.

Figure 3 shows the corresponding semi-integrals of the current, the hysteresis of which proves that the reduction step is not fast. But the cathodic limiting plateau proves that mass transfer is limited by diffusion. From the limiting value of the semi-integral, the value of the diffusion coefficient of $HfCl_6^{2-}$.could be estimated as 2.4 10^{-5} cm^2 s^{-1}. The

D.H. Kerridge and E.G. Polyakov (eds.), Refractory Metals in Molten Salts, 61-72.

cathodic process was furher analysed assuming quasireversible kinetics, and the results in terms of standard kinetic constant was : $k° = 4\ 10^{-3} \pm 2\ 10^{-3}$ cm s^{-1}.

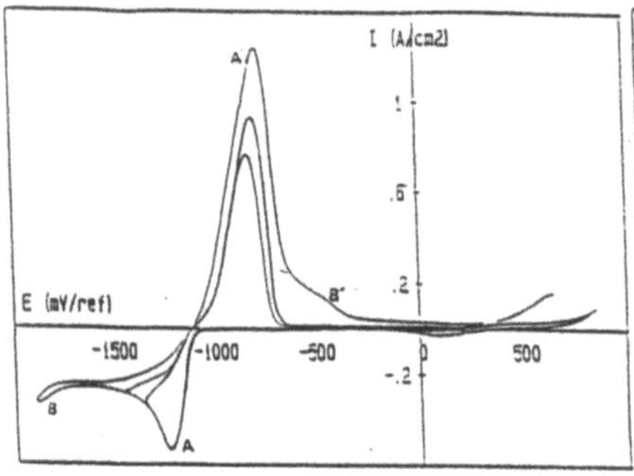

Figure 1 HfCl4 added to the electrolyte
T = 700°C, C = 10^{-2} molL-1
Working electrode : W; scan rate 0.5 Vs^{-1}

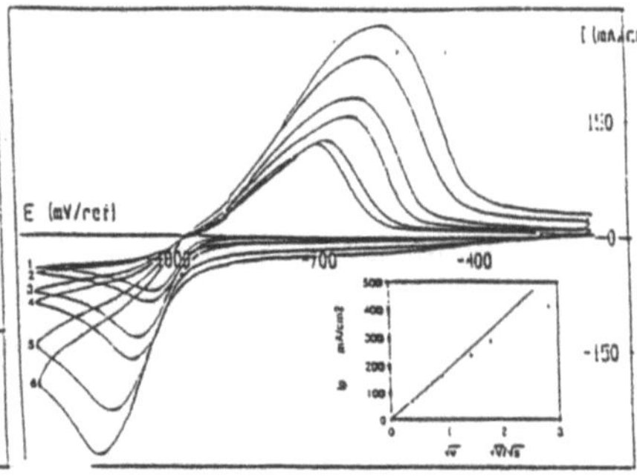

Figure 2 A series of voltammograms for various scan rates : 1, 0.125; 2, 0.2; 3, 0.5; 4, 0.8 5, 2; 6, 3.2 Vs^{-1}. and corresponding curve ip vs $v^{1/2}$

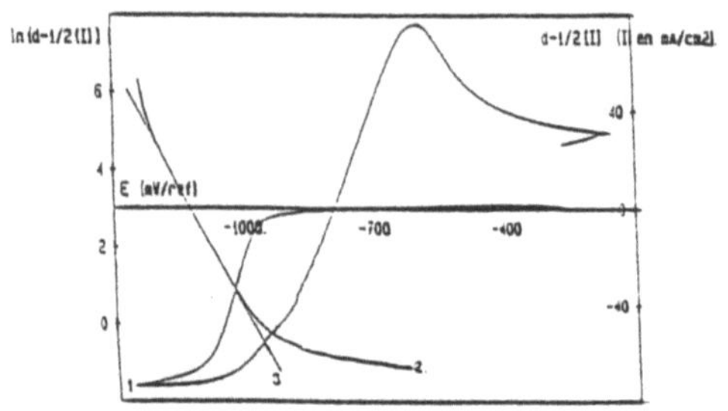

Figure 3 1, semi-integral curve related to the voltammogram of fig 4 at 0.5 V s^{-1}; 2, logarithmic analysis related to the voltammogram of fig. 4 at 3.2 Vs-1; 3, its linear interpolation.

1.2. INFLUENCE OF TEMPERATURE ON THE RAPIDITY OF THE SYSTEM

From 700 °C to 900°C, a single reduction step was observed, with a limiting plateau observed on the semi-integrals. At 800°C, the process was still found quasireversible, with $\alpha = 0.42$ and $k° = 10^{-2} \pm 2\ 10^{-3}$ cm s^{-1}. At 900°C, the process became reversible, as shown on the semi-integral of *figure 4*.

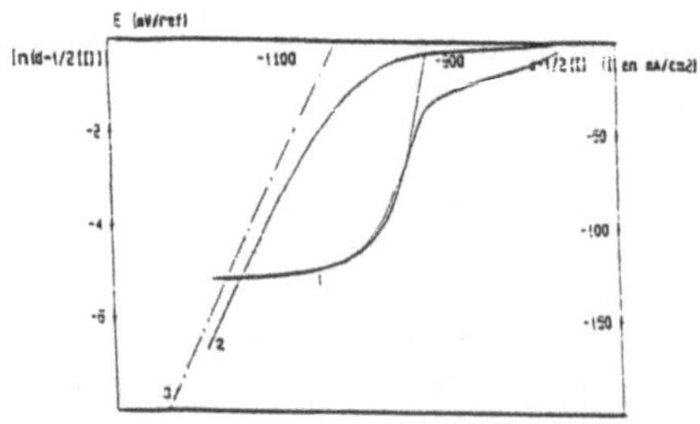

Figure 4 1, semi-integral of a voltammogram plotted at 900°C and v = 0.5 Vs^{-1}; 2, logarithmic analysis of the semi-integral; 3, theoretical slope for a 4-electron exchange.

The nernstian analysis of this semiintegral confirms that the number of electrons exchanged is 4. The peak intensity is proportional to the square root of the sweep rate, which allows the determination of the diffusion coefficient of the $HfCl_6^{2-}$ species : $D = 6.8\ 10^{-5}\ cm^2s^{-1}$.

1.3. INFLUENCE OF THE ADDITION OF FLUORIDE IONS ON THE PROCESS

Various quantities of sodium fluoride were added to the melt, and we studied the effect on the voltammograms of Rf, i.e. the ratio of the concentration of added fluorides to that of hafnium in solution in the electrolyte.

Figure 5 shows voltammograms at 700°C for Rf = 0, 2.5, 4.5, and 6.5. In the absence of fluoride ions, only one reduction step is visible, but after F^- addition a second reduction peak appears at lower potentials. This peak intensity grows at the expense of the first one as more F^- ions are added. For Rf values over 6, the first peak has totally disappeared and only one peak remains at -1.3 V/ref. The explanation fore these observations is the existence of an equilibrium

$HfCl_6^{2-} + 6\ F^- \Leftrightarrow HfF_6^{2-} + 6\ Cl^-$

On the reoxidation side, the coexistence of various complexes leads to two reoxidation peaks. for values of Rf between 0 and 6. These peaks are ascribed to the reactions

$Hf + 6\ Cl^- \Leftrightarrow HfCl_6^{2-} + 4\ e^-$

$Hf + 6\ F^- \Leftrightarrow HfF_6^- + 4\ e^-$

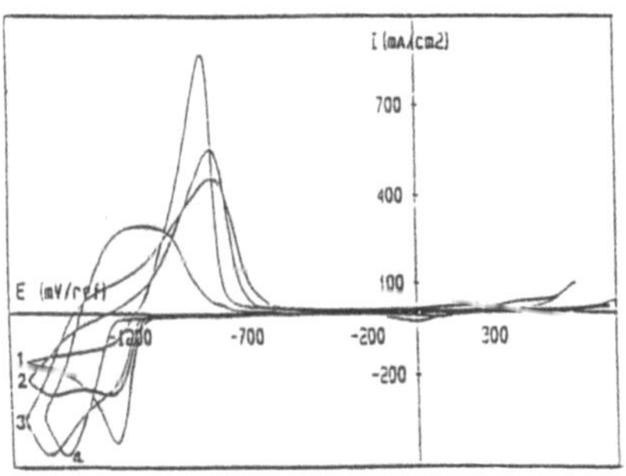

Figure 5 Effect of fluoride additions on voltammograms plotted at 700°C.
1, $R_F = 0$; 2, $R_F = 2.5$; 3, $R_F = 4.5$; 4, $R_F = 6.5$

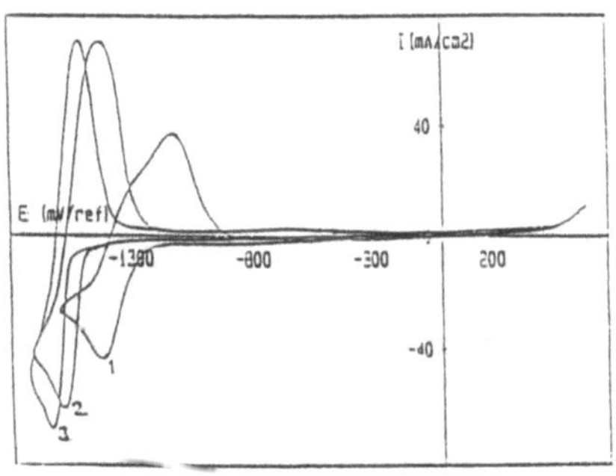

Figure 6 Voltammograms plotted for different R_F values at 700°C and 0.5 Vs^{-1}; 1, $R_F = 7$; 2, $R_F = 12$; 3, $R_F = 20$

1.4. KINETICS OF THE HfF_6^- / Hf EXCHANGE

Figure 6 shows that at 700°C a single reduction step for which mass transfer is limited by diffusion (see semi-integral curve on *figure 7*). Since the peak intensity is proportional to the square root of the sweep rate, the reaction can be either reversible or irreversible, but by no means quasireversible. A complete study shows that this reaction can be considered as reversible.

When the temperature is raised, the reaction becomes more clearly reversible, as shown on *figure 8*, which presents semiintegral curves for Rf = 7, at the temperatures of 700, 800 and 900 °C.

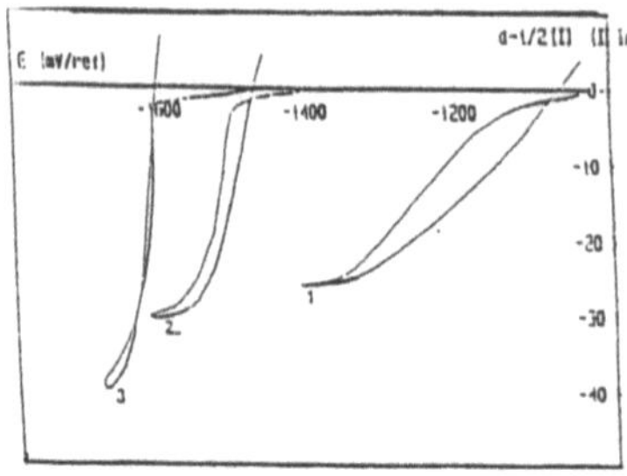

Figure 7 Semi-integral curves related to the voltammograms of fig 8
1, $R_F = 2$; 2, $R_F = 3$; 3, $R_F = 20$

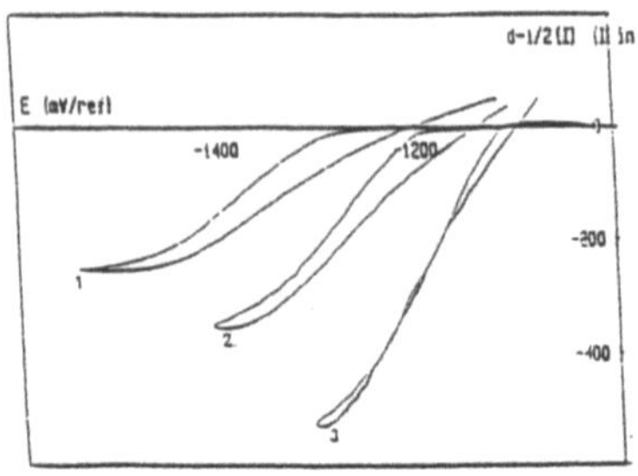

Figure 8 Semi-i tegral curve plotted for $R_F = 7$
1, T = 700°C; 2, T = 800°C; 3, T = 900°C

The limiting plateau at 900°C is not very clear, due to the fact that at the higher temperatures sodium reduction interferes severely. But it is possible to take this effect into account by subtracting the semi-integral curve due to sodium reduction only. Classical nernstian treatment of the semi-integral curves confirm that the number of electrons exchanged is 4.

1.5. QUALITY OF THE DEPOSITS OBTAINED

Hafnium deposits have been prepared on W or Ni substrates, either potentiostatically or intensiostatically, at 700°C and 750°C. Pure metal was obtained, but the coherence of the deposited metal decreased as Rf increased. Increasing the fluoride content of the melt resulted in finer metallic grains. At the higher values of Rf, metal loss was observed due to poor adherence on the substrate.
To sum up, hafnium was electrodeposited from molten chloride and chlorofluoride solutions of Hf(IV) in alkali chlorides between 700 and 900°C. Hf(IV) ions reduction occurred via a single step reaction : $HfCl_6^{2-}$ + 4e --> Hf + 6 Cl^-. This reaction remained slow up to 800 °C, and became reversible at 900 °C. Addition of fluorides ions to the melts caused the reversibility of the reaction to increase.

2. Electrochemistry of molybdenum

2.1. INTRODUCTION

The aim of this work was to deposit molybdenum on a steel substrate at temperatures sufficiently low (< 650°C) in order to avoid structural modifications of the substrate. This is why the LiCl-KCl eutectic was selected as the solvent. A literature survey shows that molybdenum deposits have already been obtained from molten chlorides, fluorides, oxides, and mixed fluoride-oxide media, and that many questions concerning the chemistry of the molybdenum solutions and the reduction mechanism remain unanswered. In this paper, we will talk about the preparation of the molybdenum salt used as a solute, then describe the electrochemical kinetic investigation performed, and finally briefly outline a practical application of the knowledge gained during this work.

The most suitable salt for obtaining Mo deposits in molten chloride media is K_3MoCl_6, which does not exist as a common commercial product and has therefore to be prepared. Its synthesis was based on the following scheme :

$MoO_4^{2-} + 8\ H^+ + 3\ e^- \Rightarrow Mo^{3+} + 4\ H_2O$

$Mo^{3+} + 6\ HCl_{gas} \Rightarrow MoCl_6^{3-} + 6\ H^+$

The K_3MoCl_6 precipitate obtained was then dehydrated under vacuum at 200°C for half an hour, analysed, then stored in a glove box.

The molten electrolyte was prepared by mixing LiCl (Strem Chemicals) and KCl (Prolabo Normapur), heating them at 300°C under vacuum during 3 hours and then melting the mixture under purified argon (containing less than 1 ppm oxygen or water). In fact, we have verified that, whatever the dehydration method used, the pO^{2-} content of the melt is controlled by the argon purity. in our case, pO^{2-} was about 6. The working electrode was vitreous carbon or gold. The counter-electrode was made of molybdenum and the reference electrode was a silver chloride/silver electrode. The K_3MoCl_6 powder was added to the melt via a lock chamber. The Mo(III) concentration in the molten electrolyte was determined at fixed intervals by sampling the solutions and chemical analysis.

The anodic behaviour of Mo was first studied.

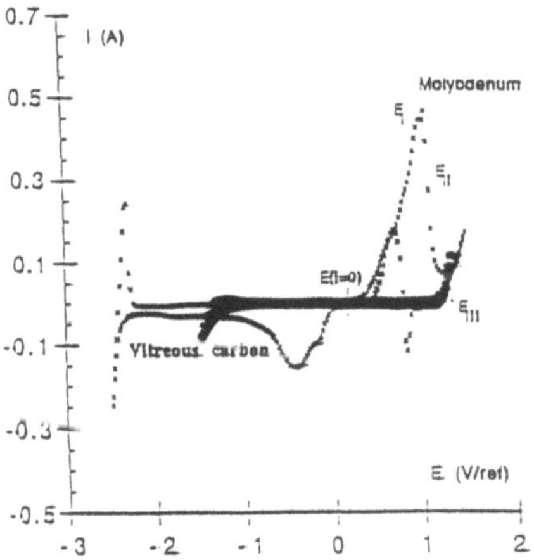

Figure 9 Cyclic voltammograms of pure LiCl-KCl at 500°C; anodic start from $E_{i=0}$; v= 200mV s-1; working electrode :– vitreous carbon; ..... Mo

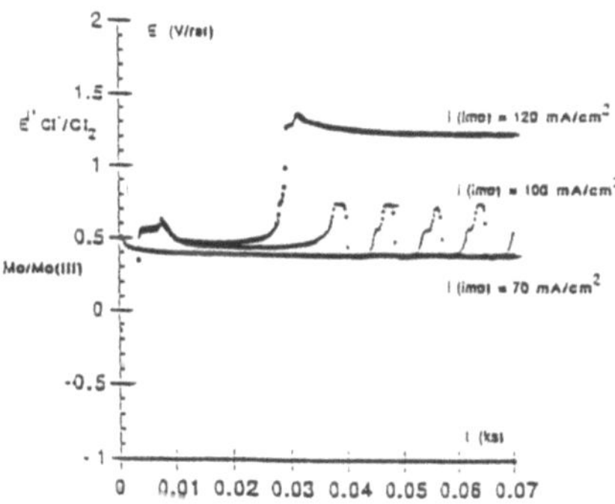

Figure 10 Chronopotentiograms related to the oxidation of a Mo electrode in Li-Cl-KCl at 500°C

Figure 9 shows a typical voltammogram, starting anodic, and showing a classical passivation curve. The anodic dissolution of the Mo electrode was further studied by chronopotentiometry at various current densities.

Figure 10 shows that when the current density applied is below 100 $mAcm^{-2}$, a stationary dissolution state is reached, for which analysis showed that the reaction is

$Mo \rightarrow Mo\ (III) + 3\ e^-$.

When the applied current density reaches the critical value of 100 $mAcm^{-2}$, the potential of the electrode swings between two stationary state values corresponding to a passivated and a non passivated state. When the applied current density is over this critical value, the electrode gets passivated because solubilisation of K_3MoCl_6 is slow. A value of 1.5 mol per litre , in agreement with estimations by Brenner, has been calculated from the chronopotentiograms.

2.2 ELECTROCHEMICAL PROPERTIES OF K_3MoCl_6 SOLUTIONS IN LiCl-KCl AT 500 °C.

Solutions were obtained either by anodic dissolution of Mo, or by adding K_3MoCl_6 powder to the LiCl-KCl melt.

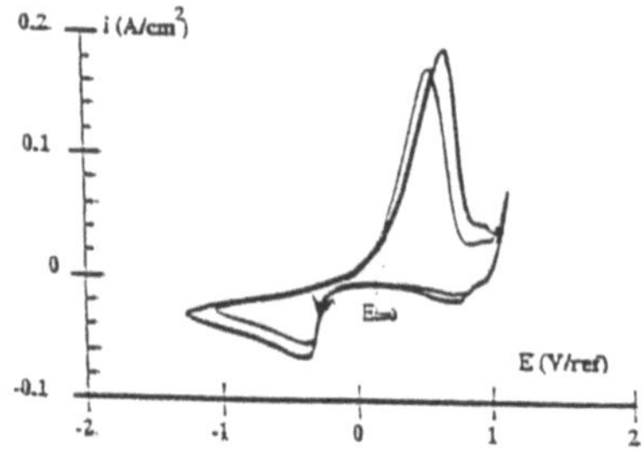

Figure 11 Typical voltammograms plotted at a vitreous carbon electrode in Mo ions solutions obtained by :
– anodic dissolution of a Mo rod; – dissolution of K_3MoCl_6; Mo(III) concentration : 0.06 M; v = 500 mV s^{-1}

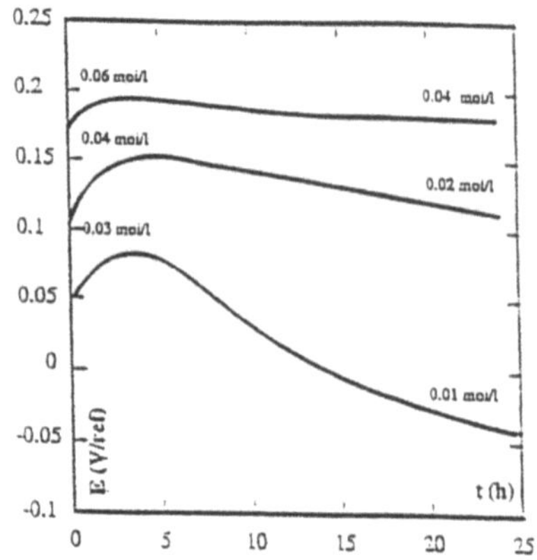

Figure 12 Evolution of the open circuit potential of a passivated electrode for various Mo(III) concentrations

Figure 11 shows that the voltammograms obtained in these differently prepared solutions are very similar. Consequently, the stable species in solution is likely to be Mo(III) in both cases. But, as has been pointed out by several authors, Mo(III) solutions decompose rapidly, and we have shown that this decomposition can be written as :

$$5\ MoCl_6^{3-} \Rightarrow 3\ MoCl_5 + 2\ Mo + 15\ Cl^-$$

When the Mo concentration was over 5 10^{-3} mol l^{-1}, decomposition was less pronounced. The existence of various complex Mo(III) ions has been put forward by authors, but at the present time new spectroscopic determinations are necessary in order to identify the species present in the solutions.

Figure 12 illustrates the slow dissolution of K_3MoCl_6 and the decomposition reaction, which results in a slow decrease of the Mo(III) concentration.

2.3. STUDY OF THE REDUCTION OF Mo(III) SOLUTIONS

Polarisation curves at gold electrodes showed that crystallisation problems were significant. Chronamperometric determinations, see *figure 13*, showed that these crystallisation phenomena involved nstantaneous nucleation ($Im^2\,tm = 0.163\ (nFC)\ 2D$).

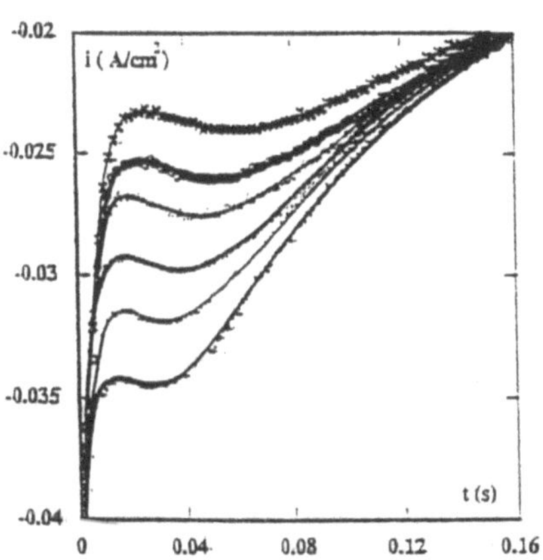

Figure 13 Study of Mo crystallization by chronoamperometry at gold electrodes; applied potentials : -300, -320, -340, -360, -380, -400 mV/ref

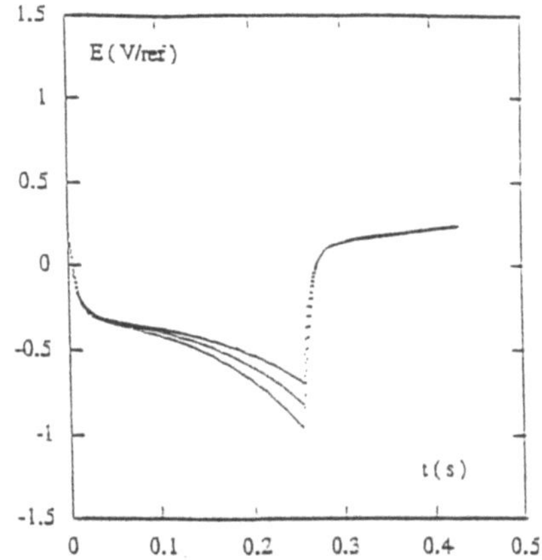

Figure 14 Typical reversed current chronopotentiograms for the reduction of K_3MoCl_6; i = 90, 100 and 110 mA cm^{-2}

We therefore covered the surface of the electrodes with a molybdenum layer prior to each determination. Chronopotentiometry was used in stabilised solutions. *Figure 14* shows a characteristic chronopotentiogram.

Contrarily to what was reported by Inman and White at high concentrations and high current densities, a single reduction wave was observed and attributed to the reduction of $MoCl_6^{3-}$. Sand's law was obeyed, and the diffusion coefficient of the $MoCl_6^{3-}$ ions was $D_{Mo} = 8 \pm 0.5\ 10^{-6}\ cm^2\ s^{-1}$. The reduction process was irreversible, since the difference between the cathodic and anodic plateaux was about 700 mV. Analysis of the chronopotentiograms yielded values of α between 0.2 and 0.3. The standard rate constant was between 10^{-8} and 10^{-9} cm s^{-1}.

These results were applied to molybdenum deposition within the pores of Y_2O_3-stabilised zirconia coatings obtained by plasma spraying on a steel substrate. Low current densities (< 10 $mAcm^{-2}$) and high Mo(III) concentration was found necessary for a good filling of the pores. When the sample was immersed in LiCl-KCl + K_3MoCl_6 melt, a mixed potential was observed (*Figure 15*) and the cathodic partial current was too high for a good deposit.

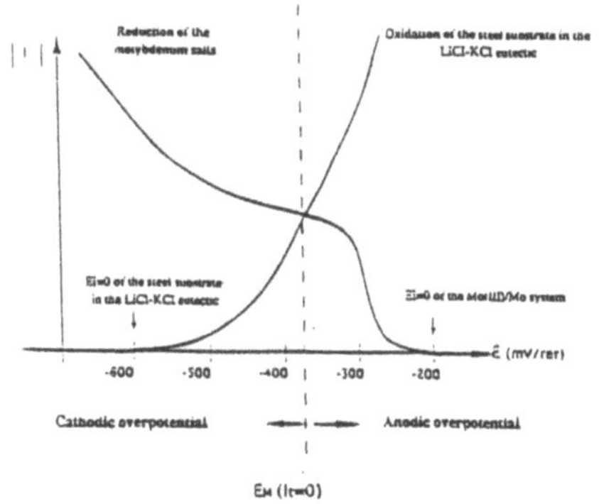

Figure 15 Principle and improvement of the cementation process of the Mo salts dissolved in LiCl-KCl at 500°C by a Fe-Co-Ni alloy. Quasi-stationary polarization curves of reduction of the Mo salts and of oxidation of the steel substrate.

In fact, the molybdenum deposit obtained by cementation was powdery and porous. In order to obtain a coherent and filling deposit, it was proved sufficient to apply an anodic overpotential of 100 mV with respect to the mixed potential.

3. Electrochemistry of chromium

3.1. INTRODUCTION

A literature survey shows that the corresponding data are rather contradictory, as far as the solubilities of $CrCl_2$ and $CrCl_3$ are concerned, and as far as the Cr(II) / Cr deposition step is concerned. In fact, the kinetics of this step has been reported to be fast, or slow, and there is some agreement to think that this step is complicated by adsorption or electrocrystallisation phenomena. In this talk, we will tackle the questions of the chromium chlorides solubility, of the Cr(III) or Cr(II) adsorption on various substrates and the kinetics of the final electron exchange.
The LiCl-KCl was prepared following the usual procedure, and once molten it ws kept under purified argon (less than 1ppm O_2 and H_2O). $CrCl_3$ and $CrCl_2$, high purity products (STREM Chemicals), were added to the molten salt via a lock chamber. The working electrode was either a vitreous carbon rod, a tungsten wire or a platinum or gold wire. The reference electrode was Ag / AgCl, and the counter-electrode was a carbon rod.

3.2. SOLUBILITY OF CHROMIUM CHLORIDES

The solubilities were measured by sampling the solutions obtained after introduction of $CrCl_2$ or $CrCl_3$ and measuring the chromium concentration at time intervals by chemical analysis.
The dissolution of $CrCl_3$ was found to be slow (see TABLES 1 and 2). The solubilisation of $CrCl_2$ was faster, part of the $CrCl_2$ initially in solution disappeared from the solution yielding fine colloidal particles but eventually resulting in a solution which remained stable for more than one week. This has not yet been explained. Nevertheless the solubility of $CrCl_2$ is higher than 0.2 mol l^{-1}.

TABLE 1 : Solubilisation of $CrCl_3$ in LiCl-KCl at 406°C
(total amount added : 0.044 mol.L^{-1})

C_{CrCl3} mol.L^{-1}	2.86 10^{-2}	3.49 10^{-2}	3.49 10^{-2}	3.89 10^{-2}
time / h	13	24	36	60
% $CrCl_3$ dissolved	64.9	79.3	79.3	86.7

TABLE 2 : Solubilisation of $CrCl_2$ in LiCl-KCl at 439°C
(total amount added : 0.027 mol.L^{-1})

C_{CrCl2} mol.L^{-1}	2.70 10^{-2}	2.22 10^{-2}	2.21 10^{-2}	2.22 10^{-2}
time / h	12	24	65	80
% $CrCl_2$ dissolved	100	81.9	81.9	81.9

3.3. ELECTROCHEMISTRY OF $CrCl_3$ SOLUTIONS

I will not focus much on this, and will just say that the reduction to Cr appeared via two well-separated steps (see Figure 16). The first step corresponds to the Cr(III) / Cr(II) couple, and the second one to the Cr(II) / Cr couple. Concerning the Cr(III) / Cr(II) couple, I will merely mention that our study has shown that this exchange was fast, with no significant kinetical complication. The second step corresponds to the Cr(II) / Cr exchange, and will be dealt with in the next section.

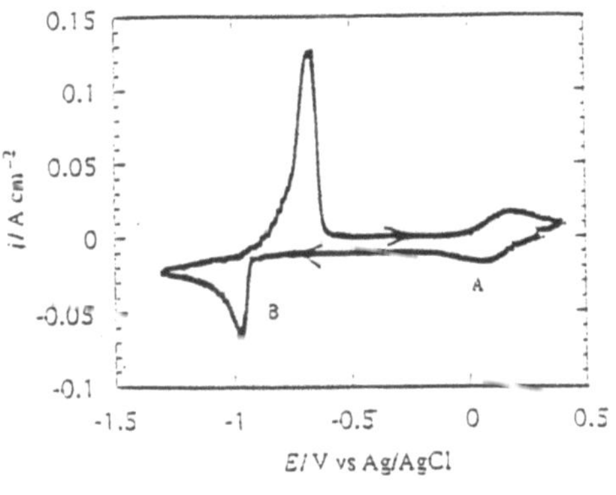

Figure 16 Cyclic voltammogram for the reduction of $CrCl_3$ in molten LiCl-KCl at 673°C; v = 0.9 V s^{-1}; C_{CrCl3} = 3.81 10^{-2} mol L^{-1}; Ref : Ag/AgCl;
A : Cr(III) / Cr(II); B : Cr(II) / Cr

3.4. ELECTROCHEMISTRY OF $CrCl_2$ SOLUTIONS

Voltammetry, chronopotentiometry and chronoamperometry were used. A series of voltammograms related to the reduction of a $CrCl_3$ solution at a vitreous carbon electrode is reported in Fig 17 with the corresponding convoluted curves.

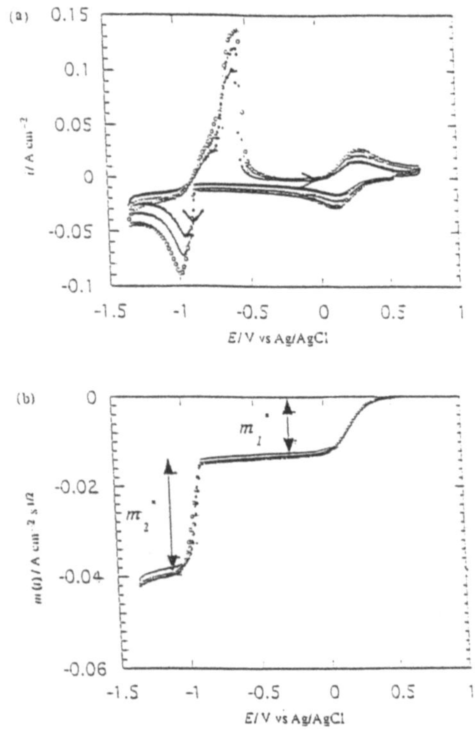

Figure 17 (a) A series of voltammograms related to the reduction of $CrCl_3$ in fused LiCl-KCl at 406°C; $C_{CrCl_3} = 3.49\ 10^{-2}$ mol L^{-1}; Working electrode : vitreous carbon; v = 0.5, 0.9 and 1.4 V s^{-1}.
(b) Corresponding semi-integral curves

Two plateaux are visible on the convoluted curves, with indicate diffusion for both steps. From the ratio of these plateaux, it can be concluded that the second one corresponds to a 2-electron exchange.

A typical chronopotentiogram related to the oxidation of Cr(II) solutions at gold or vitreous carbon electrodes is given in figure 18.

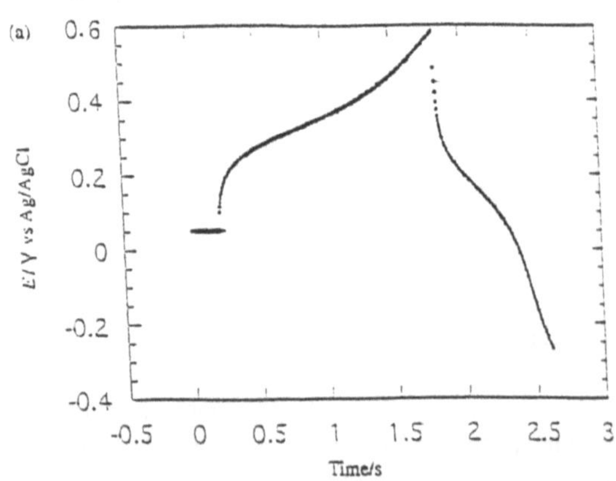

Figure 18 Typical chronopotentiogram for the oxidation of Cr(II) ions

Sand's law was obeyed, with no evidence of an adsorption phenomenon. Calculation of the diffusion coefficient of Cr(II) using Sand's law yielded $D_{Cr(II)} = 0.62 \pm 0.37\ 10^{-5}\ cm^2\ s^{-1}$, in agreement with previous data. But if the gold electrode was immersed a long time in advance, the results were different and adsorption of Cr(II) ions could be postulated. Adsorption of Cr(II) on copper and nickel electrodes was also noticed, since the $t^{1/2}$ versus I^{-1} plots yielded a positive intercept with the $t^{1/2}$ axis.

Finally, a thin chromium deposit was plated in situ on a tungsten substrate, then chronopotentiograms were recorded, the analysis of which showed that Cr(II) species was adsorbed, and that adsorbed and dissolved Cr(II) species participated in the reduction process. No adsorption was demonstrated on a vitreous carbon electrode. Cyclic voltammetry was then used at gold, copper, nickel, vitreous carbon and tungsten electrodes (Figure 19).

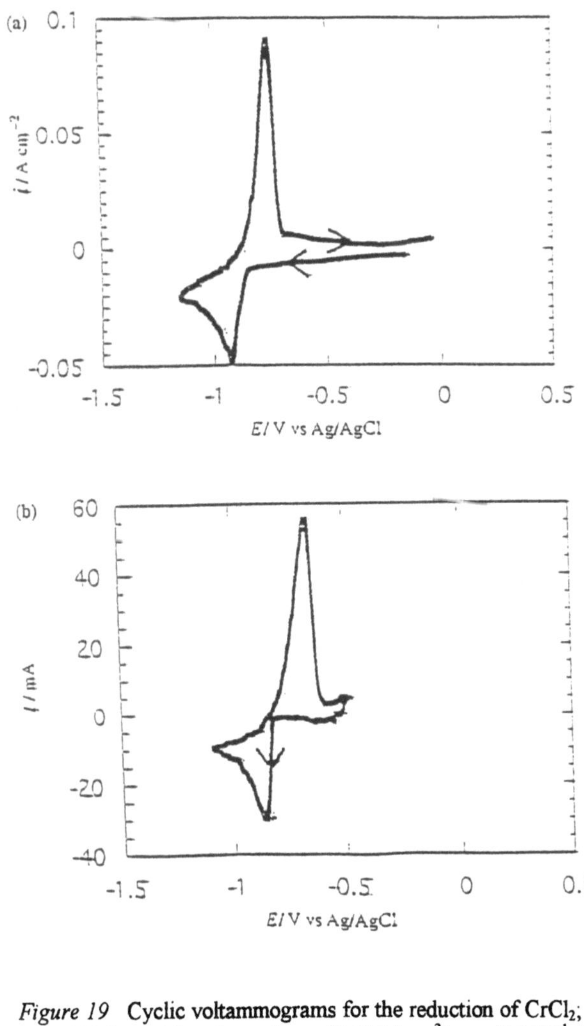

Figure 19 Cyclic voltammograms for the reduction of $CrCl_2$; (a) : $C_{CrCl2} = 2.71\ 10^{-2}$ mol L^{-1}; T = 489°C; working electrode : gold (0.14 cm^2). v = 0.5 Vs^{-1} b) : $C_{CrCl2} = 4.13\ 10^{-2}$ mol L^{-1}; T = 461 °C; working electrode : Cu (0.32 cm^2); v = 1 Vs^{-1}

The curves show that an overpotential was needed for the formation of the first Cr nuclei and that the reverse potential had an influence on the nucleation crossover effect in the reverse sweep. Chronoamperometry was the used to characterise the chromium nucleation effect (Figure 20).

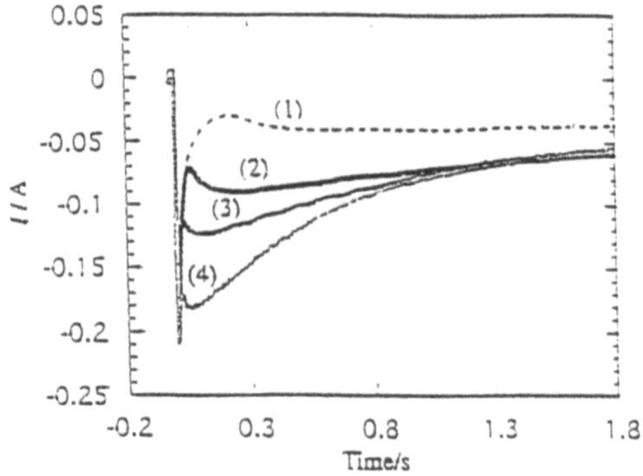

Figure 20 A series of chronoamperograms for the reduction of CrC_2 1.76 10^{-1} mol L^{-1}; T = 414 °C working electrode : vitreous carbon (1.4 cm^2); (1) : E = -0.860 V; (2) : E = -0.910 V; (3) : E = -0.970 V; (4) : E = -1.0 V vs Ag/AgCl

Considering the value of the $Imax^2$ tmax product, instantaneous nucleation was proved on a gold electrode, and progressive nucleation on a vitreous carbon electrode.

From these results, we have designed chromium plating procedures which yield compact and adherent deposits. Next step is to scale up the process, and to test the use of a soluble chromium anode.

DEPOSITION OF TITANIUM, ZIRCONIUM AND HAFNIUM DIBORIDE COATINGS BY HIGH-TEMPERATURE ELECTROCHEMICAL SYNTHESIS FROM CHLORO-FLUORIDE MELTS

S.V. DEVYATKIN, G. KAPTAY
Dept. of Physical Chemistry, University of Miskolc, Miskolc - Hungary
V.I. SHAPOVAL, I.V. ZARUTSKII, V.P. LUGOVOI
Inst. of General and Inorg. Chemistry 32-34 Palladin pr., Kiev - Ukraine
S.A. KUZNETSOV
Institute of Chemistry, Kola SC RAS, Apatity, 184200 - Russia

1. Introduction

The objective of this work was to study and to develop a method for high-temperature electrochemical synthesis (HTES) of transition metal diboride coating from chloro-fluoride melts.
Electrodeposition of titanium diboride coatings from the system FLiNaK-K_2TiF_6-KBF_4 was studied by H. Wendt, K. Reuhl, and V. Schwarz [1], also from $LiBO_2$, $NaBO_2$, Li_2TiO_3,Na_2TiO_3, and TiO_2 by D. Schlain, F.X. McCawley, and G.R. Smith [2]. Electrodeposition of zirconium diboride coatings from the system FLiNaK-K_2ZrF_6-KBF_4 (B_2O_3) was studied by G.W. Mellors and S. Senderoff [3], and from cryolite-ZrO_4-B_2O_3 melt by E.J. Frazer, K.E. Anthony, and B.J. Welsh [4].
The melts chosen for this study were NaCl-KCl(50-50mol%)-NaF-KBF_4-K_2MeF_6 (Me-Ti, Zr, Hf). The chloro-fluoride melt was chosen, as the pure fluoride melt is more aggressive, and, on the other hand, the oxo-fluoride melt gives a lower current efficiency. Also, the cathodic product is easier to wash from the quenched salt if the chloro-fluoride melt is used for the synthesis.
The possibility of synthesis of diborides was studied by thermodynamic calculations, by S.V. Devyatkin, and G. Kaptay [5]. Conditions of high-temperature electrochemical synthesis of TiB_2, ZrB_2, and HfB_2 from chloro-fluoride melts were determined on the basis of our calculation, the synthesis taking place in one step at a more positive potential than its components.

2. Experimental

Linear voltammetry on the wide range of sweep rates (from 10 mV/s to 10 V/s) was applied at 973 K. Experiments were carried out under atmosphere of purified argon using a Ag/KCl-NaCl-AgCl (2 w%) and Pb(Mo)/KCl-NaCl-$PbCl_2$ (2 w%) reference electrodes. Glassy carbon was used as the crucible, which also served as the counter-electrode. Cylindrical glassy carbon with a surface area of S=0.5-0.7 cm^2 was used as the working electrode.

D.H. Kerridge and E.G. Polyakov (eds.), Refractory Metals in Molten Salts, 73-80.

Electrolysis experiments in laboratory scale were performed in steel containers at 973 K under an atmosphere of purified argon. The graphite crucible served as a container of melt. Graphite, glassy carbon, Ni, Mo, and W was used as cathode materials. Cathodic products were analysed by X-ray and SEM.
To prevent decomposition of boron fluoride complexes at 973 K, accompanied by formation of gaseous BF_3 [6], to the melt were added F-ions (NaF, KF).

3. Voltammetric Experiments

Voltammetric experiments in the system NaCl-KCl-K_2TiF_6-KBF_4-NaF have been performed [6, 7], also in the systems NaCl-KCl-K_2ZrF_6-KBF_4-NaF [7, 8], and NaCl-KCl-K_2HfF_6-KBF_4-NaF [5, 9].
This paper presents a summary of these voltammetric experiments and the kinetic parameters of refractory metals ions, boron ions and electrochemically active species from electrochemical synthesis in fused salts.
In the system NaCl-KCl-K_2TiF_6 two processes more electropositive than the discharge of the alkali metals have been observed (Figure 1.). The first process corresponds to a one electron reversible change transfer with formation of soluble product:

$$\mathrm{Ti(IV)+e \leftrightarrow Ti(III)} \quad (1)$$

The second process corresponds to a three-electron irreversible change transfer with formation of insoluble product:

$$\mathrm{Ti(III)+3e \rightarrow Ti(0)} \quad (2)$$

In the system NaCl-KCl-KBF_4-NaF one process, more electropositive than the discharge of the alkali metals has been observed (Figure 1.). This process corresponds to a three electron irreversible change transfer with formation of insoluble product:

$$\mathrm{B(III)+3e \rightarrow B(0)} \quad (3)$$

In the system NaCl-KCl-K_2TiF_6-KBF_4-NaF a new process, more electropositive than the change transfer Ti(III)/Ti(0) and the deposition of boron has been observed (Figure 1.). The process corresponds to a 9-electron reversible change transfer with formation of insoluble product (TiB_2):

$$\mathrm{Ti(III)+2B(III)+9e \leftrightarrow TiB_2} \quad (4)$$

In the system NaCl-KCl-K_2ZrF_6-NaF one process more electropositive than the discharge of the alkali metals has been observed (Figure 2.). This process corresponds to a four electron irreversible change transfer with formation of insoluble product:

$$\mathrm{Zr(IV)+4e \rightarrow Zr(0)} \quad (5)$$

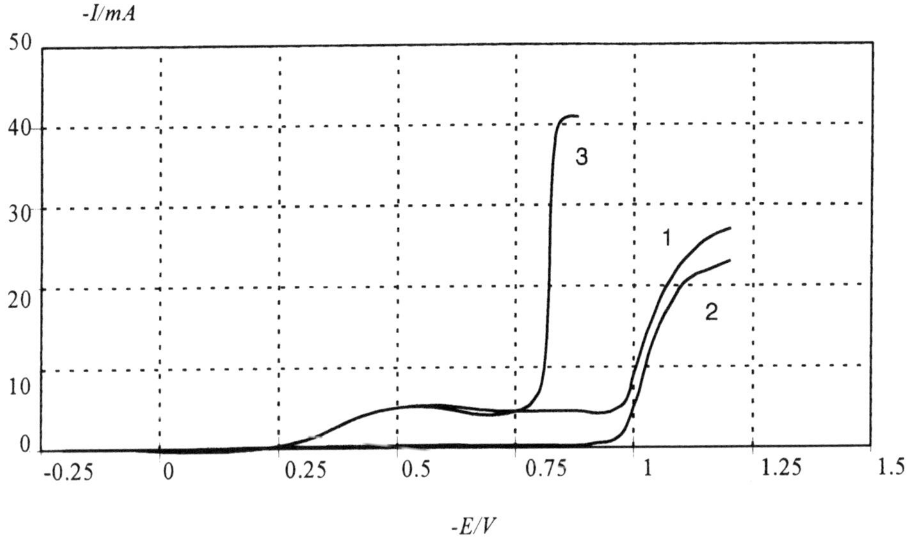

Figure 1. Voltammetric curves at the scan rate of 0.1 V/s at 973 K, reference electrode Ag/KCl-NaCl-AgCl (2 w%) in the systems:

1. NaCl-KCl-K_2TiF_6(6×10^{-5} mol/cm^3)-NaF
2. NaCl-KCl- KBF_4(8×10^{-5} mol/cm^3)-NaF
3. NaCl-KCl-K_2TiF_6(6.8×10^{-5} mol/cm^3)-KBF(7.6×10^{-5} mol/cm^3)-NaF

In the system NaCl-KCl-K_2ZrF_6-KBF_4-NaF a new process, more electropositive than the discharge of the zirconium and boron has been observed (Figure 2.). The process corresponds to a 10-electron irreversible change transfer with formation of insoluble product (ZrB_2):

$$Zr(IV)+2B(III)+10e\rightarrow ZrB_2 \tag{6}$$

In the system NaCl-KCl-K_2HfF_6-NaF one process, more electropositive than the discharge of the alkali metals has been observed (Figure 3.). This process corresponds to a four electron irreversible change transfer with formation of insoluble product:

$$Hf(IV)+4e\rightarrow Hf(0) \tag{7}$$

In the system NaCl-KCl-K_2HfF_6-KBF_4-NaF a new process, more electropositive than the discharge of the hafnium and boron has been observed ("Figure 3."). The process corresponds to a 10-electron irreversible change transfer with formation of insoluble product (HfB_2):

$$Hf(IV)+2B(III)+10e\rightarrow HfB_2 \tag{8}$$

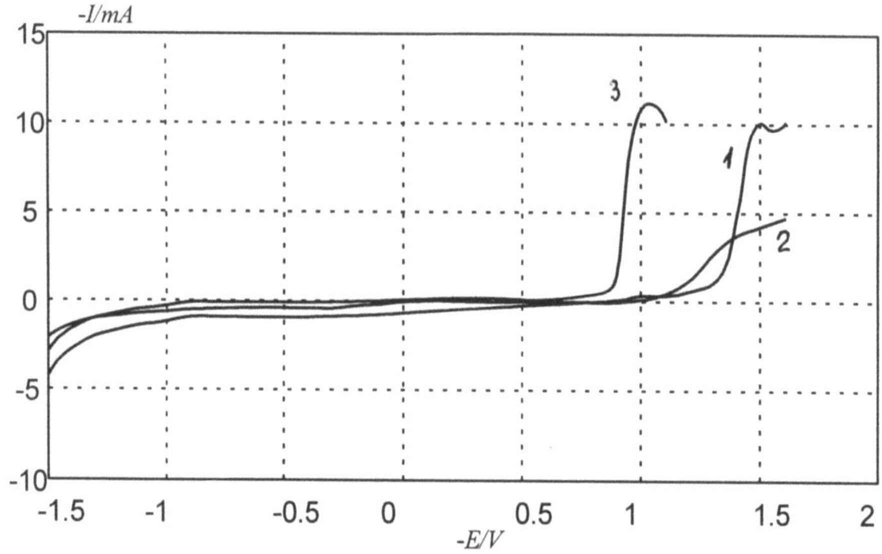

Figure 2. Voltammetric curves at the scan rate of 0.1 V/s at 973 K at reference electrode Pb(Mo)/KCl-NaCl-$PbCl_2$ (2 w%) in the systems:

1. NaCl-KCl-K_2ZrF_6(7.4×10^{-5} mol/cm^3)-NaF
2. NaCl-KCl- KBF_4(8.5×10^{-5} mol/cm^3)-NaF
3. NaCl-KCl-K_2ZrF_6(6.3×10^{-5} mol/cm^3)-KBF(9.3×10^{-5} mol/cm^3)-NaF

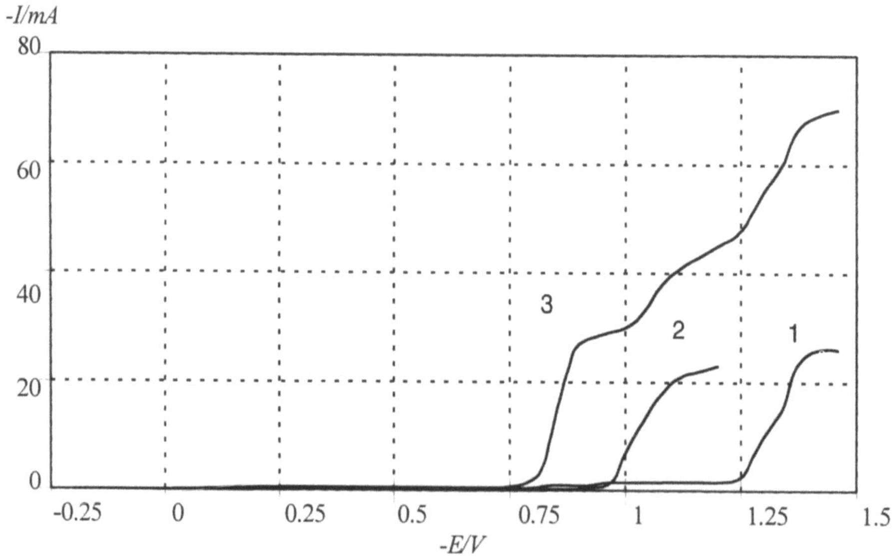

Figure 3. Voltammetric curves at the scan rate of 0.1 V/s at 973 K at reference electrode Ag/KCl-NaCl-AgCl (2 w%) in the systems:

1. NaCl-KCl-K_2HfF_6(7×10^{-5} mol/cm^3)-NaF
2. NaCl-KCl- KBF_4(9×10^{-5} mol/cm^3)-NaF
3. NaCl-KCl-K_2HfF_6(6×10^{-5} mol/cm^3)-KBF(8×10^{-5} mol/cm^3)-NaF

TABLE 1. Kinetic parameters of electrochemically active species in melts

Melts	Cations	αn_α	D (10^{-5} cm^2/s)
NaCl-KCl-KBF_4-NaF	B(III)	1.3±0.02*	3.2±0.05*
NaCl-KCl-K_2TiF_6-NaF	Ti(III)	0.95±0.02*	3.0±0.05*
NaCl-KCl-K_2TiF_6-KBF_4-NaF	Ti(III)+2B(III)	n≈9**	0.89±0.05**
NaCl-KCl-K_2ZrF_6-NaF	Zr(IV)	1.3±0.02*	1.66±0.05*
NaCl-KCl-K_2ZrF_6-KBF_4-NaF	Zr(IV)+2B(III)	2.0±0.02*	0.99±0.05*
NaCl-KCl-K_2HfF_6 -NaF	Hf(IV)	1.4±0.02*	1.36±0.05*
NaCl-KCl-K_2HfF_6-KBF_4-NaF	Hf(IV)+2B(III)	1.55±0.02*	0.4±0.05*

*- irreversible change transfer: αn_α (α is the transfer coefficient and n_α is number of electrons taking part in irreversible electrochemical reaction) were calculated by the equation of Matsuda-Ayabe [10] and D (the diffusion coefficient) were determined by Delahay's equation [11]
**- reversible change transfer with insoluble product: n (the number of electrons) were calculated using the Mamontov-Manning-Dale equation [12] and D was determined utilizing of the Berzins-Delahay equation [11]

From Table 1. it can be seen that the diffusion coefficients of electrochemically active species in melts, used for synthesis, are approximately 3 times less than in melts containing only ions of boron or ions of the transitional metals. Therefore, the assumption about formation of heteronuclear electroactive species receives some additional support.

4. Electrolysis Experiments

In the system NaCl-KCl-K_2TiF_6-KBF_4-NaF of a one-electron reversible change transfer Ti(IV)/Ti(III) occurs more positive than the HTES of TiB_2. Therefore in order to obtain a TiB_2 coating, metallic Ti was used as an anode to produce low-valent titanium ions in the melt:

$$Ti+3Ti(IV)=4Ti(III) \tag{9}$$

TiB_2 coatings have been obtained on W-, Ni-, graphite and glassy carbon cathodes at current densities in the range of 0.05 and 0.5 A/cm^2 (Figure 4.) and a very good quality TiB_2 coating was deposited on the Ni-cathode.
In the system NaCl-KCl-K_2ZrF_6-KBF_4-NaF it was impossible to obtain dense coatings. In our opinion, this was due to passivation of the cathodic surface by low-valent, poorly-soluble zirconium compounds, like $KZrF_4$ [13], or ZrF [14].
In order to obtain a ZrB_2 cathodic coating, 5 w% of B_2O_3 was added into the melt. The B_2O_3 additive depassivated the cathode (probably by dissolving the ZrF_{4-X} compound), and a ZrB_2 cathodic coating on a glassy carbon-cathode has been obtained at current densities in a range of 0.05 and 0.5 A/cm^2. The layered structure depended probably on the influence of the passivating ZrF_{4-X} compound (Figure 5.).

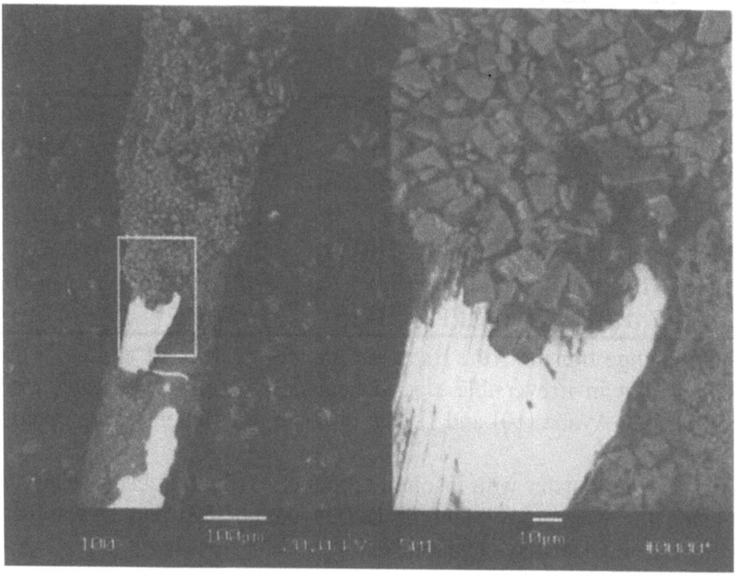

Figure 4. SEM micrograph showing of TiB_2 coating deposited onto a W cathode.

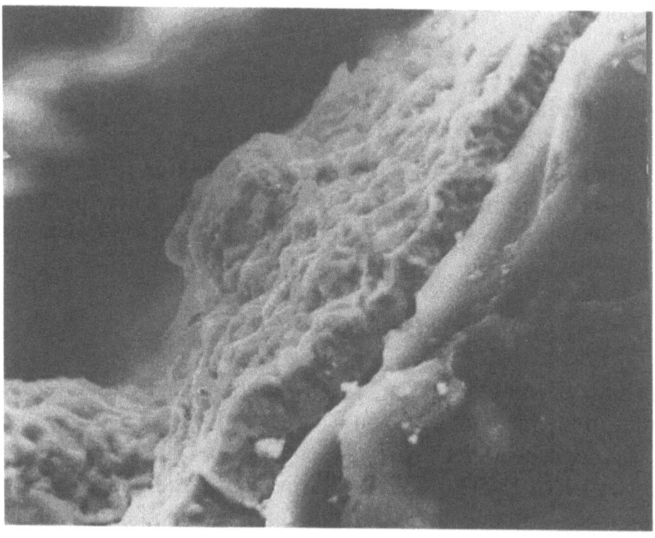

Figure 5. SEM micrograph showing of ZrB_2 coating deposited onto a glassy carbon cathode.

In the system NaCl-KCl-K_2HfF_6-KBF_4-NaF the HTES of HfB_2 proceeds unobstructedly. HfB_2 coherent coatings have been obtained on Mo-cathodes at current densities in a range of 0.005 and 0.05 A/cm^2 (Figure 6.).

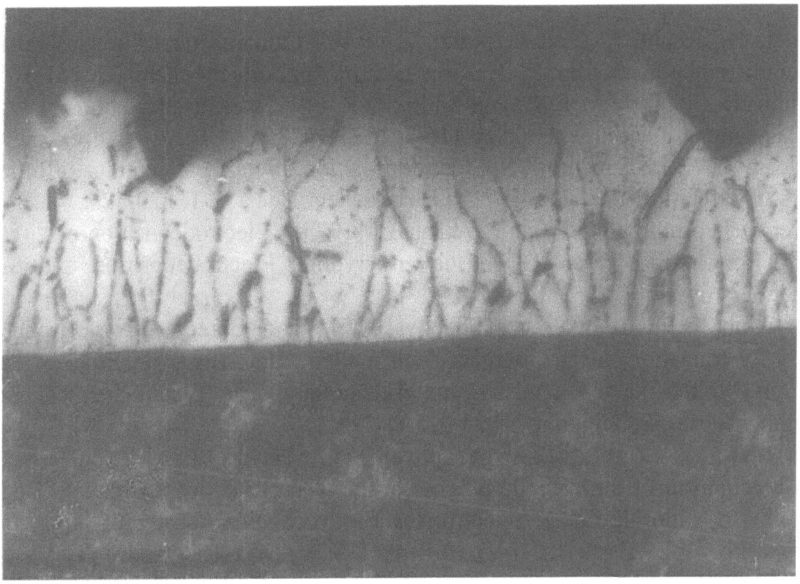

Figure 6. SEM micrograph showing of HfB_2 coating deposited onto a Mo cathode.

The deposition of diboride coating was possible only on a glassy carbon or graphite cathode of cylindrical form (max diameter 2 mm). The problem was, that coefficients of linear expansion of carbon and transition metal diboride differed greatly (Table 2.). After electrolysis at 973 K the graphite (glassy carbon) and diboride coating were cooled to room temperature. The graphite contracted to a larger extent than the diboride, coating which failed under compression.

TABLE 2. Coefficient of linear expansion at 25 ^{0}C [15, 16, 17].

	TiB_2	ZrB_2	HfB_2	graphite	Mo	W	Ni
$\alpha(10^6\ K^{-1})$	4.6	5.9	6.3	27.2	4.8	4.5	13.4

5. Conclusion

The HTES of all the Titanium-group transition metal diborides from chloro-fluoride melts was possible at a wide range of current densities, using more positive potentials than those for deposition of boron, or transition metal from the same melt.

Necessary requirements for producing coatings whit the HTES method are:

(a) one-step HTES;

(b) absence of more positive electrochemical reactions;

(c) similarity of coefficient of linear expansion of coatings and material of cathode.

(d) desirable formation of an interim layer between coating and material of cathode.

6. References

1. Wendt, H., Reuhl, K., and Schwarz, V. (1992) Cathodic deposition of refractory intermetallic compounds from FLiNaK melts, J. of Appl. Electrochem. **22**, 161-165.
2. Schlain, D., McCawley, F.X., and Smith, G.R. (1969) Electrodeposition if titanium diboride coatings, J.Electrochem. Soc **116-9**, 1227-1228.
3. Mellors, G.W. and Senderoff, S. Electrodeposition of coherent coatings of refractory metals VII. Zirconium diboride (1971) J.Electrochem. Soc **118-1**, 220-225.
4. Frazer, E.J., Anthony, K.E., and Welsh, B.J (1975) . Electrodeposition of zirconium diboride from oxides dissolved in molten cryolite, Electrodeposition and Surface Treatment **3**, 169-177.
5. Devyatkin, S.V., Kaptay, G., Shapoval, V.I., and Berecz, E. Electrochemical synthesis of transition metal diborides from molten salts (1993), in Proc. of the Int. Symp. on Molten Salts Chemistry and Technology, The Electrochemical Society Proceedings Series, Pennington, NJ **93-9**, 584-599,
6. Taranenko, V.I., Zarutskii, I.V., Shapoval, V.I., Makyta, M., and Matiasovsky, K. (1992) Mechanism of the cathode process in the electrochemical synthesis of TiB_2 in molten salts-II. Chloride-fluoride electrolytes, Electrochimica Acta **37**, 263-268.
7. Taranenko, V.I., Zarutskii, I.V., Bidenko, V.A., and Devyatkin, S. V. (1991) Electrochemical synthesis of titanium and zirconium diborides from chloro-fluoride melts, in: Borides (in Russian), IPM AN UkrSSR, Kiev 55-61.
8. Lugovoi, V.P., Devyatkin, S.V., Kaptay, G., and Kuznetsov, S.A. (1996) High temperature electrochemical synthesis of zirconium diboride from chloro-fluoride melts, in Proc. of the Int. Symp. Molten Salts X, The Electrochemical Society Inc., NJ **96-7**, 303-311.
9. Kuznetsov, S.A., Devyatkin, S.V., Glagolevskoya, A.L., Taranenko, V.I., Stangrit, P.T., and Shapoval, V.I. (1992) Electrochemical synthesis of hafnium diboride from chloro-fluoride melts, Rasplavy (in Russian) **2**, 67-70.
10. Matsuda, H.and Ayabe, Y. (1955) Zur theorie der Randles-Sevcikschen kathodenstrahl-polarographie, Z. fur Elektrochemie **59-6**, 494-503.
11. Delahay, P. (1954) New Instrumental Methods in Electrochemistry, Interscience Publ., New York.
12. Mamontov, G., Manning, D.L., and Dale, J.M. (1965) Reversieble deposition of metals on solid electrodes by voltammetry with linearly varying potential, J. of Electroanal. Chem. **9**, 253-259.
13. Ivanovskii, L.E. and Petenev, O.S.(1961) Processes in cathodic deposition of zirconium from chloro-fluoride melts, Trudy Inst. Elektrokhimii UF AN USSR (in Russian) **2**, 71-78.
14. Polyakova, L.P., Polyakov, E.G., and Stangrit, P.T. (1991) Electrochemical behavior of zirconium in chloro-fluoride melts, Rasplavy (in Russian) **1**, 54-61.
15. Lide, D.R. (1994) Handbook of Chemistry and Physics, CRS Press, Boca Raton (74th Edition).
16. Dric, M.E. (1985) Svoistva elementov (in Russian), Metallurgiya, Moscow.
17. Samsonov, G.V. and Vinickii, I.M. (1976) Tugoplavkie soedineniya (in Russian), Metallurgiya, Moscow.

THE SYNTHESIS OF BORIDES, CARBIDES AND SILICIDES OF REFRACTORY METALS IN IONIC-ELECTRONIC MELTS

N.I. SHUROV, A.I. ANFINOGENOV, V.V. CHEBYKIN,
L.P. KLEVTSOV, E.G. KAZANSKII
Institute of High-Temperature Electrochemistry, Ural Branch of RAS, Yekaterinburg, Russia

1. Introduction

When refractory compounds are produced by the solid-phase synthesis from powder components, the direct contact area of the particles is small and therefore the synthesis is performed at elevated temperatures to enhance the mutual diffusion. If any of the components is transferred to the solution by means of transport reactions, the effective contact area is increased many-fold and the synthesis temperature and time are reduced considerably. This, in turn, allows producing fine particles of refractory compounds.

Previously, we used the phenomenon of an oriented isothermal transport of metals and nonmetals in ionic melts for the medium-temperature synthesis of powder alloys, inorganic compounds and diffusion coatings [1]. Recently we have detected an isothermal transport of metals in the LiCl-Li ionic-electronic melt, the transport being associated with the formation of diffusion alloys – coatings.

Advancing in this direction, we studied the possibility of synthesising carbides, borides and silicides of titanium, zirconium, niobium and tantalum in Li-LiCl and Ca-$CaCl_2$ ionic-electronic melts. These melts are referred to as the ionic-electronic ones, because dissolved lithium and calcium dissociate to cations and delocalized electrons:

$$Me_{soln.} \leftrightarrow Me^{b+} + be^{-}. \quad (1)$$

Such melts are intermediate between ionic and metallic ones and transport reactions should take place in these melts too.

2. Experimental Technique

The experiments were performed in an argon atmosphere in a gas-tight reactor made of a heat-resistant steel. The mixture of the metal and nonmetal powders being in the assigned stoichiometric ratio with a small surplus of the nonmetal was placed in a tube made of a low-carbon steel. Lumps of the remolten lithium chloride (calcium chloride) and lumps of lithium (calcium) were charged on top in the amount 3- to 5-fold larger than their

D.H. Kerridge and E.G. Polyakov (eds.), Refractory Metals in Molten Salts, 81-86.

solubility in the salt melt. The filled tube was placed in the reactor, which was sunk vertically in a shaft furnace. After evacuation and heating to 200° C the reactor was filled with pure argon. Subsequently the reactor was heated in the electric furnace up to 800-1000° C and was allowed to stand for 5-10 hours.

After the end of the experiment the reactor was withdrawn from the furnace, cooled in air to room temperature, and taken apart. The tube was taken out of the reactor and the salt melt was washed with water. The precipitate of the refractory compound was collected and washed on a filter. Then the precipitate was dried, removed from the filter, weighed, and underwent the X-ray phase and chemical analyses.

3. Results and Discussion

3.1. SYNTHESIS OF CARBIDES

Fine carbon (black), which we used to synthesise carbides, is poorly wetted with LiCl and $CaCl_2$ melts and therefore it comes to the surface when the salt melts at the beginning of the experiment, whereas the metal powders are left on the tube bottom. The density of the molten lithium and calcium is lower than the density of their chloride melts and therefore they come to the melt surface too. Thus, at the very beginning of the process the components are spatially separated and the question arises as to the mechanism of the carbide synthesis under these conditions.

The solubility of lithium in its (own) chloride at 900° C is 1.67 mole % [3], while the solubility of calcium in its own chloride equals 3.26 mole % at 900° C [4]. Carbon does not dissolve in molten LiCl and $CaCl_2$. However, lithium and calcium form ionic-type carbides Li_2C_2 and CaC_2. Thus, the first stage of the synthesis is the formation of lithium and calcium carbides, which dissolve in the salt melts and dissociate as was shown by Morris [5]:

$$Li_2C_2 \leftrightarrow 2Li^+ + C_2^{2-} \leftrightarrow 2Li_{(soln)} + 2C_{(s)}\,, \qquad (2)$$
$$CaC_2 \rightarrow Ca^{2+} + C_2^{2-} \leftrightarrow Ca_{(soln)} + 2C_{(s)}\,. \qquad (3)$$

The second stage of the synthesis is the transport of carbon by the carbide anions through the salt melt to the metal surface. The motive force of the carbon transport is the gradient of the carbon chemical potential at the boundaries between the solid phases and the salt melt:

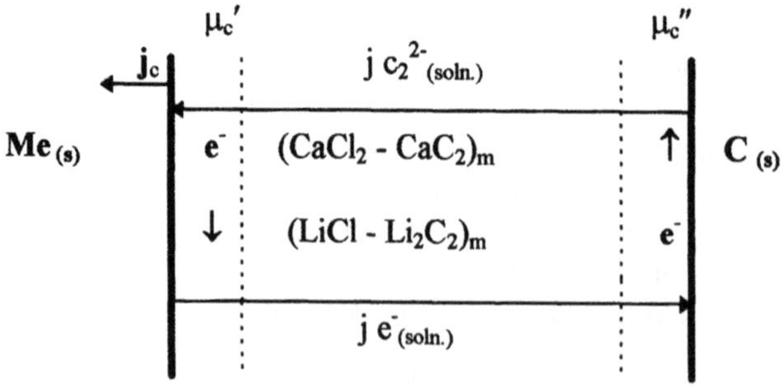

(4)

The reactions at the phase boundaries are as follows:
r.h. boundary: $2C_{(s)} + 2e^-_{(soln.)} \rightarrow C_2^{2-}{}_{(solv.)}$;
l.h. boundary: $2Me_{(s)} + 2C_2^{2-}{}_{(soln.)} \rightarrow 2MeC_s + 2e^-_{(solv.)}$.
The total transport reaction is

$$2Me_{(s)} + 2C_{(s)} \rightarrow 2MeC_{(s)} \qquad \Delta_f G^0 << 0. \qquad (5)$$

The slowest stage of the synthesis is the diffusion of carbon into the refractory metal accompanied by the formation of the growing carbide layer.

The thermodynamic estimation confirms a high probability of these reactions:

Reaction	$\Delta_r G^0_{(1200K)}$, kJ
$CaC_2 + 2Ti \rightarrow 2TiC + Ca$	−250.3
$CaC_2 + 2Zr \rightarrow 2ZrC + Ca$	−280.5
$CaC_2 + 2Nb \rightarrow 2NbC + Ca$	−177.3
$CaC_2 + 2Ta \rightarrow 2TaC + Ca$	−191.0

Considering that the Gibbs free energy of the formation of the Li_2C_2 carbide (−43.0 kJ/mole) is twice as low as that of CaC_2 (−90.7 kJ/mole), the same reactions with the lithium carbide are still more probable.

The X-ray phase analysis confirmed that the synthesis products are TiC, ZrC, NbC and TaC carbides, which do not contain free carbon. Fineness of the carbides is determined by the grain size of the metal powder. The synthesis time depends on the diffusion rate of carbon into the volume of the grain or the crystallite. We used oxides of refractory metals as the starting materials for production of fine (1-5 μm) carbides. In this case the mass of the metallic calcium was increased taking into account its consumption for the calcium thermal reduction of the oxides. The presence of a higher amount of the calcium oxide in the salt melt did not incur large difficulties in the synthesis of the carbides and their washing with water to remove salts, because the calcium oxide easily dissolves in acidified water. This method of the carbide synthesis has been covered by a Russian Federation patent [6].

3.2. SYNTHESIS OF BORIDES

Like the carbon black, fine amorphous boron is poorly wetted with salt melts and emerges to the surface of the salt immediately after the salt is molten, while the refractory metal powder is left on the bottom. Thus, the first stage of the synthesis is the formation of lithium or calcium borides.

There has been no state diagram of the lithium-boron system, but according to certain data, several borides – Li_2B, Li_7B_6 or LiB-based solid solution, LiB_3, LiB_4 and LiB_6 and a solid solution of lithium in boron $LiB_{10.85\pm0.35}$ – are formed. The higher lithium borides are refractory compounds. Boron atoms in these compounds are coupled by strong covalent bonds and therefore these borides dissolve, as the elementary boron does, little if at all in ionic melts. The structure of the low-melting-point borides LiB_2 and

LiB has been so far unknown. We think they probably have a polar-covalent type of the chemical bonds, which is similar to the ionic type. In the LiCl melt, which contains a certain amount of free delocalized electrons, the said borides may dissociate to lithium cations and boron anions having the hypothetical composition B^{m-}. The latter may serve as the boron carriers from pure boron or the higher lithium borides to the metal particles. As a result, thermodynamically stronger borides of refractory metals are formed through diffusion.

Thus, the synthesis of Ti, Zr, Nb and Ta borides may be represented by a hypothetical scheme of the boron transport through the ionic-electronic melt:

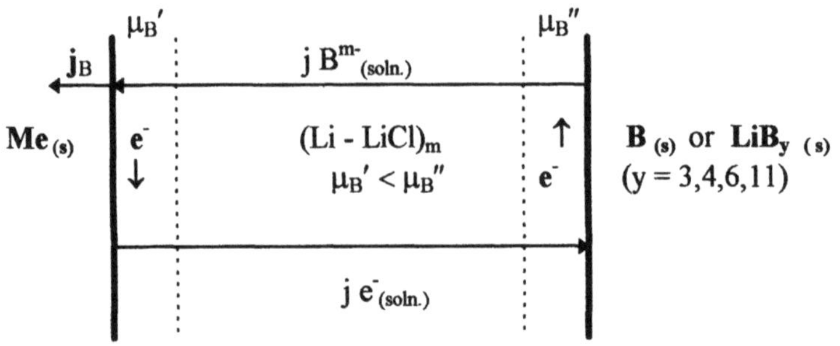

(6)

The reactions at the boundaries between the solid and liquid phases are as follows:

r.h. boundary: $2B\ (\text{from}\ LiB_{y(s)}) + 2me^- \rightarrow 2B^{m-}{}_{(soln.)}$;

l.h. boundary: $2B^{m-}{}_{(soln.)} + Me_s \rightarrow MeB_2 + 2me^-$.

The total reaction is

$$Me_{(s)} + 2B\ (\text{from}\ LiB_{y(s)}) \rightarrow MeB_{2(s)} \quad (7)$$

The synthesis reaction continues until all the boron passes to the refractory boride. The X-ray phase analysis of the products of the synthesis performed at 900° C for 10 hours revealed the presence of the TiB_2, ZrB_2, NbB_2 and TaB_2 phases. As fine metal powders as possible should be used if one needs to produce single-phase powders.

A somewhat different picture is observed for the Ca-$CaCl_2$ melt. Two strong refractory borides CaB_4 and CaB_6, which are insoluble in the ionic melt, are formed in the Ca-B system. The density of these borides is higher than the density of the $CaCl_2$ melt and therefore they should subside to the tube bottom and may contact the refractory metal powder. We have been unaware of the thermodynamic data on CaB_4. The standard values of the CaB_6 formation are $\Delta_f H^0_{298} = -58.0$ kJ/mole and $\Delta_f S^0_{298} = 76.89$ J/K·mole. Using a rough formula $\Delta_f G^0_{(1200K)} = \Delta_f H^0_{298} - 1200 \cdot \Delta_f S^0_{298}$, for CaB_6 we have $\Delta_f G^0_{1200} = -150$ kJ/mole. For comparison, $\Delta_f G^0_{1200}(TiB_2) = -304.5$ kJ/mole.

Thus, the thermodynamic reaction

$$CaB_6 + 3Ti \rightarrow 3TiB_2 + Ca \quad (8)$$

is quite probable. In actual fact, this reaction is very slow when CaB_6 and Ti are separated by the salt melt. Specifically, after a 10-hour exposure at 900° C the X-ray phase and chemical analyses of the solid products of the synthesis revealed the presence of the Ti and CaB_6 phases and a small amount of TiB_2, the last compound being probably in the form of a very thin diffusion layer on Ti particles. The reason for a very slow transfer of boron from CaB_6 to titanium in the Ca-CaCl melt (as compared to the Li-LiCl melt) is, in our opinion, a very small concentration of boron dissolved in the salt melt. The boron atoms in the CaB_4 and CaB_6 compounds are coupled by strong covalent bonds. The dissociation of these compounds to calcium cations and boron anions is probably infinitesimally small, that is, the limiting stage of the synthesis process is an extremely slow transport of boron through the salt melt by the boron anions.

3.3. SYNTHESIS OF SILICIDES

Crystalline silicon, which was ground to a powder having the grain size less than 160 μm and used for the synthesis, had a density of 2330 kg/m^3. Therefore, in the molten salts it subsided to the tube bottom, where silicon could contact the refractory metal powder. However, the direct contact area of the particles is very small as the particles are separated by a layer of the salt melt. Liquid lithium and calcium, which are soluble in their own chloride melts, can diffuse from the surface to the bulk of the salt melt and form silicides.

Two silicides – Li_4Si (t_{melt} = 635° C) and $Li_{14}Si_6$ (t_{melt} = 706° C) – exist in the Li-Si system. The Ca-Si system has four silicides: Ca_2Si, which melts incongruently at 910° C; CaSi (t_{melt} = 1245° C); Ca_3Si_4, which melts incongruently at 1020° C, and $CaSi_2$ (t_{melt} = 1000° C) [8].

By analogy with lower lithium and calcium borides, lower lithium (Li_4Si, $Li_{14}Si_6$) and calcium (Ca_2Si) silicides may, in our opinion, partially dissociate in ionic-electronic melts and form silicon anions of the hypothetical composition Si^{n-}. These anions may serve as the silicon carriers from the silicon powder or higher silicides to the metal. This is followed by the formation of refractory silicides of *d*-transition metals of the 4th and 5th groups. The hypothetical scheme of the silicon transfer through the salt melt is similar to the scheme given above for boron:

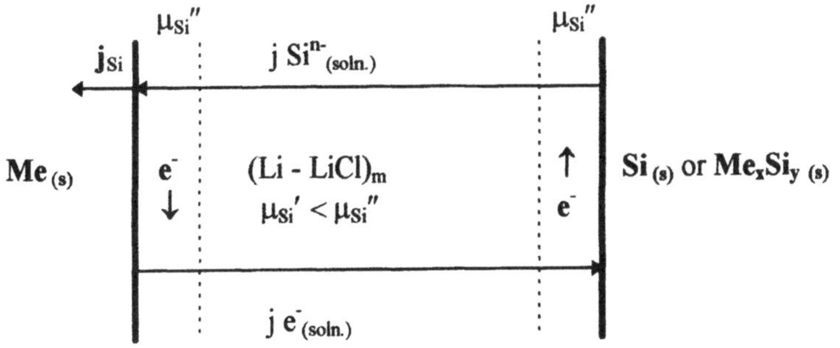

(9)

Thus, Ti_5Si_3, TiSi, $TiSi_2$, ZrSi, $ZrSi_2$, Nb_5Si_3, $NbSi_2$, Ta_5Si_3 and $TaSi_2$ powders were synthesised in the Li-LiCl and Ca-$CaCl_2$ melts and identified by the X-ray phase analysis.

4. Conclusions

1. It is shown experimentally that powdered carbides, borides and silicides of Ti, Zr, Nb and Ta can be synthesised in the Li-LiCl and Ca-$CaCl_2$ melts.
2. A probable mechanism of the refractory compounds synthesis is discussed. Hypothetical schemes of the carbon, boron and silicon transport in ionic-electronic melts are proposed.
3. It is for the first time hypothesised that boron and silicon are present in ionic-electronic melts in the form of anions. The role played by these anions in the transport reactions involved in the synthesis of the refractory compounds is discussed.

5. References

1. Ilyushchenko, N.G., Anfinogenov, A.I., and Shyrov, N.I. (1991) *Interaction of Metals in Ion Melts*, Nauka Publishers, Moscow.
2. Ilyushchenko, N.G., Anfinogenov, A.I., Shyrov, N.I., Chebykin, V.V., and Zyryanov V.G. (1995). *Zh. Prikl. Khim.* **68**(6), 1027-1029.
3. Dvorkin, A.S., Bronstein, H.R., Bredig, M.A. (1962) J.Phys.Chem. **66**(3), 572-573.
4. Sharma, R.A. (1970) J.Phys.Chem. **74**(22), 3896-3900.
5. Morris, D.R. (1982) Ionic Carbides, in D.G. Lovering (ed.), *Molten Salt Technology*, Plenum Press, New-York–London, p.p.153-183.
6. RF Patent No. 2043967 B22F 9/16. The Composition for Synthesis of Carbides of Refractory Metals. N.G. Ilyushchenko, A.I. Anfinogenov, V.V. Chebykin, Ya.B. Chernov, Yu.A. Ryaposov, A.I. Dobrynin, A.V. Gorshkov, and A.V. Chub (Russia). In: *Bull. Isobr.* **26** (1995).
7. *Alloys of Alkali and Alkali-Earth Metals*, Handbook (1986), M.E. Drits and L.L. Zusman (eds), Metallurgiya Publishers, Moscow.
8. *The Structure and Properties of Binary Metallic Systems*, Handbook, v.IV (1979), A.E. Vol and I.K. Kagan (eds), Nauka Publishers, Moscow.

PHYSICOCHEMICAL PROPERTIES OF THE MOLTEN SYSTEM $KF—KCl—KBF_4—K_2TiF_6$

M. CHRENKOVÁ, V. DANĚK, A. SILNÝ, D. K. NGUYEN
Institute of Inorganic Chemistry, Slovak Academy of Sciences, Dúbravska cesta 9, 842 36 Bratislava, Slovakia

Abstract

On the basis of the measurements of the phase equilibria, density and viscosity of the melts of the system $KF–KCl–KBF_4–K_2TiF_6$, using the complex thermodynamic and physicochemical analysis, the structure or ionic composition was studied. It was found that in this electrolyte complex anions TiF_7^{3-} and TiF_6Cl^{3-} with lowered symmetry of co-ordination sphere and different stability are present. The electrodeposition of titanium and titanium diboride from the K_2TiF_6 based electrolytes is enhanced by the formation of these anions.

1. Introduction

The quaternary system $KF–KCl–KBF_4–K_2TiF_6$ is suitable electrolyte in the electrochemical synthesis of TiB_2, especially when coherent coatings on metallic bases have to be prepared [1]. Besides of the mentioned application the molten system $KF–KCl–K_2TiF_6$ is promising electrolyte for the deposition of titanium and $KF–KCl–KBF_4$ is interesting also as possible electrolyte for electrochemical boriding of steels. From the theoretical point of view, on the basis of the physicochemical properties considerations on the structure, i.e. the ionic composition of the melts, can be made.

The knowledge of the structure of these electrolytes is needed for the understanding of the mechanism of the electrochemical processes. The interaction of the components and the chemical reactions taking place in the melt effect the ionic composition, determining the kind of the electroactive species. The suitable choice of the electrolyte composition may suppress the formation of volatile compounds which leads to undesirable exhalation and lowers the efficiency of the process.

The aim of the present work was to determine some physicochemical properties, like phase equilibria, density and viscosity of the electrolyte and to elucidate the structure, i.e. the actual ionic composition of the melts using the complex thermodynamic and physicochemical analysis of the melts as well as the X-ray diffraction and spectroscopic analysis of quenched melts.

D.H. Kerridge and E.G. Polyakov (eds.), Refractory Metals in Molten Salts, 87-101.

2. Computational Procedure

For the whole concentration region of the investigated system two different approaches may be used. In the first approach the structure *(i.e.* the ionic composition) is determined by the thermodynamic equilibrium composition, after all the chemical reactions taking place in the system are over. Then the ideal mixing of components after reaching the equilibrium is supposed. If the obtained standard deviation of the calculated and measured property for the given chemical reactions is in the range of the experimental error, we may assume that the structure of the electrolyte is given by the equilibrium composition determined by the calculated equilibrium constants. Besides, information on *e.g.* the thermal stability and the Gibbs energy of present compounds may be obtained. The task is solved by means of the material balance and using the thermodynamic relations valid for ideal solutions. In general, this approach may be used in the evaluation of those properties for which the ideal behaviour of the system is physically defined, *e.g.* for Gibbs energy of mixing and molar volume.

In the second approach, which finds application in real systems, the validity of the general Redlich-Kister's type equation for the excess property is supposed. For the description of the composition dependence of the given property Y in the system the following equation is then used (*e.g.* in a ternary system)

$$Y = \sum_{i=1}^{3} A_i . x_i + \sum_{\substack{i,j=1 \\ i \neq j}}^{3} x_i . x_j \sum_{n=0}^{n} B_{nij} . x_j^n + \sum_{a,b,c}^{m} C_m . x_1^a . x_2^b . x_3^c \qquad (1)$$

The first term represents the additive behaviour, the second one the binary interactions and the third one the interaction of all three components. For the excess Gibbs energy of mixing in the real solution the following equation may be supposed (*e.g.* for a ternary system)

$$\Delta G^{ex} = \sum_{\substack{i,j=1 \\ i \neq j}}^{3} \sum_{n=1}^{5} A_{ijn} . x_i . x_j^n + \sum_{\substack{i,j,k=1 \\ i \neq j \neq k}}^{3} B_{ijk} . x_i^a . x_j^b . x_k^c \qquad (2)$$

In case of transport properties, like viscosity, the ideal behaviour is not physically defined, since we deal with scalar quantities which do not possess the total derivative. Thus the simple additivity rule may not be used. However, these properties are thermally activated and the additivity of activation energies is allowed. Based on this idea the additivity of logarithms of these properties is often accepted as "ideal" behaviour. For the description of the viscosity in a three component system the following equation has been suggested

$$\eta = \eta_1^{x_1} . \eta_2^{x_2} . \eta_3^{x_3} + \sum_{\substack{i,j=1 \\ i \neq j}}^{3} x_i . x_j \sum_{n=0}^{k} A_{nij} . x_j^n + B . x_1^l . x_2^m . x_3^n \qquad (3)$$

Coefficients of the regression equations (1), (2) and (3) are calculated using the multiple linear regression analysis. Omitting the statistical non-important terms on the chosen confidence level and minimising the number of relevant terms we effort to get a solution, which describes the concentration dependence of the investigated property with a standard deviation of the fit being in the same order as the experimental error. For statistically important binary and ternary interactions we look for appropriate chemical reactions and check their thermodynamic probability calculating their standard reaction Gibbs energies. The reaction products are then identified using the X-ray phase analysis and IR spectroscopy of quenched melts. Again we assume that the high temperature composition is after quenching at least qualitatively conserved.

Interactions are mostly considered as chemical reactions between components. However, Van der Waals bonds and the formation of associates may not be excluded as interactions, in spite they cannot be detected by spectroscopic measurements.

3. Results

3.1. Electrodeposition of titanium

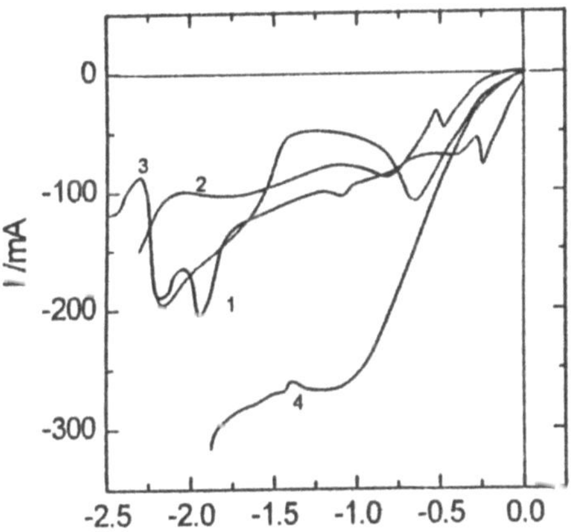

Fig. 1. Voltammetric curves in the system KCl–K_2TiF_6. 1 - K_2TiF_6; 2 - K_2TiF_6 + 5 mole % KCl; 3 - K_2TiF_6 + 50 mole % KCl; 4 - K_2TiF_6 + 90 % KCl.

The influence of the melt composition on the deposition of titanium from the K_2TiF_6 melt was studied by Makyta et al. [2]. The essential results of this work is summarised in Fig. 1. From the comparison of the cathodic part of the voltammetric curves obtained in the pure K_2TiF_6 (curve 1) and those obtained in the $KF–K_2TiF_6$ and $KCl–K_2TiF_6$ mixtures (curves 2 - 4) it follows that substantial changes occur when KF or KCl is added to the pure K_2TiF_6. The reduction of Ti(IV) to titanium metal in the all-fluoride and chloride-fluoride melts proceeds in two electrochemical consecutive reduction steps, Ti(IV) → Ti(III) → Ti(0). These two reduction steps are clearly visible on all recorded curves, the first one starting at approximately -0.2 V and the second one starting at approximately -1.5 V. It can be seen from the figure that the current values of the peaks, representing the above mentioned titanium reduction steps, increase with increasing content of potassium chloride in the melt. Similar changes were observed also on voltammetric curves with the KF addition to K_2TiF_6. Even when titanium may be deposited also from pure K_2TiF_6, most probably due to the presence of KF impurities, the addition of either KF or KCl significantly influences the peaks current corresponding to the titanium reduction.

On the basis of these results it can be assumed that the addition of KF or KCl causes most probably changes in the ionic composition of the electrolyte, *i.e.* the kind of the electroactive species, which enhances the electroreduction of titanium. The changes in the ionic composition were studied using the physicochemical and thermodynamic analysis.

3.2. Phase diagram

The temperature of the primary crystallization was measured using the thermal analysis method recording the cooling and heating curve of the studied samples. A computer-controlled automatic thermal analysis device with a subsequent statistical processing of the output data was used. The reproducibility of the temperature of the primary crystallization measurement was ± 2 K.

The phase diagram of the ternary system KF–KCl–KBF_4 was measured in [3]. It was found that all three binary systems and the ternary system are simple eutectic ones. The calculated co-ordinates of the eutectic point in the ternary system is: 19.2 mole % KF, 18.4 mole % KCl, 61.4 mole % KBF_4 and the temperature of the eutectic crystallization 422 °C. The probable inaccuracy in the calculated ternary phase diagram is ± 6.9 °C.

For the dependence of the molar excess Gibbs energy of mixing on composition in the ternary system, which is needed for the calculation of the activity coefficients of components, the following relation was considered

$$\begin{aligned}\Delta G^{E} / \mathrm{J\,mol^{-1}} &= x_{KF}x_{KCl}\left(1997 - 7854x_{KCl} + 8722x_{KCl}^{2}\right) + \\ &+ x_{KCl}x_{KBF_4}\left(1437 + 1124x_{KBF_4}^{2} - 721x_{KBF_4}^{3}\right) + \\ &+ x_{KBF_4}x_{KF}\left(3014 + 6760x_{KF}^{2} + 394x_{KF}^{3}\right) - \\ &= 10601x_{KF}x_{KCl}x_{KBF_4}\end{aligned} \tag{4}$$

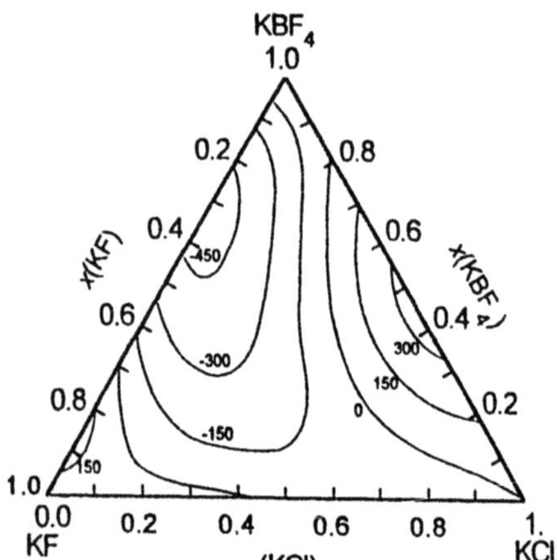

Fig. 2. Excess molar Gibbs energy of mixing (J mol⁻¹) in the system KF–KCl–KBF₄.

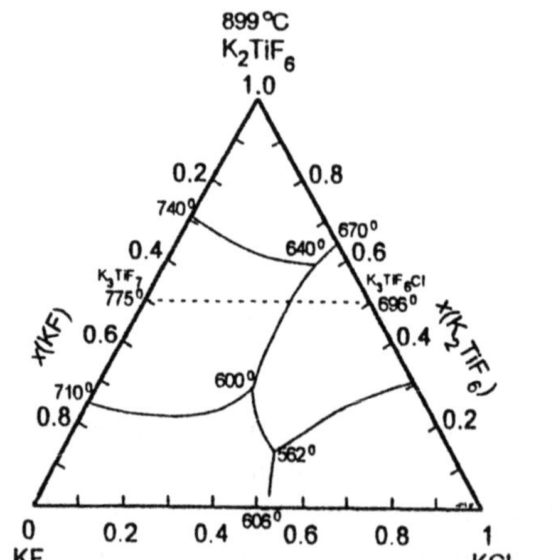

Fig. 3. Phase diagram of the system KF–KCl–K₂TiF₆. Isotherm values in °C.

The excess molar Gibbs energy of mixing of the molten system KF–KCl–KBF_4 is shown in Fig. 2.

The phase diagram of the ternary system KF–KCl–K_2TiF_6 has been studied in [4]. Two congruently melting compounds, K_3TiF_7 and K_3TiF_6Cl, are formed in this system (Fig. 3).

The measurements of the phase equilibria of the system KF–KBF_4–K_2TiF_6 was performed in [5]. The phase diagram of this system, constructed using the coupled analysis of the thermodynamic and measured phase diagram data, is shown in Fig. 4. Four crystallisation fields are present in the phase diagram corresponding to the primary crystallisation of KF, KBF_4, K_2TiF_6 and the intermediate compound K_3TiF_7. The calculated co-ordinates of the two ternary eutectic points are:

e_1: 26 mole % KF, 69 mole % KBF_4, 5 mole % K_2TiF_6, t_{e1} = 448 °C

e_2: 4 mole % KF, 69 mole % KBF_4, 27 mole % K_2TiF_6, t_{e2} = 440 °C.

The probable inaccuracy in the calculated ternary phase diagram is ± 11.5 °C.

For the molar excess Gibbs energy of mixing the following equation was obtained

$$\begin{aligned}\Delta G^{E} / \mathrm{J\,mol^{-1}} = {} & x_{KF}x_{KBF_4}\left(3569 - 15930x_{KBF_4} + 8347x_{KBF_4}^2\right) - \\ & - x_{KF}x_{K_2TiF_6}\left(13319 + 14895x_{K_2TiF_6} - 26989x_{K_2TiF_6}^2\right) - \\ & - x_{KBF_4}x_{K_2TiF_6}\left(3688 - 11864x_{K_2TiF_6}\right) - \\ & - 26897x_{KF}x_{KBF_4}x_{K_2TiF_6}\end{aligned} \qquad (5)$$

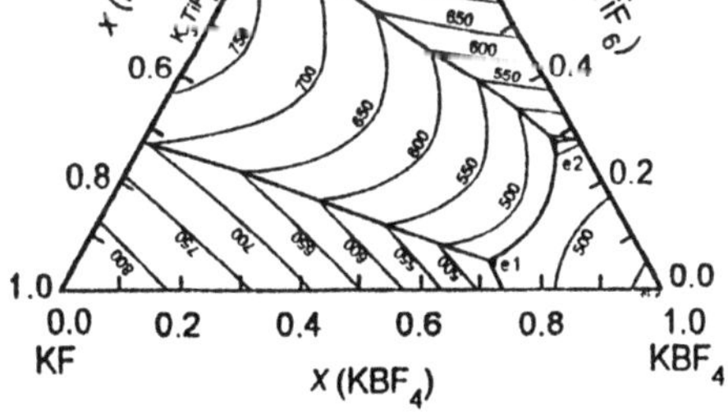

Fig. 4. Phase diagram of the system KF–KBF_4–K_2TiF_6. Isotherm values in °C.

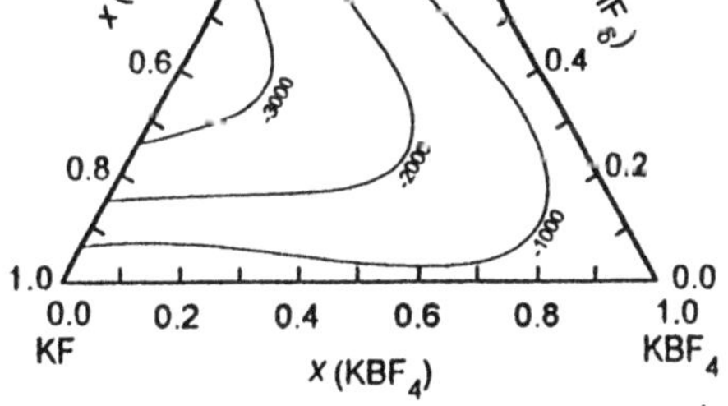

Fig. 5. Excess molar Gibbs energy of mixing (J mol^{-1}) in the system KF–KBF_4–K_2TiF_6.

The molar excess Gibbs energy of mixing in the ternary system KF–KBF_4–K_2TiF_6 is shown in Fig. 5.

The phase diagram of the ternary system KCl–KBF_4–K_2TiF_6 was determined in [6] and it is shown in Fig. 6. Four crystallisation fields are present in the phase diagram corresponding to the primary crystallisation of KCl, KBF_4, K_2TiF_6 and the intermediate compound K_3TiF_6Cl. The calculated co-ordinates of the two ternary eutectic points are as follows:

e$_1$: 24.1 mole % KCl, 62.1 mole % KBF_4, 13.8 mole % K_2TiF_6, t_{e1} = 447.1 °C,

e$_2$: 6.5 mole % KCl, 62.5 mole % KBF_4, 31.0 mole % K_2TiF_6, t_{e2} = 414.5 °C.

The probable inaccuracy in the calculated ternary phase diagram is ± 17.1 °C.

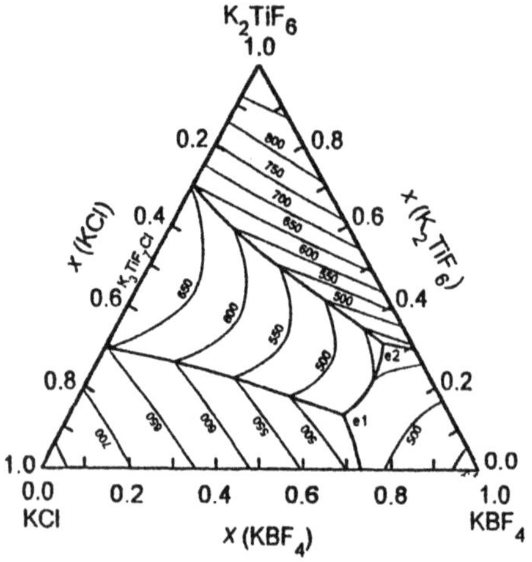

Fig. 6. Phase diagram of the system KCl–KBF$_4$–K$_2$TiF$_6$. Isotherm values in °C

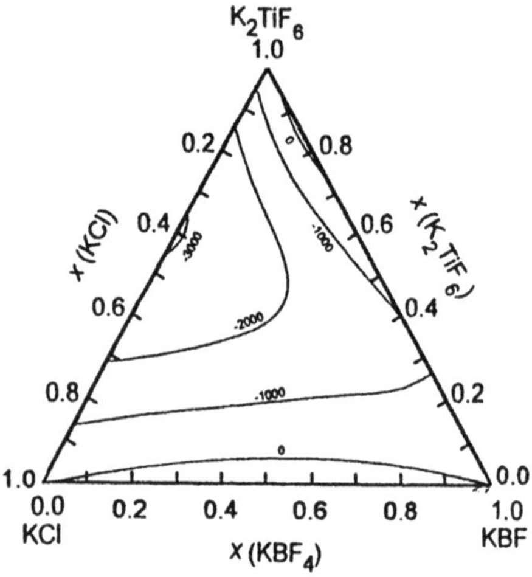

Fig. 7. Excess molar Gibbs energy of mixing (J mol^{-1}) of the system KCl–KBF$_4$–K$_2$TiF$_6$.

For the molar excess Gibbs energy of mixing the following equation was obtained

$$\Delta G^{E} / \mathrm{J\,mol^{-1}} = 1745 x_{KCl} x_{KBF_4} - x_{KCl} x_{K_2TiF_6}\left(6906 + 10055 x_{K_2TiF_6}\right) - \\ - x_{KBF_4} x_{K_2TiF_6}\left(6092 - 14202 x_{K_2TiF_6}^2\right) - 36401 x_{KCl}^2 x_{KBF_4} x_{K_2TiF_6} \qquad (6)$$

The molar excess Gibbs energy of mixing in the ternary system KCl–KBF_4–K_2TiF_6 is shown in Fig. 7.

3.3. Density

The density of the investigated melts was measured using the Archimedean method. The platinum sphere with the diameter of 20 mm, suspended using the platinum wire on the automatic analytical balance. For the measuring device control and the evaluation of experimental data the on-line PC XT computer was used. The experimental error of the density measurement was less then ± 0.4 %.

The density of the system KF–KCl–KBF_4 was measured in [7]. On the basis of the density data the values of the molar volume and excess molar volume were calculated. The concentration dependence of the molar volume at 1100 K was described by the equation

$$V\,/\,\mathrm{cm^3mol^{-1}} = 30.11x_{KF} + 49.84x_{KCl} + 75.29x_{KBF_4} + \\ + 0.91x_{KF}x_{KCl} + 4.38x_{KF}x_{KBF_4} + 2.63x_{KCl}x_{KBF_4} - \\ - 14.63x_{KF}x_{KCl}x_{KBF_4} \qquad (7)$$

with the standard deviation of the fit $\sigma = 0.17$ cm^{3}mol^{-1}. The excess molar volume of the molten system KF–KCl–KBF_4 is shown in Fig. 8.

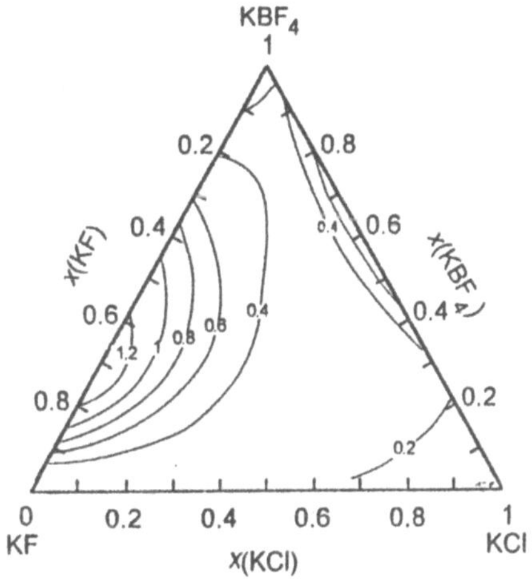

Fig. 8. Excess molar volume (cm^3 mol^{-1}) of the molten system KF–KCl–KBF_4 at the temperature of 1100 K.

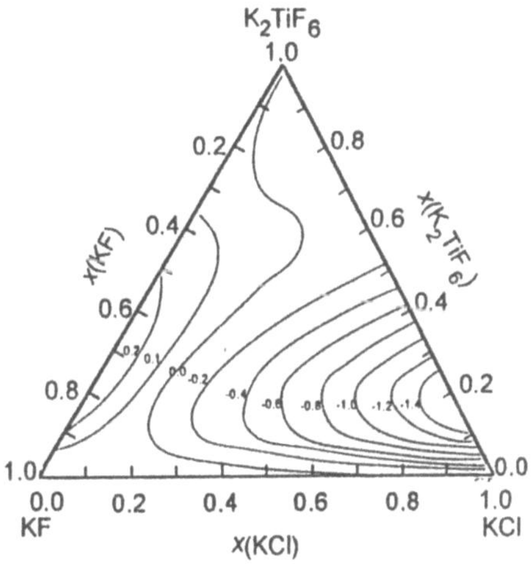

Fig. 9. Excess molar volume (cm^3 mol^{-1}) of the molten system KF–KCl–K_2TiF_6 at the temperature of 1100 K.

The density of the KF–KCl–K_2TiF_6 melts [8] is mostly influenced by the KCl component, which strongly decreases the density of the melts. For the dependence of the excess molar volume on composition the following equation for the temperature 1100 K was obtained

$$V^E\,/\,\mathrm{cm^3\,mol^{-1}} = 0.763x_{KF}x_{KCl} + x_{KF}x_{K_2TiF_6}\left(2.236 - 2.956x_{K_2TiF_6}\right) - \\ - x_{KCl}x_{K_2TiF_6}\left(21.817 - 68.506x_{K_2TiF_6} + 54.552x_{K_2TiF_6}^2\right) \qquad (8)$$

with the standard deviation of approximation $\sigma = 0.047$ cm^{3}mol^{-1}.

In Fig. 9 the excess molar volume of the KF–KCl–K_2TiF_6 melts at the temperature 1100 K is shown. No ternary interaction was found.

The volume properties of the melts of the ternary system KF–KBF_4–K_2TiF_6 were studied in [9]. For the molar volume of this system at 1100 K the following equation has been calculated

$$V / \mathrm{cm^3mol^{-1}} = 30.57x_{\mathrm{KF}} + 75.12x_{\mathrm{KBF_4}} + 114.62x_{\mathrm{K_2TiF_6}} + 5.41x_{\mathrm{KF}}x_{\mathrm{KBF_4}} + \\ + x_{\mathrm{KBF_4}}x_{\mathrm{K_2TiF_6}}\left(18.44 - 28.29x_{\mathrm{K_2TiF_6}}\right) - 152.50x_{\mathrm{KF}}x_{\mathrm{KBF_4}}x_{\mathrm{K_2TiF_6}} \quad (9)$$

with the standard deviation of the fit $\sigma = 0.34\ \mathrm{cm^3mol^{-1}}$. The excess molar volume of this system is shown in Fig. 10. From the figure it follows, that two different regions are present in the ternary system: the region of volume expansion with the maximum at approx. 25 mole % K_2TiF_6 and 75 mole % KBF_4 and the region of volume contraction with the maximum at approx. 60 mole % K_2TiF_6 and 20 mole % KBF_4.

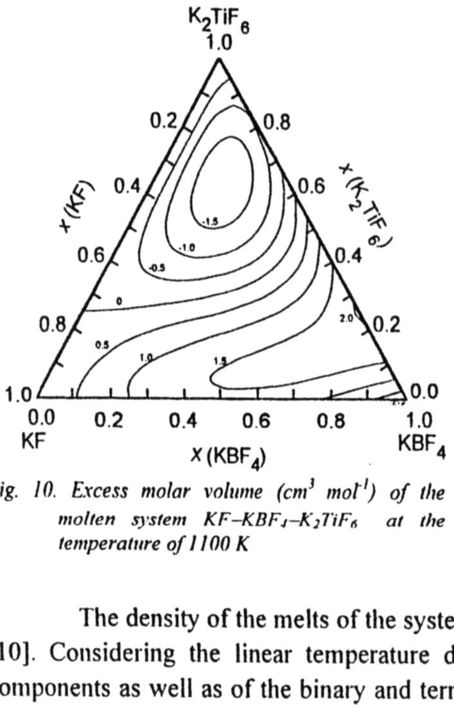

Fig. 10. Excess molar volume (cm³ mol⁻¹) of the molten system KF–KBF₄–K₂TiF₆ at the temperature of 1100 K

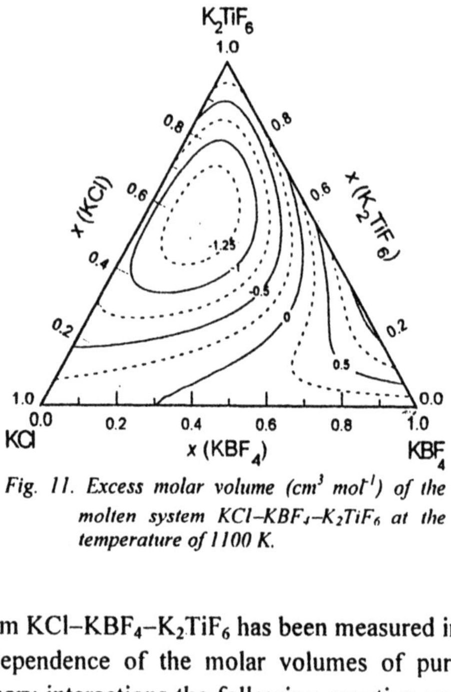

Fig. 11. Excess molar volume (cm³ mol⁻¹) of the molten system KCl–KBF₄–K₂TiF₆ at the temperature of 1100 K.

The density of the melts of the system $KCl–KBF_4–K_2TiF_6$ has been measured in [10]. Considering the linear temperature dependence of the molar volumes of pure components as well as of the binary and ternary interactions the following equation was obtained for the molar volume of the ternary system

$$\begin{aligned} V / \mathrm{cm^3mol^{-1}} = {} & x_{\mathrm{KCl}}\left(34.450 + 1.879 \times 10^{-2}\, T/\mathrm{K}\right) + \\ & + x_{\mathrm{KBF_4}}\left(50.760 + 2.982 \times 10^{-2}\, T/\mathrm{K}\right) + \\ & + x_{\mathrm{K_2TiF_6}}\left(86.149 + 3.383 \times 10^{-2}\, T/\mathrm{K}\right) + 3.985x_{\mathrm{KCl}}x_{\mathrm{KBF_4}}^2 + \\ & + x_{\mathrm{KCl}}x_{\mathrm{K_2TiF_6}}\left(8.883x_{\mathrm{K_2TiF_6}} - 7.624 \times 10^{-3}\, T/\mathrm{K}\right) - \\ & - x_{\mathrm{KBF_4}}x_{\mathrm{K_2TiF_6}}\left(14.776 - 1.447 \times 10^{-2}\, T/\mathrm{K}\right) - \\ & - 88.661x_{\mathrm{KCl}}x_{\mathrm{KBF_4}}x_{\mathrm{K_2TiF_6}}^2 \end{aligned} \quad (10)$$

The standard deviation of approximation of eqn (10) is $\sigma = 0.343$ cm3mol-1.

The excess molar volume of the ternary system $KCl-KBF_4-K_2TiF_6$ is shown in Fig. 11. Like in the ternary system $KF-KBF_4-K_2TiF_6$ [9], two different regions are present in the investigated system: the region of volume expansion with the maximum in the binary system $KBF_4-K_2TiF_6$ at approx. 25 mole % K_2TiF_6 and 75 mole % KBF_4 and the region of volume contraction with the maximum at approx. 60 mole % K_2TiF_6, 20 mole % KCl and 20 mole % KBF_4.

The density of the melts of the system $KF-KCl-KBF_4-K_2TiF_6$ was measured in [11]. The concentration dependence of the molar volume of the quaternary system may be described by the general equation

$$V = \sum_{i=1}^{4} A_i x_i + \sum_{i \neq j}^{4} x_i x_j \left(B_{ij} + C_{ij} x_j\right) + \sum_{i \neq j \neq k}^{4} x_i x_j x_k \left(D_{ijk} + E_{ijk} x_k\right) + F x_1 x_2 x_3 x_4 \qquad (11)$$

The first term represents the ideal behaviour of the system (A_i being the molar volumes of the pure components), the second one expresses the interactions in the binary boundary systems, the third term describes the interactions in the ternary systems and the fourth one describes the interaction of all four components. The sum of the second, third and fourth terms represents the excess molar volume of the system. The values of the individual constants were calculated using the multiple linear regression analysis. Omitting the statistically non-important terms on the 0.95 confidence level, the following equation for the temperature 1100 K was obtained

$$\begin{aligned} V\,/\,\mathrm{cm^3 mol^{-1}} = {} & 30.13 x_{KF} + 49.93 x_{KCl} + 75.76 x_{KBF_4} + 114.33 x_{K_2TiF_6} + \\ & + 2.87 x_{KF} x_{KBF_4} - 3{:}74 x_{KCl} x_{K_2TiF_6} + 4.31 x_{KBF_4} x_{K_2TiF_6} + \\ & + x_{KF} x_{KBF_4} x_{K_2TiF_6} \left(43.47 - 231.80 x_{K_2TiF_6}\right) - 74.66 x_{KCl} x_{KBF_4} x_{K_2TiF_6}^2 + \\ & + 1456.38 x_{KF} x_{KCl} x_{KBF_4} x_{K_2TiF_6}^2 \end{aligned} \qquad (12)$$

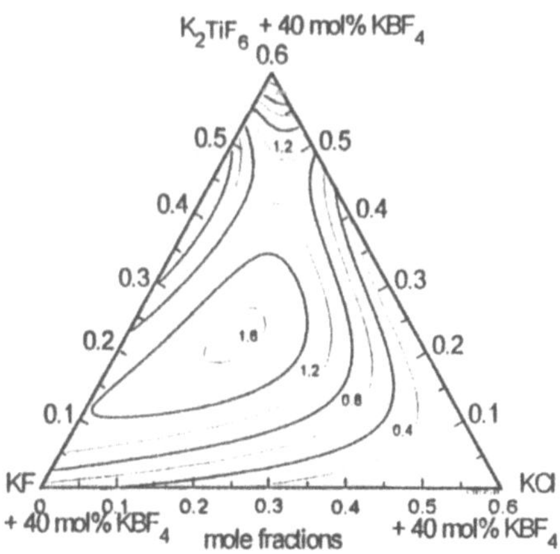

Fig. 12. Iso-lines of excess molar volume of the system $KF-KCl-KBF_4-K_2TiF_6$ with constant content 40 mole % of KBF_4 at 1100 K. Values are in $cm^3\,mol^{-1}$.

The standard deviation of approximation is $0.375\ cm^3\,mol^{-1}$.

The course of the excess molar volume of the $KF-KCl-KBF_4-K_2TiF_6$, calculated according to the eqn (12), is represented in the cross-section with constant content of 40 mole % KBF_4 (Fig. 12).

3.4. Viscosity

The torsion pendulum method was used for the viscosity measurement The platinum cylinder with the diameter of 15 mm and the height 20 mm was used as the measuring body. The oscillation of the pendulum system were followed by means of two phototransistors, placed in the path of a light beam reflected from a mirror attached to the pendulum. The experimental error in the viscosity measurement did not exceed ± 1 %.

The viscosity of the melts of the system KF–KCl–KBF_4 was measured in [12] Using the multiple linear regression analysis of data published in [12] the following final equation at 1100 K was obtained

$$\eta/(\mathrm{mPa\,s}) = 1.431^{x_{KF}}0.953^{x_{KCl}}0.638^{x_{KBF_4}} - \\ - 0.562x_{KF}x_{KCl} - 0.501x_{KF}x_{KBF_4} - 0.541x_{KCl}x_{KBF_4} - \\ - 4.576x_{KF}x_{KCl}x_{KBF_4} \qquad (13)$$

The standard deviation of the fit is 0.019 mPa.s. The excess viscosity of the investigated system is shown in Fig. 13.

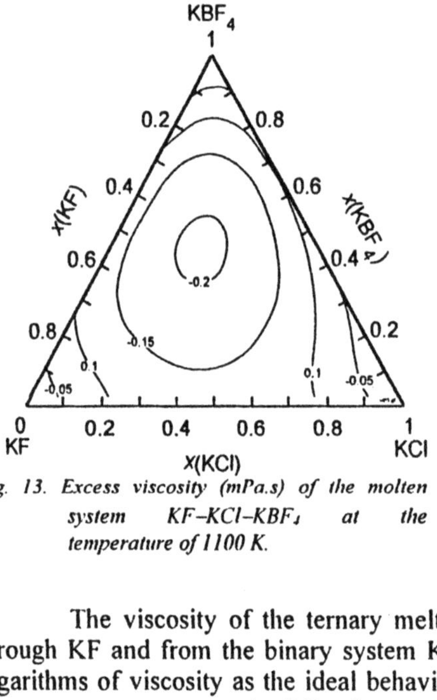

Fig. 13. Excess viscosity (mPa.s) of the molten system KF–KCl–KBF₄ at the temperature of 1100 K.

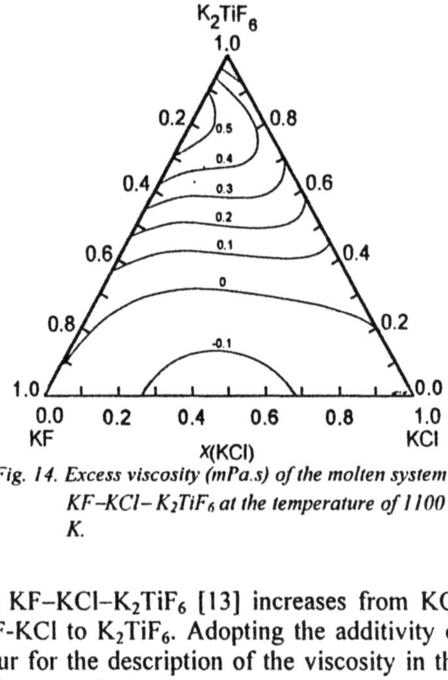

Fig. 14. Excess viscosity (mPa.s) of the molten system KF–KCl–K₂TiF₆ at the temperature of 1100 K.

The viscosity of the ternary melts KF–KCl–K_2TiF_6 [13] increases from KCl through KF and from the binary system KF-KCl to K_2TiF_6. Adopting the additivity of logarithms of viscosity as the ideal behaviour for the description of the viscosity in the three-component system the following equation was obtained

$$\ln\eta = \Big[-2.438x_{KF} - 2.777x_{KCl} - 1.948x_{K_2TiF_6} + \\ + \frac{1}{T}\left(3068x_{KF} + 3002x_{KCl} + 3641x_{K_2TiF_6}\right)\Big] - \\ - \Big[\left(1.000x_{KF}x_{KCl} + 5.728x_{KF}x^3_{K_2TiF_6} + 11.20x_{KCl}x^3_{K_2TiF_6}\right) + \\ + \frac{1}{T}\left(620.1x_{KF}x_{KCl} + 8072x_{KF}x^3_{K_2TiF_6} + 13390x_{KCl}x^3_{K_2TiF_6}\right)\Big] + \\ + 21.20x_{KF}x^2_{KCl}x^3_{K_2TiF_6} \qquad (14)$$

with the standard deviation of the fit σ = 7.1×10-4. The validity of eqn (14) is in the temperature interval 700–1300 K. The course of the excess viscosity in the binary systems is shown in Fig. 14.

The viscosity of the melts of the system KF– KBF_4–K_2TiF_6 was measured in [14]. The viscosity of the ternary melts increases from KBF4 through KF to K_2TiF_6. Surprisingly low seems to be the viscosity of KBF4, however, there is a substantial overheating of the KBF_4 melt at 1100 K compared with that of KF.

For the viscosity of this system at 1100 K the following equation has been calculated

$$\begin{aligned}\eta / \text{mPa.s} = {} & 1.421^{x_{KF}} \times 0.638^{x_{KBF_4}} \times 4.247^{x_{K_2TiF_6}} - 0.489 x_{KF} x_{KBF_4} + \\ & + 4.795 x_{KF} x_{K_2TiF_6}^4 + 4.385 x_{KBF_4} x_{K_2TiF_6}^4 + \\ & + 0.079 x_{KF} x_{KBF_4}^5 x_{K_2TiF_6}^3 \end{aligned} \qquad (15)$$

with the standard deviation of the fit $\sigma = 0.011$ mPa.s. The excess viscosity of the ternary system KF– KBF_4–K_2TiF_6 is shown in Fig. 15.

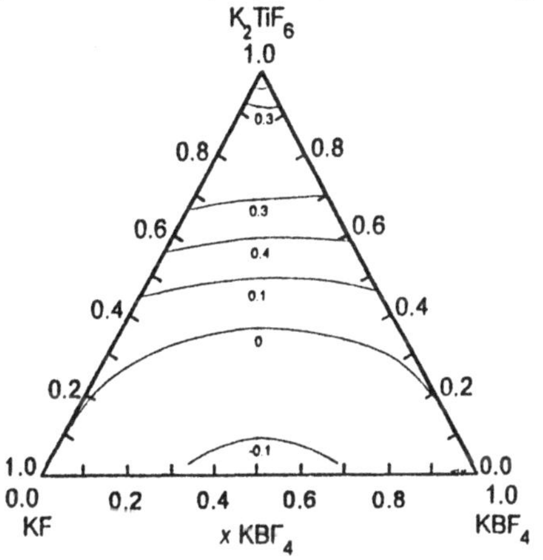

Fig. 15. Excess viscosity (mPa.s) of the molten system KF– KBF_4–K_2TiF_6 at the temperature of 1100 K.

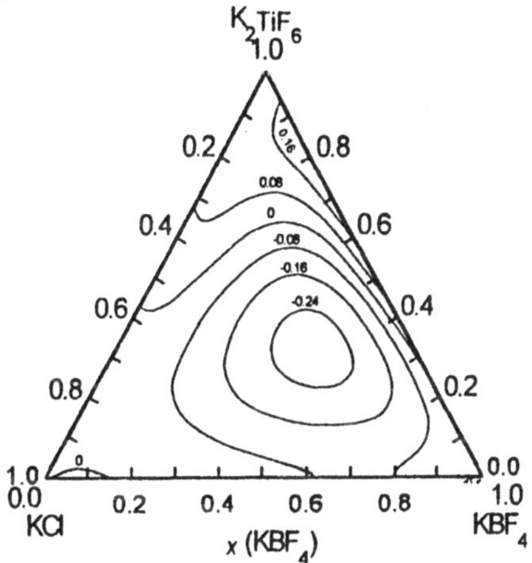

Fig. 16. Excess viscosity (mPa.s) of the molten system KCl KBF_4–K_2TiF_6 at the temperature of 1100 K.

The viscosity of the ternary system KCl–KBF_4–K_2TiF_6 was measured in [15]. For the viscosity of the system KCl–KBF_4–K_2TiF_6 at the temperature of 1100 K the following equation was obtained

$$\begin{aligned}\eta / \text{mPa.s} = {} & 0.960^{x_{KCl}} \times 0.631^{x_{KBF_4}} \times 4.342^{x_{K_2TiF_6}} - \\ & - 0.644 x_{KCl} x_{KBF_4}^2 - 2.427 x_{KCl} x_{K_2TiF_6} - \\ & - 2.549 x_{KBF_4} x_{K_2TiF_6} - 63.784 x_{KCl} x_{KBF_4}^2 x_{K_2TiF_6}^2 \end{aligned} \qquad (16)$$

with the standard deviation of approximation $\sigma = 0.033$ mPa.s. The excess viscosity of the investigated system at 1100 K based on eqn (16) is shown in Fig. 16. It can be seen from the figure that there are two different regions in the ternary system. The first region with low concentration of K_2TiF_6 shows negative deviations from the viscosity, while the second region with higher K_2TiF_6 concentration exhibits positive excess viscosity.

4. Discussion

From the results of the physicochemical analysis it follows that deviations from the ideal behaviour were observed in all the boundary binaries as well as in the ternary system. With regard to the fact that the investigated system has a common cation, the observed deviations from the ideal behaviour must be a consequence of the anionic interaction only. The interaction of components found may be of different origin. First of all a different character of interaction has to be considered in the boundary binaries KF–KBF_4 and KCl–KBF_4.

In the pure KF–KBF_4 melt BF_4^- tetrahedra tend to link, forming relatively weak B-F-B bonds. The strength of this bond depends strongly on temperature. Introducing F^- ions into the KBF_4 melt by addition of KF, the B-F-B bridges break off, which leads to the lowering of the viscosity and the negative deviation from the ideal behaviour in the KF–KBF_4 system. Besides, the mixing of small anions F^- with relatively large BF_4^- ones takes place. In systems of this type the deviation from additivity is proportional to the fractional difference in the radii of the different anions [16]. Therefore a relatively large deviation from ideality may be achieved, which was confirmed either by the density [7], conductivity [17], and viscosity [12] measurements.

An opposite interaction effect takes place in the system KCl–KBF_4. There is the mixing of two relatively large and more polarizable anions BF_4^- and Cl^-. Like in the systems of alkali metal chlorides, bromides and iodides with a common cation, no important deviations from additivity were found in these systems [18, 16]. Such behaviour was confirmed also in the system KCl–KBF_4 by the density [7], conductivity [17], and viscosity [12] measurements. On the other hand, introducing Cl^- ions into the KBF_4. melt, the exchange of fluoride atoms in the BF_4^- tetrahedra by the chloride ones according to the general scheme

$$BF_4^- + nCl^- \Leftrightarrow \left[BF_{4-n}Cl_n\right]^- + nF^- \qquad (17)$$

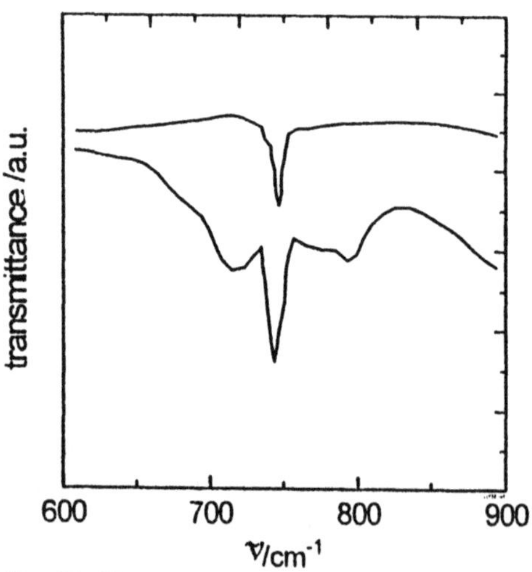

Fig. 17. IR spectra of KBF_4 and of the quenched molten KCl–KBF_4 (1:1 mole ratio) mixture.

and the possible presence of the $[BF_{4-n}Cl_n]^-$ mixed anions may be achieved. Consequently, the lower stability of the B-Cl-B bridges and the lower concentration of the B-F-B ones lead to the negative deviations of the properties in the KCl–KBF_4 system. This explanation is supported also by the asymmetric course of the excess viscosity which is due to the reaction (17) shifted to the right side in the region of high concentration of KBF_4.

The anionic interaction according to the reaction (17) was confirmed by means of the infrared spectroscopy method [19]. The IR spectra of KBF_4 and that of the quenched molten KCl–KBF_4 (1:1 mole ratio) mixture are shown in Fig. 17.

With except of the 600-900 cm^{-1} region, the mid infrared spectra of both samples are almost identical. In agreement with the earlier study of the alkali metal tetrafluoroborates [20], common vibrations can be assigned to crystalline KBF_4. Significant differences have been observed only in the 600–900 cm^{-1} region. It can be seen that beside the ν(1) vibration corresponding to KBF_4, the quenched molten $KCl–KBF_4$ mixture produces two additional peaks at 760 and 796 cm^{-1} with a shoulder at 770 cm^{-1}. It may be assumed, that these peaks arise due to the different B-F and B-Cl valence vibrations in the $[BF_{4-n}Cl_n]^-$ anions. The substitution of the fluorine atom by the chlorine one in the co-ordination sphere of the BF_4^- anion in the molten $KF–KCl–KBF_4$ mixtures was thus confirmed. However, the type of the mixed anion had not be determined neither by physicochemical analysis nor by spectroscopic method.

In the binary systems $KF–K_2TiF_6$ and $KCl–K_2TiF_6$ the compounds K_3TiF_7 and K_3TiF_6Cl are formed according to the reactions

$$K_2TiF_6(l) + KF(l) = K_3TiF_7(l) \qquad (18)$$
$$\Delta_r G^\circ_{1100\,K} = 0.963\,kJ\,mol^{-1}$$

$$K_2TiF_6(l) + KCl(l) = K_3TiF_6Cl(l) \qquad (19)$$
$$\Delta_r G^\circ_{1100\,K} = 0.587\,kJ\,mol^{-1}$$

The formation of the K_3TiF_7 and K_3TiF_6Cl compounds does not affect substantially the volume properties. This indicates that either these compounds dissociate in the melt in a considerable degree, or the volumes of the anions TiF_6^{2-}, TiF_7^{3-} and TiF_6Cl^{3-} are fairly similar. However, the formation of both compounds affects the course of the volume properties differently. While the formation of K_3TiF_7 is connected with small volume expansion, the formation of K_3TiF_6Cl is connected with substantially higher volume contraction. The difference is probably coherent with different stereochemistry of both complex anions and with its relations to the other present anions.

The thermodynamic analysis of the phase equilibria and of the volume properties of the system $KF–K_2TiF_6$ and $KCl–K_2TiF_6$ were used for the calculation of the dissociation degree of the K_3TiF_7 and K_3TiF_6Cl compounds. The dissociation degree of K_3TiF_7 calculated on the basis of the experimentally determined phase diagram is α_0 = 0.64 [97], which is in a very good accordance with the values obtained on the basis of the density data, α_0(1000 K) = 0.6 and α_0(1100 K) = 0.7 [8]. The corresponding data for K_3TiF_6Cl are α_0 = 0.78 from the phase equilibria, α_0(1000 K) = 0.72 and α_0 (1100 K) = 0.81 [8] from the density data. Based on the temperature dependence of the equilibrium dissociation constant the values of dissociation enthalpies ΔH°(dis,K_3TiF_7) = 52.4 kJ mol^{-1} and ΔH°(dis,K_3TiF_6Cl) = 46.6 kJ mol^{-1}, which is in both cases a major part enthalpy of fusion [21,22].

In the binary system $KBF_4–K_2TiF_6$ at 1100 K the partial molar volume of K_2TiF_6 for $x(KBF_4) \rightarrow 1$ is $V(K_2TiF_6)$ = 129.69 cm^3mol^{-1}. This value is higher than the molar volume of pure K_2TiF_6 (V° = 114.62 cm^3mol^{-1}), which indicates the formation of larger complex ions, *e.g.* TiF_7^{2-}. The partial molar volume of KBF_4 for $x(K_2TiF_6) \rightarrow 1$ is

$V(KBF_4)$ = 68.08 cm^{3}mol^{-1}. This value is lover than the molar volume of pure KBF_4 (V^o = 75.12 cm^{3}mol^{-1}) The reason for this lowering is the decomposition of KBF_4 and escape from the melt of gaseous BF_3, which was actually observed. On the basis of the given findings the following chemical reaction may be considered to take place in the binary system KBF_4–K_2TiF_6

$$KBF_4(l) + K_2TiF_6(l) = K_3TiF_7(l) + BF_3(g) \qquad (20)$$
$$\Delta_r G''_{1100K} = 10.71\,\text{kJ mol}^{-1}$$

The relatively low positive value of the reaction Gibbs energy, as well as the observed escape of gaseous BF_3, indicate the probability of reaction (20) in the melt.

In quaternary system KF–KCl–KBF_4–K_2TiF_6 the next chemical reaction is possible

$$KCl(l) + 2K_2TiF_6(l) + KBF_4(l) = K_3Ti_7F(l) + K_3TiF_6Cl(l) + BF_3(g) \qquad (21)$$
$$\Delta_r G^o_{1100K} = 12.05\,\text{kJ mol}^{-1}$$

The originating additive compounds dissociate thermally according to reactions (18, 19). The Gibbs energy of reaction (20, 21) was calculated on the basis of the Gibbs energies of dissociation reactions of K_3TiF_7 and K_3TiF_6Cl [8], as well as of the Gibbs energies of formation of KF, KCl and KBF_4 and BF_3 given in [23].

5. Conclusion

From the physicochemical properties of the melts of the system KF–KCl–KBF_4–K_2TiF_6 it follows that the most characteristic feature of these melts is the formation in the melt of the thermodynamically less stable ionic species TiF_7^{3-} and TiF_6Cl^{3-} with lowered symmetry of the co-ordination sphere. The presence of these anions in the melt facilitates most probably the electroreduction of titanium and thus the formation of titanium diboride on the cathode.

References

1. K. Matiašovský, K. Grjotheim, M. Makyta, *Metall* **42**, 1192 (1988).
2. M. Makyta, P. Zatko and V. Daněk, *Influence of the Melt Structure on the Deposition of Molybdenum, Titanium and Aluminium.* Proc. IX. Int. Symp. Molten Salts, San Francisco, May 1994.
3. O. Patarák and V. Daněk, *Chem. Papers, 46* (1992) 91.
4. R. V. Chernov and P. M. Ermolenko, *Zh. Neorg. Khim., 18* (1973) 1372.
5. M. Chrenková, O. Patarák and V. Daněk, *Thermochim. Acta, 273* (1996) 157.
6. M. Chrenková, O. Patarák and V. Daněk, *Chem. Papers, 49* (1995) 167.
7. M. Chrenková and V. Daněk, *Chem. Papers, 45* (1991) 213.
8. V. Daněk, J. Šiška and K. Matiašovský, *Chem. Papers, 42* (1988) 753.
9. M. Chrenková, V. Daněk and D. K. Nguyen, *Z. physik. Chemie, 194,* (1996) 51.

10. M. Chrenková, V. Daněk and A. Silný, *Ber. Bunsenges. Phys. Chem.*, in press.
11. M. Chrenková, V. Daněk, A. Silný, to be published
12. V. Daněk and D. K. Nguyen, *Chem. Papers, 49* (1995) 64.
13. D. K. Nguyen and V. Daněk , *Chem. Papers, 50* (1996) 4.
14. A. Silný, V. Daněk and D. K. Nguyen, *Ber. Bunsenges. Phys. Chem., 99* (1995) 985.
15. D. K. Nguyen and V. Daněk, *Chem. Papers,*.51 (1997) 73.
16. M. Chrenková and V. Daněk, *Chem. Papers, 44* (1990) 329.
17. M. Chrenková, M. Hura and V. Daněk, *Chem. Papers, 45* (1991) 739.
18. J. Sangster and A.D. Pelton, *J. Phys. Chem. Ref. Data, 16* (1987) 506.
19. V. Daněk, M. Chrenková, D. K. Nguyen, K. Putyera, *Chem Papers, 51* (1997) 69
20. J. B. Bates and A. S. Quist, *Spectrochim. Acta, 31(A)* (1975) 1317.
21. K. Adamkovičová, L. Kosa, I. Nerád, I. Proks, *Thermochim. Acta,* 262 (1995) 83
22. K. Adamkovičová, L. Kosa, I. Nerád, I. Proks, *Thermochim. Acta*, 287 (1996) 1
23. D.P. Stuhl, H. Prophet, *JANAF Thermochemical Tables*, 2nd Edition, NSRDS, Nat. Bur. Stand. Washington 1971.

STUDY OF ELECTRODE PROCESSES IN FLINAK-K_2TaF_7 - KBF_4 MELT

L.P. Polyakova[1] , E.G. Polyakov[1], G.A. Bukatova[1] A.A. Shevyryov[1], O.V. Makarova [1] , N.J.Bjerrum[2]

1- Institute of Chemistry KSC RAS, Apatity, Russia
2 - The Danish Technical University, Lyngby, Denmark

1. Introduction

Electrochemical synthesis in melts is one of several promising methods of producing refractory metal borides. To develop this method one needs to be aware of the mechanism of the processes occurring both in the melt bulk and on the surface of electrodes. The evidence available is mainly concerned with production of titanium and zirconium borides. The data on electrochemical synthesis of tantalum borides are contradictory and limited by the area of chloride-fluoride melts [1-5].

Rameau [1] believes that TaB is formed in the KCl-K_2TaF_7-KBF_4 melt in the course of two processes occurring simultaneously at the cathode, i.e. Ta(V) and B(III) reduction. TaB_2 is formed in melts with a low tantalum concentration owing to boriding of the previously formed monoboride.

According to Devyatkin et al. [2], one cathodic and one anodic wave in the NaCl-KCl-K_2TaF_7 melt are associated with three processes at coincident potentials, i.e. Ta(V) and B(III) reduction with the formation of TaB_2 and, individually - to B(III) reduction. However, the presented voltammograms contradict both this explanation and the data of other workers [6,7].

Kuznetsov et al. [4,5] observed a single cathodic peak in NaCl-KCl and NaCl-KCl-NaF melts containing K_2TaF_7 and KBF_4. This peak was attributed to the formation of heteronuclear tantalum and boron complexes with the formation of TaB_2. In the opinion of the authors, these complexes are formed as the result of a chemical reaction preceding the electrochemical stage. However, the experimental data cited in this work do not corroborate the suggested pattern of the process of synthesis.

The aim of this work is studying the electrochemical behaviour of tantalum and boron complexes in fluoride melts containing both components.

2. Experimental

In this work the methods of linear voltammetry (LVA) and reversive chronopotentiometry were used. The experiment was described in detail earlier [8-10]. The working

D.H. Kerridge and E.G. Polyakov (eds.), Refractory Metals in Molten Salts, 103-108.

electrodes were made of silver and glassy carbon, the glassy carbon vessel for the electrolyte was used as an auxiliary electrode. The potentials were measured relatively to a glassy carbon quasireference electrode (GC-QRE). A potential sweep rate to the extent of silver dissolution made possible a more rigid control of the measured E values. Oxygen content in the melt measured by the LVA method did not exceed $(1\text{-}3)\cdot 10^{-3}$ m/o. The studies were conducted in the temperature interval 700-750°C, the K_2TaF_7 concentration was varied from 0.03 to 0.75 m/o, and that of KBF_4 - from 0.05 to 2.0 m/o.

3. Results and discussion

Tantalum in FLINAK-K_2TaF_7 melt was earlier shown to reduce at $V<0.5\ V\cdot s^{-1}$ in one quasireversible stage [9]. According to [10], the B(III) reduction in a fluoride melt is a more complicated process and depends on the KBF_4 concentration. When the latter is lower than $5{,}7\cdot 10^{-2}$ m/o, the cathodic process $B(III) + 3e \rightarrow B(0)$ is irreversible. At higher boron concentrations its reduction occurs under surface ohmic control. A typical voltammogram for the latter case is shown in Fig.1 (curve 1). Peak R_4 corresponds to the fluoride complex discharge with the formation of elemental boron, peak R_5 may be attributed to the formation of an alkali metal boride.

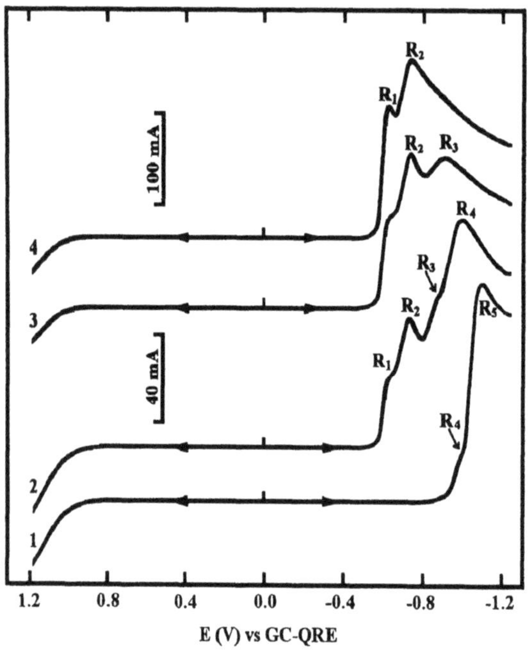

Figure 1. Change in the shape of voltammograms of the FLINAK-KBF_4 (0.594 m/o) melt with change in K_2TaF_7 concentration. T =710°C, scan rate 0.25 $V\cdot s^1$, A_{Ag}=0.24 cm^2, $C_{K_2TaF_7}$ (m/o) : 1- 0; 2 -0.075; 3 - 0.298; 4 -0.593

The addition of K_2TaF_7 to this melt modelled the case of a simultaneous presence of boron and tantalum in the fluoride melt. To avoid the possible errors two ways of titration were used: tantalum addition to the boron containing melt and boron addition to the tantalum containing one.

The first additions of K_2TaF_7 to the FLINAK - KBF_4 melt resulted in the appearance of new peaks R_1, R_2 and R_3 (Fig. 1, curve 2) and disappearance of peak R_5. The peak R_1 potential corresponded to that of tantalum discharge in the fluoride melt. Further additions of K_2TaF_7 resulted in decrease of the number of cathodic peaks. Only R_1 and R_2 were seen when the molar ratio B/Ta was equal to 1 (Fig. 1, curve 4).

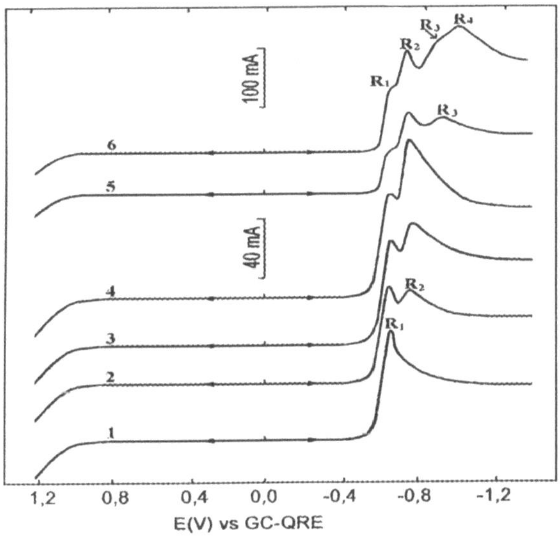

Figure 2. Change in the shape of voltammograms of the FLINAK-K_2TaF_7-KBF_4 melt with change in B/Ta molar ratio. 710^0C, $A_{Ag} = 0.36\ cm^2$, scan rate $0.25\ V{\cdot}s^1$, $C_{K_2TaF_7} = 0.222$ m/o, B/Ta molar ratio: 1 - 0; 2 - 0.25; 3 - 0.5; 4 - 1.0; 5 - 2.0; 6 - 4.0

The voltammograms observed on titration of tantalum containing melts with fluoroborate had the same shape when the molar ratio B/Ta reached unity (Fig. 2, curve 4). At first glance it was possible to suggest that peak R_1 in Figures 1 and 2 (curves 2-6) corresponded to Ta(V) reduction. However, the cyclic voltammograms are at variance with this supposition. The anodic part of the voltammograms is considerably more complicated than the cathodic one in the peak R_1 region and the number of peaks on it depends on the reverse potential.

Three anodic peaks in Fig. 3, curves 2′, and 3″, point to the complex character of the boron and tantalum reduction in the narrow region overlapped by the cathodic peak

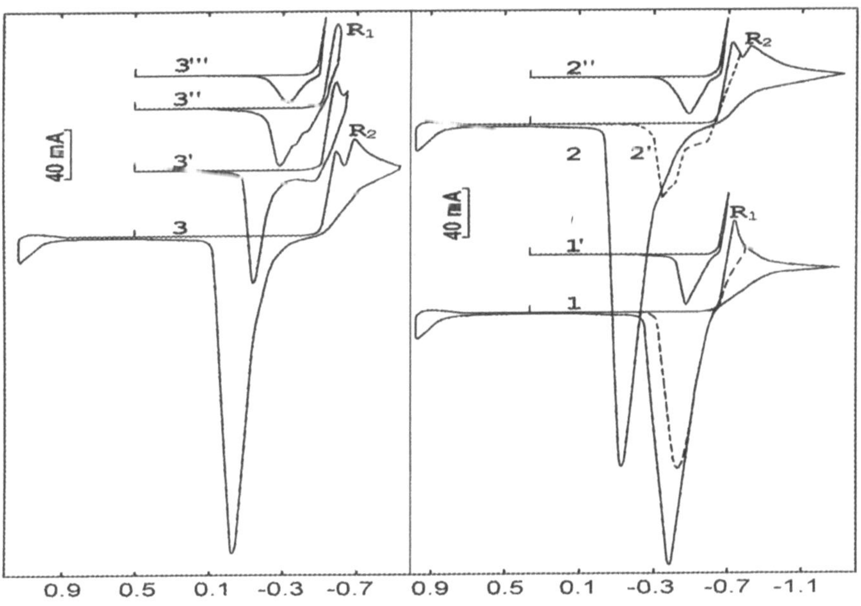

Figure 3. Cyclic voltammogramms of FLINAK - K_2TaF_7 (1,1′) and FLINAK-K_2TaF_7-KBF_4 melts (2-2″, 3-3‴). $A_{Ag}=0.36\ cm^2$, $V= 0.25\ V{\cdot}s^{-1}$, $T=710^0C$, $C_{K_2TaF_7} = 0.222$ m/o, C_{KBF_4} (m/o): 1 -0; 2 -0.055; 3 - 0,111

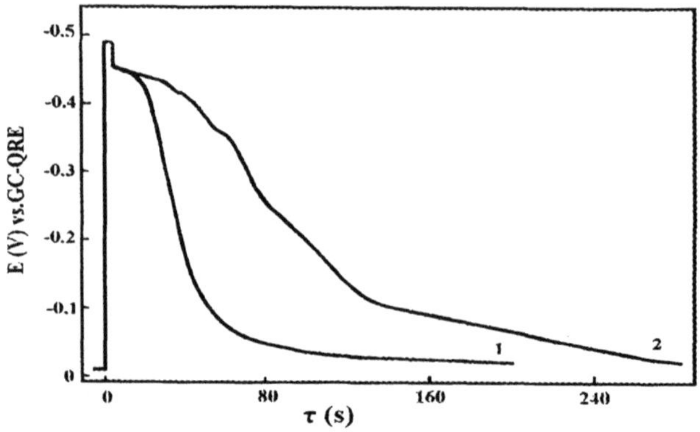

Figure 4. Open-circuit potential decay FLINAK- K_2TaF_7 (1) and FLINAK- K_2TaF_7- KBF_4 (2) melts, i_c=5 mA, τ_c=5 s, T = 710°C, A_{Ag}= 0.36 cm^2, $C_{K_2TaF_7}$ =0.222m/o, B/Ta molar ratio: 1- 0; 2 - 0.25

R_1. Fig. 4 shows the E-τ curves obtained under similar conditions in melts either undoped with KBF_4 (curve 1) and doped ones with the molar ratio B/Ta=0.25 (curve 2). The three characteristic plateau within the range -0.42 ÷ -0.1V on the curve are in this case the evidence of the formation of three B-Ta compounds. The plateau common for both curves around -0.45V corresponds to tantalum dissolution.

Electrolysis at a constant potential was conducted in melts where the molar ratio B/Ta=1. According to the XRD data, at -0.555V the cathodic deposit is represented by cubic tantalum, but at -0.570V - by its tetragonal form. This fact can be explained as restructuring of the crystalline lattice due to increase of the boron concentration in tantalum as a solid solution. The deposits obtained at -0.590 V(R_1) and -0.680 V(R_2) are multiphase with boron contents of 0.95 and 1.93 w/o, respectively.

Increasing the B/Ta molar ratio to more than unity in the melt caused the appearance of a more negative peak R_3 (Fig.2, curve 5). When the above ratio became higher than three, an extra peak R_4 appeared (Fig. 2, curves 6). There was practically no change in peaks R_1, R_2 and R_3. Electrolysis in the melt with the B/Ta ratio greater than 4 at a potential -0.555V also yielded tetragonal tantalum. The latter is evidence of the increased contamination of tantalum with boron in melts with a high KBF_4 concentration. At peak potentials R_3 and R_4 of a cathodic deposit TaB_2 was obtained. The precise XRD analysis of the products obtained at higher positive potentials is impossible due to overlapping of the spectra of numerous borides in the Ta-B system [11].

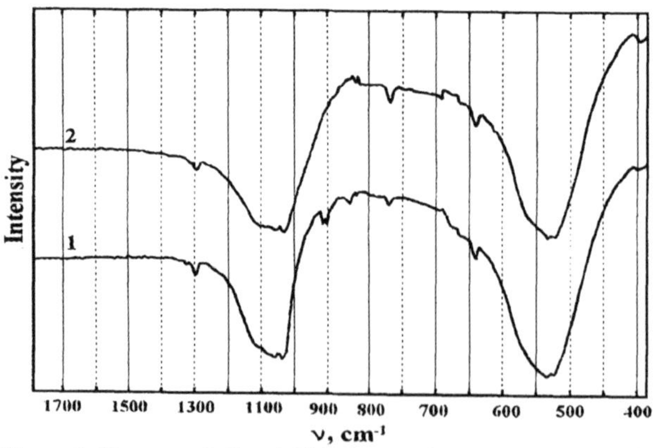

Figure 5. IR spectra of alloyed (1) and mechanical (unmelted)(2) K_2TaF_7- KBF_4 mixtures, molar ratio B/Ta =1

In this work an attempt was made to find in the bulk of the melt in question the heteronuclear complexes whose existence has recently been extensively discussed in the literature [4, 5, 12-14]. For this purpose the KBF_4 and K_2TaF_7 mixtures with the molar ratio 1/1 and 2/1 were melted, held in the molten state and speedily cooled. In Fig. 5 the IR spec-

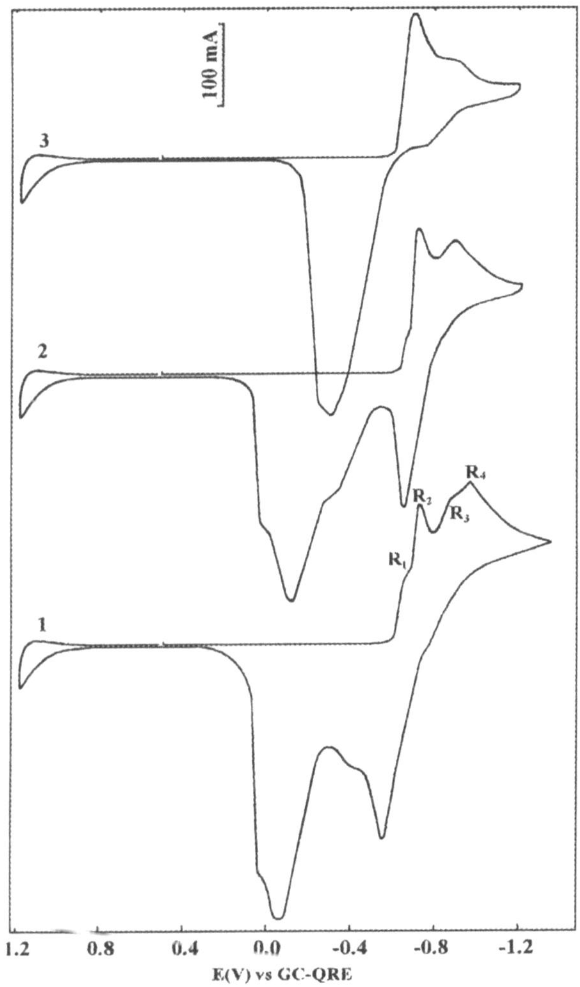

Figure 6. Change in the shape of voltammograms of the FLINAK- K_2TaF_7-KBF_4 melt with change in oxide-ion concentration. V=0.25 V·s[1], A_{Ag}=0.36 cm[2], T=710[0]C, $C_{K_2TaF_7}$ =0.218 m/o, C_{KBF_4} =0.594 m/o, C_{Na_2O} (m/o): 1 - 0.033; 2 - 0.139; 3 - 0.545

trum of an alloyed sample consisting of 50 m/o K_2TaF_7 and 50 m/o KBF_4 is shown in comparison with a mechanical (unmelted) mixture of the same composition. The coincidence is indicative of the absence of interaction between the melt components. This conclusion is corroborated by the XRD data as well. The results obtained for the samples studied by us as well as the data pertaining to investigation of the physico-chemical properties of the fluoride melts doped with K_2TiF_6 and KBF_4 [15, 16] do not indicate the existence of heteropolynuclear complexes in them.

The shape of the FLINAK-K_2TaF_7-KBF_4 melt voltammograms changes radically on introduction of oxygen in it (Fig. 6). As the O^{2-} concentration increases, one cathodic and one anodic peak began to predominate on both branches. This circumstance is probably the reason for the contradictory results for the melts of the same type.

4. Conclusions

The mechanism of electrode reactions in FLINAK- K_2TaF_7-KBF_4 melt was shown to change on varying the B/Ta ratio in the electrolyte. The cathodic deposit composition depends on the electrode potential value. It consisted initially of cubic and then of tetragonal tantalum, but that the increasing boron concentration made it possible to form individual phases of tantalum borides. Our findings suggest that tantalum and boron in the above melts discharge independently in a narrow range of potentials. The positive shift of the boron discharge potential is caused by depolarization due to formation of chemical compounds with a strong bond between tantalum and boron. There was not even indirect evidence that heteropolynuclear complexes exist in FLINAK- K_2TaF_7-KBF_4 melts.

5. Acknowledgements

The authors[1] wish to acknowledge their thanks to the Russian Foundation for Basic Research for the financial support of this work (grant 96-03-32175).

6. References

1. Rameau, J.J. (1971) Contribution à l'etude de la formation du tantale et des borures de tantale par électrolyse ignée, *Rev. Int. Hautes Temp. Réfract.*, **8**,.59-69.
2. Deviatkin, S.V., Taranenko, V.I., Kushkhov, Kh.B., and Shapoval V.I. (1992) High-temperature electrochemical synthesis of tantalum diboride in chloride-fluoride melts, *Rasplavy (Russ.)*, **2**, 70-73.
3. Deviatkin, S.V., Kaptay, G., Shapoval, V.I., and Berecz, E. (1993) Electrochemical synthesis of transition metal diboride from molten salts, *Proc.- Electrochem. Soc., (Molten Salt Chemistry and Technology 1993)*, **93-9**, 584-599.
4. Kuznetsov, S.A., Glagolevskaya, A.L., and Belyaevskii, A.T. (1994) Electrochemical preparation of tantalum boride coatings in salt melts, *Zh. Prikl.Khim. (Russ)*. **67**, 1093-1099.
5. Kuznetsov, S.A., Glagolevskaya, A.L. (1996) Cathodic processes involved in the electrosynthesis of tantalum borides, *Electrokhimiya (Russ.)*, **32**. . 344-351.
6. Polyakova, L.P., Kononova, Z.A., Polyakov, E.G., and Kremenetsky, V.G. (1996) Effect of caesium chloride on the diffusion coefficient value of tantalum complexes in NaCl-KCl-K_2TaF_7 melts, *Zh. Prikl.Khim. (Russ.)*, **69**, 1307-1313.
7. Lantelme, F., Barhoun, A., and Li, G. (1992) Electrodeposition of tantalum in NaCl-KCl-K_2TaF_7 melts, *J. Electrochem. Soc.*, **139**, 1249-1255.
8. Polyakova, L.P., Polyakov, E.G., Sorokin, A.I., and Stangrit, P.T. (1992) Secondary processes during Ta electrodeposition in molten salts, *J. Appl. Electrochem.*, **22**, 628-637.
9. Polyakova, L.P., Polyakov, E.G., Matthiesen, F., Christensen, E., and Bjerrum, N.J. (1994) Electrochemical study of tantalum in fluoride and oxofluoride melts, *J. Electrochem. Soc*, **141**, 2982-2988.
10. Polyakova, L.P., Bukatova, G.A., Polyakov, E.G., Christensen, E., Barner, J.H., and Bjerrum, N.J. (1996) Electrochemical behaviour of boron in LiF-NaF-KF melts, *J. Electrochem. Soc.*, **143**, 3178-3186.
11. Samsonov, G.V., and Vinitskij, I, M. (1976) *Tugoplavkie soedinenija (Refractory compounds)*, Metallurgiya, Moscow.
12. Taranenko, V.I., Zarutskii, I.V., Shapoval, V.I., Makita, M., and Matiašovský, K. (1992) Mechanism of cathode process in the electrochemical synthesis of titanium diboride, *Electrochim. Acta.*, **37**, 263-268.
13. Kuznetsov, S.A., Glagolevskaya, A.L. (1995) Stabilization of Cr(III) in the formation of heteronuclear complexes of chromium, aluminium and electrodeposition of alloys from these complexes, *Electrokhimiya (Russ.)*, **31**, 1389-1392.
14. Volkov, S.V. (1990) Coordinative compounds in high-temperature and laserochemical synthesis, *Zhur. Vsesoyuzn. Khim. ob-va im. D.I. Mendeleeva (Russ) [Mendeleev Chemistry Journal]*, **35**, 686-696.
15. Chrenkova, M., Daněk, V., and Silny, A. (1996) Structure of melts of the system KCl-KBF_4-K_2TaF_7, *EUCHEM Conference on Molten Salts. Smolenice. Slovakia*, A-21.
16. Nguen, K.D., Daněk, V., and Silný, A. (1996) Viscosity of subsystems of the molten system KF-KCl-KBF_4-K_2TaF_7, *EUCHEM Conference on Molten Salts. Smolenice. Slovakia*, A-38.

THE COMMON REGULARITIES OF EPITAXIAL GROWTH OF MONOCRYSTALLINE MOLYBDENUM AND TUNGSTEN ELECTRODEPOSITED FROM MOLTEN SALTS

N.O. ESINA, A.M. MOLCHANOV, Z.I. VALEEV,
A.N. BARABOSHKIN
Institute of High-Temperature Electrochemistry, Ural Branch of RAS, Ekaterinburg, Russia

1. Introduction

The formation and growth of an electrodeposited phase is a complex process and many methods have been used to study it. The main feature of our laboratory scientific investigations is intensive study of the structure of poly- and monocrystalline refractory metal deposits, such as tungsten, molybdenum, rhenium, iridium, ruthenium and etc. We have shown that the structure of electrodeposited layers directly depends on the conditions of electrolysis [1]. Moreover this structures a most sensitive instrument to study the properties of our electrolyte, including purity of the melt, ion composition, dissipative ability, and kinetic parameters of electrodeposition. It was established that the reduction of oxygen in a chloride melt changed the direction of molybdenum growth texture from <111> to <001> direction with a fine structure of epitaxial layers [2].

The present report is devoted to the epitaxial growth of molybdenum and tungsten, electrodeposited from oxide and halide melts onto monocrystalline molybdenum and tungsten substrates with four orientations (<001>; <011>; <112>; <111>) . In particular to study the structure of deposits, the relative growth rates of monocrystalline planes with different oientations and the process of destruction of epitaxial growth.

2. Experimental

Electrodeposition was carried out from a molten mixture of sodium, potassium and molybdenum (tungsten) chlorides (chloride-fluoride) using a sealed quartz electrolyser with a pure argon atmosphere. In addition the electrodeposition was performed from polytungstate molten salts in air, or in a gas-tight reactor made of a heat- resistant steel in an argon atmosphere.

D.H. Kerridge and E.G. Polyakov (eds.), Refractory Metals in Molten Salts, 109-116.

Deposition was under galvanostatic conditions at temperatures from 700 to 900°C with various cathodic current density within the range 0.01 -0.20 A/cm^2. The thickness of the deposits was from 0.01 to 1.00 mm [3,4].

The structural perfection of the deposits was investigated by metallographic, X-ray diffraction and X-ray topography methods [5]. The surface morphology was analysed with an optical microscope "Neophot-32" and by SEM "Camebax" and opticogoniometric methods.

3. Results and Discussion

3.1. SURFACE MORPHOLOGY

It was established that epitaxial growth proceeded layer by layer by a two-dimensional mechanism (Fig.1a). Hence there are crystallization difficulties, which are conditioned by two-dimensional nucleus formation. The main morphological features of single crystal molybdenum or tungsten deposits are the growth pits with a symmetry corresponding to the substrate . We also observed pyramids, or hillocks , and growth steps. It is assumed, that the appearence of pits results from crystal defects. The defect regions have a higher free energy, and therefore the thermodynamic activity of the metal in them is more than unity. This results in lowering of the local overpotential and decreasing of the growth rate of these regions. This phenomena is accompanied by facing of lateral sides of the pits. As seen from the opticogoniometric data, the lateral sides of the pits are faced by planes, which are not the ones with the least free surface energy for b.c.c. metals. This can be explaned assuming, either that the concave form must be faced by the most fast growing planes [6], or that the lateral sides of pits are, in reality, formed by the steps of close packed planes such as (001); (110) and (112). In latter case the step sizes must be less then the visible light vave length.

For example the surfaces of molybdenum and tungsten deposits were electrodeposited from halide melts on the (111) plane were fully covered with three-sided pits (piramids) faced by (113) planes (Fig.1b).

In contrast the surface of deposits on the (001) plane were smooth with separate square stepped pyramids (Fig.1c), which are typical for growth on screw dislocation. However square pits appeared on the deposits, if the current density was high enough. Pits were faced by (012) or (013) planes (Fig.1d).

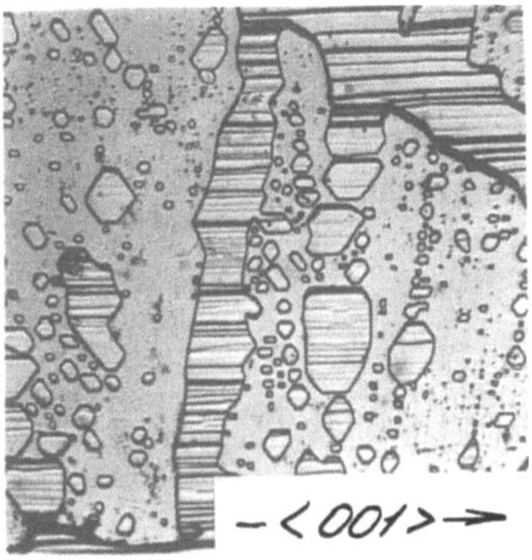

Figure 1a. The initial stage of molybdenum growth onto (011) plane of molybdenum substrate. x200

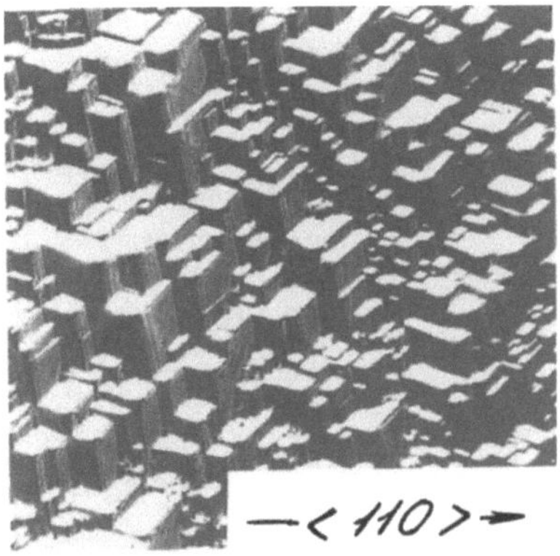

Figure 1b. Surface structure of single crystal molybdenum deposit on (111) plane of molybdenum. x200

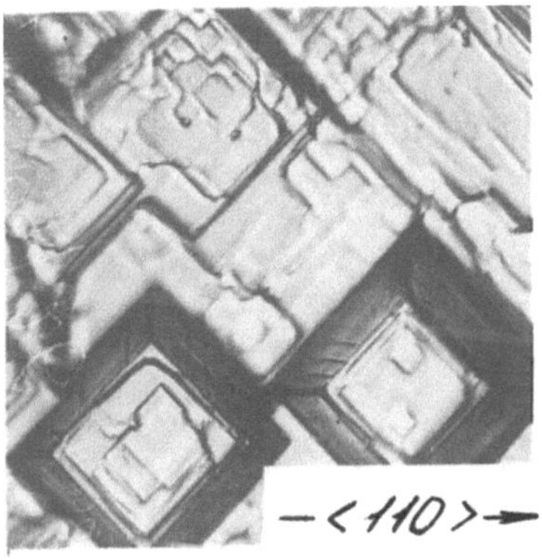

Figure 1c. Surface structure of tungsten deposit on (001) plane of molybdenum substrate. Cathodic current density: $0.01 A/sm^2$. x200

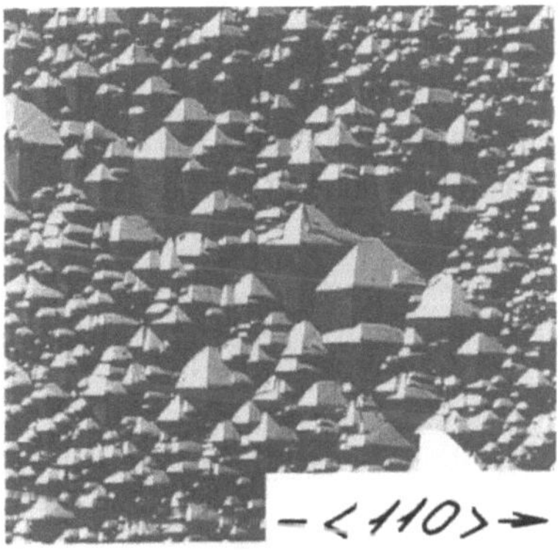

Figure 1d. Surface structure of molybdenum deposit onto (001) plane of molybdenum substrate. $i_c = 0.1 A/sm^2$. x200

The deposits on a (112) plane obtained at low current density had a lamellar structure with zigzag edges of layers (Fig.2a). At high current density the surface of deposits was formed with pits faced by (113) planes (Fig.2b).

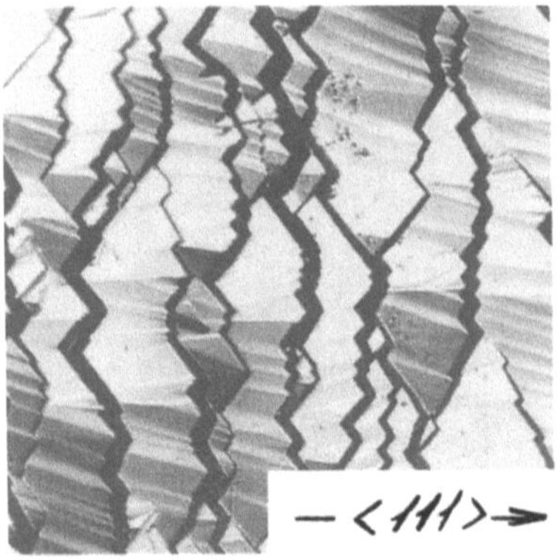

Figure 2a. Surface structure of molybdenum deposit on (112) plane of molybdenum substrate. i_c =0.01A/sm^2 . x200

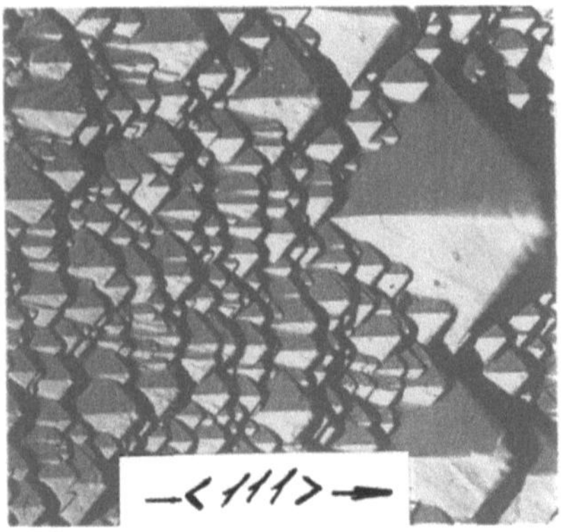

Figure 2b. Surface substrate of molybdenum deposit on (112) plane of molybdenum substrate. i_c = 0.1A/sm^2. x200

The surfaces of monocrystalline deposits on a (011) plane were covered by grooves, which were parallel to the <001> direction (Fig.2c). The grooves were faceted with (012) planes. Moreover the deposits grew smooth, if the current density was high enough (Fig.2d).

Figure 2c. Surface structure of molybdenum deposit on (011) plane of molybdenum substrate. i_c = 0.01A/sm ^{2}. x200

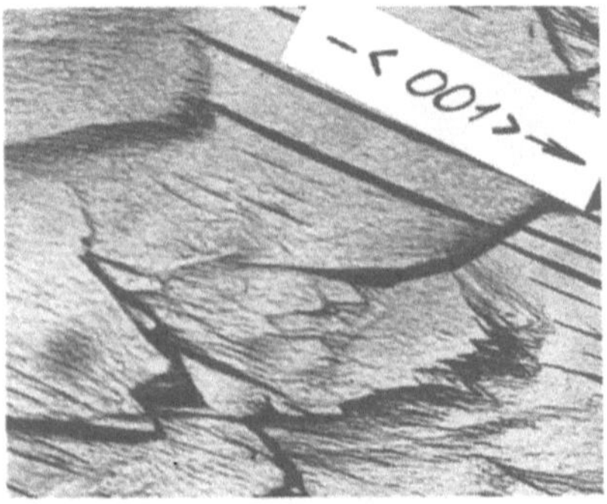

Figure 2d. Surface structure of molybdenum deposit on (011) plane of molybdenum substrate. i_c = 0.1A/sm^2 . x200

The surface of deposits formed from molten salts with low concentrations of discharge ions were covered by pyramids (but not pits) with symmetry corresponding to the symmetry of substrate plane.

It was established that transition from pits to pyramids of growth was conditioned by the slowest stage of the electrocrystallization process and was dependent on the concentration of the discharge ions. If the process of crystallization was the slowest stage (with a relatively high concentration of discharge ions- 9 mass% Me), then the surface morphology was conditioned the energy nonhomogeneus nature of the substrate and growth pits of were formed.

3.2 GROWTH RATES of SINGLE CRYSTAL

The relative growth rates of single crystal molybdenum and tungsten were measured at various conditions of electrolysis (current density, oxychemical impurities, concentration of discharge ions and temperature). The growth rate of monocrystalline deposits was always more than the polycrystalline. It was established that growth rate anisotropy of different single crystal planes for molybdenum and tungsten took place (Fig.3a,b).

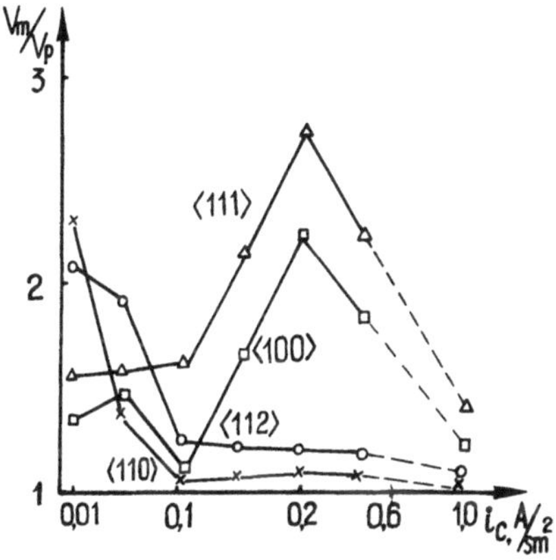

Figure 3a. Plots of the relative growth rates of monocrystalline molybdenum as a function of cathode current density.
NaCl-$MoCl_3$
(9mass.% of Mo)

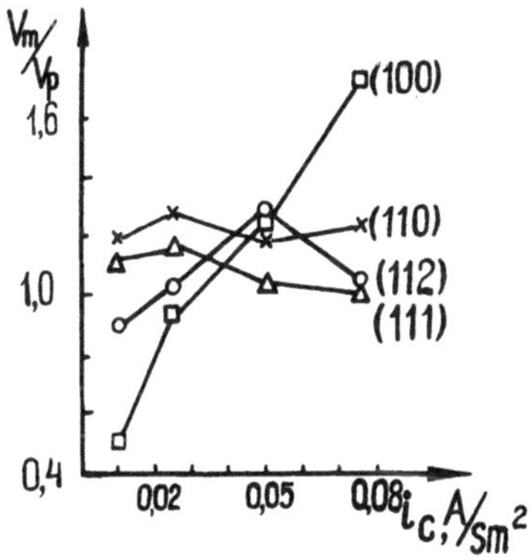

Figure 3b. Plots of the relative growth rates of monocrystalline molybdenum as a function of cathode current density.Polytungstate melt. 1.5 mass.% of Mo.

V_m - single crystall growth rate; V p - polycrystalline growth rate.

The degree of anisotropy depended on the purity of melt and a limited stage of the electrocrystallization process. The maximum value of growth rate anisotropy of the various monocrystalline planes corresponded to minimum impurity contained in the melt and growth layer, and the slowest stage of electrolysis is the act of crystallization.

Additions of the oxygen containing compounds to the melt lead to change the relative growth rates of single crystal molybdenum and tungsten planes.

Introduction into this electrolyte of an oxygen acceptor (Zr) caused the relative growth rates of planes to return their to initial values (Fig.4).
The crystallization overpotentials are small and not exceed 5 -10 mV (Fig.5).

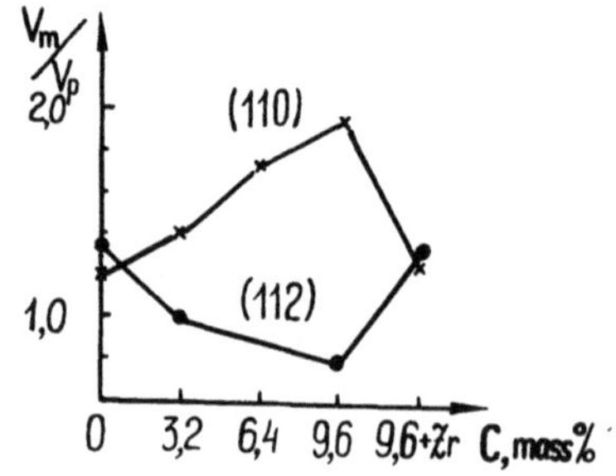

Figure 4. Plots of the relative growth rates of single crystal molybdenum as a function of concentration Na $_2$MoO$_4$ in the melt NaCl-KCl-MoCl $_3$.

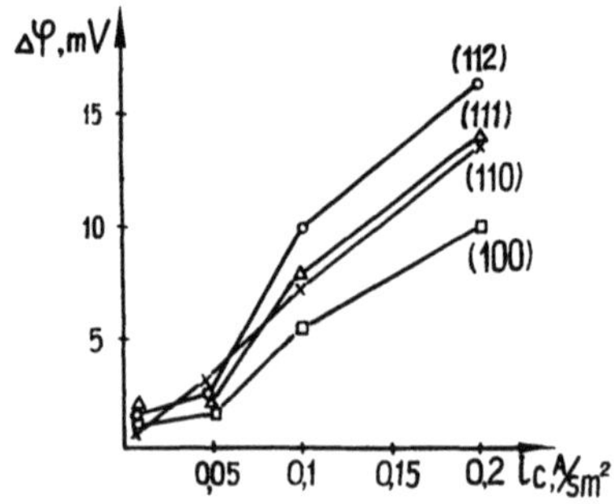

Figure 5. Overpotential of molybdenum electrodeposition onto the monocrystalline planes as a function of cathode current density.
NaCl-MoCl (9mass.%Mo).

3.3. EPITAXIAL GROWTH DESTRUCTION

The destruction of monocrystalline epitaxial growth occurs by accumulation of crystal defects as dislocations and than twin formation (Fig.6).
The grains in a twin position grow faster than the single crystal matrix and fill in all the surface of the deposit. After that these twin grains gradually incline to the direction of the growth texture axis. Multiple twinning, which can be seen in deposits from aqueous solutions, does not occur. In halide melts the destruction of monocrystalline epitaxial growth with formation of twins occurs only on the (001) plane for molybdenum and for tungsten deposits. The thickness of the monocrystalline layers on the (001) substrate plane depended on the presence of oxychemical impurities in the melt [7,8].

In oxide melts the twinning takes place on the substrates with <001>; <111>; <112> orientations, whereas on the sustrate with <011> orientation twinning was not observed. The formation of twins on (111) and (112) planes took place at low current density and high temperature. No twins were formed on the (001) plane at low current density but begin to form with increasing current density.

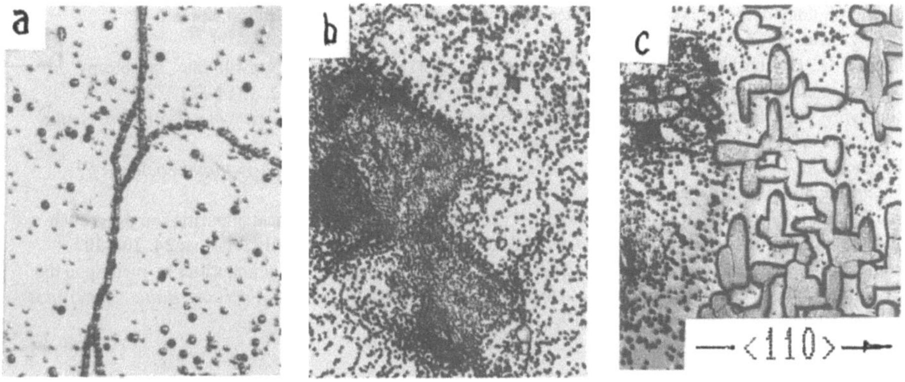

Figure 6. **Etch pits of dislocations with increasing thickness of molybdenum deposit on the (001) plane of molybdenum substrate (a,b); c- the twin defect formation.** x200
a - substrate, b - molybdenum deposit (thickness - 0.05mm)
c - molybdenum deposit (thickness - 0.1mm).

It was shown that epitaxial growth of molybdenum and tungsten onto bent monocrystalline molybdenum substrates with different orientations and radius of curvature occured during electrodeposition from oxide and halide melts. The morphology of bent growth surface corresponded to the morphology of platelike growth surface.

4. Conclusions

On the basis of these investigations the scientific foundations for technological production of monocrystalline specimens of different forms and orientations with high grade perfect structures (i.e. poly- and monofaced tubes, crucibles, boat containers, foils) were developed.

5. Acknowledgments

The NATO Organising Committee is acknowledged for its kind invitation. The authors wish to thank Dr. D.H. Kerridge for his help in editing the present manuscript

6. References

1.. Baraboshkin A.N. (1976) *Electrocrystallization of metals from molten salt*, Nauka Publishers, Moscow
2. Esina N.O., Valeev Z.I., Pankratov A.A. (1996) Structure of molybdenum electrodeposited from chloride melt with various contents of oxygen impurity in the melt. *Extend Abstracts, Baltic Conference of Interface*

Electrochemistry, Tartu 1996, 73-74.

3. Esina N.O., Tarasova K.P., Baraboshkin A.N. (1987) The effect of tungsten trioxide concentration on the morphology of monocrystalline tungsten layers electrodeposited from polytungstate melt, *Elektrokhimiya* **23**, 116-120.
4. Esina N.O., Baraboshkin A.N., Talanova M.I. and etc. (1981) Structure and growth rate anisotropy of monocrystalline molybdenum layers, *Physica metallov i metallovedenie* **52**, 1314-1317.
5. Gundyrev V.M., Belova N.V., Esin V.O.(1973) *Single crystal refractory and rare metals growth*, Nauka Publishers, Moscow.
6. Baraboshkin A.N., Esina N.O. (1981), *Extended Abstracts, 30th Meeting international Society of Electrochemistry, Yugoslavia* **1**, 349-352.
7. Baraboshkin A.N, Esina N.O., Talanova M.I.(1988) The effect of oxychemical impurities on the growth and dissolution rates of monocrystalline molybdenum facets, *Vysokochistye Veschestva* **24**, 206-207
8. Esina N.O., Molchanov A.M., Kalinin M.G.(1996) The effect of oxychemical impurities in melt upon the structure of tungsten electrodeposited onto the plane of molybdenum, *Extend Abstracts, Baltic Conference of Interface Electrochemistry, Tartu* 1996, 71-72.

THE ELECTRODEPOSITION OF ALUMINUM-NIOBIUM ALLOYS FROM CHLOROALUMINATE ELECTROLYTES

G.R. STAFFORD
Materials Science and Engineering Laboratory
National Institute of Standards and Technology
Gaithersburg, Maryland 20899

G.M. HAARBERG
Department of Electrochemistry
Norwegian University of Science and Technology
N-7034 Trondheim, Norway

1. Abstract

The electrochemical behavior of niobium in a 52:48 mole ratio $AlCl_3$:NaCl molten salt electrolyte was examined by linear sweep voltammetry, chronoamperometry and chronopotentiometry at 190 °C. Niobium oxidation begins at about 1.10 V vs. Al but appears to be limited to a charge of about 4 mC/cm^2 before the electrode becomes inhibited. This inhibition is likely due to limited solubility of the niobium oxidation product. The inhibition region extends from about 1.20 V to 1.40 V. At higher oxidation potentials, the current can be continuously increased up to the anodic limit of the chloroaluminate electrolyte. Gravimetric analysis during oxidation indicates that the dissolution of Nb at 1.50 V is a two-electron process. Aluminum-niobium alloys can be electrodeposited from the above electrolyte following Nb dissolution. Alloys containing a Nb atomic fraction of up to 13.5 % have been electrodeposited. X-ray diffraction and transmission electron microscopy examinations of these electrodeposits indicate that the low Nb concentration deposits (< 5 % Nb) are face centered cubic (fcc) aluminum. As the Nb content increases, an amorphous phase is introduced into the structure.

2. Introduction

The susceptibility of conventional aluminum alloys to chloride-induced pitting corrosion is a major impediment to their general use. During the last several years, considerable progress has been made towards enhancing the corrosion resistance by alloying aluminum with a variety of solute elements, including Cr, Ta, Mo, W, Ti, Nb, Zr, Si, Cu and V (1-11). For these alloying additions to be effective, they must

D.H. Kerridge and E.G. Polyakov (eds.), Refractory Metals in Molten Salts, 117-130.

typically be in solid solution at concentrations exceeding the equilibrium solubility. Consequently, nonequilibrium alloying methods are required to obtain single phase material. Processing schemes such as melt spinning (11), sputter deposition (1-7) and ion implantation (8-10) have been implemented. Electrodeposition is an excellent method for producing metastable alloys since undesirable compositional inhomogeneities are very limited in scale and grain sizes are typically very small. However, the electrodeposition of aluminum-refractory metal alloys has yet to be adequately explored.

Pure aluminum can be electrodeposited from chloroaluminate electrolytes (12-22). When an equimolar mixture of $AlCl_3$ and NaCl melts, an ionic liquid composed exclusively of Na^+ and $AlCl_4^-$ is produced. When the relative concentrations deviate from equimolar, additional ionic species are introduced. Cl^- is present in melts containing excess NaCl while higher order aluminum complexes, such as $Al_2Cl_7^-$, are present with excess $AlCl_3$. The chemical equilibria operative in $AlCl_3$-NaCl melts under a wide range of $AlCl_3$ concentrations above the equimolar point is well known (23-26). This melt is often considered as an acid-base system where the acid ($Al_2Cl_7^-$) is defined as a chloride ion acceptor and the base ($AlCl_4^-$) is defined as a chloride ion donor,

$$2AlCl_4^- \longleftrightarrow Al_2Cl_7^- + Cl^- \tag{1}$$

The equilibrium constant for Equation (1) is reported to be 1.06×10^{-7} at 175 °C (23). In the 52:48 $AlCl_3$:NaCl melt the $AlCl_4^-$ and $Al_2Cl_7^-$ concentrations are about 7.68 mol/L and 0.662 mol/L, respectively. Aluminum can be electrodeposited from either one of these electroactive species. The reduction of $AlCl_4^-$ occurs at potentials more negative than that required for $Al_2Cl_7^-$ reduction and becomes prominent as the acidity ($AlCl_3$ content) of the melt is reduced. The aluminum deposition reaction, involving $Al_2Cl_7^-$ as the electroactive species, is normally expressed by equation (2). One can see that aluminum deposition results in a localized acidity decrease near the electrode surface since $Al_2Cl_7^-$ is consumed and $AlCl_4^-$ is produced.

$$4Al_2Cl_7^- + 3e^- \longrightarrow Al + 7AlCl_4^- \tag{2}$$

The electrochemical behavior of niobium in alkali metal chlorides, fluorides, and mixtures of the two is well documented (27-42). Pure niobium can be electrodeposited from a variety of electroactive niobium species, although much of the work has focused on $NbCl_5$ and K_2NbF_7. Some limited studies involving niobium have been conducted in chloroaluminate electrolytes. The structure and electrochemical behavior of $NbCl_5$ have been examined in alkali metal chloride based chloroaluminates (43-48) as well as the room temperature organic chloride based chloroaluminates (49). The general behavior reported is the reversible step-wise reduction to Nb(III). The formation of lower oxidation states results in dimerization and cluster formation. Metal-metal-bonded chloride clusters having a $\{Nb_6Cl_{12}\}^{z+}$ (z=1-4) core have in fact been synthesized and examined in both basic and acidic $AlCl_3$:MeEtimCl room temperature melts (50). No evidence for the deposition of niobium metal has been reported from any of these systems. It was recently reported, however, that Nb-Al alloys could be electrodeposited from a $NaCl-AlCl_3$ melt containing either Nb_3Cl_8 or anodically dissolved niobium (48). The present paper examines the anodic dissolution of niobium into a slightly acidic $NaCl-AlCl_3$ electrolyte and the electrodeposition of Al-Nb alloys from this melt.

3. Experimental Procedure

The electrochemical cell, shown in Figure 1, was constructed of Pyrex* glass. It was placed in a muffle furnace and its temperature was maintained at 190 °C. The reference and counter compartments were separated from the working compartment by a fine frit, having an average pore size of 4 μm. Each compartment contained a 2 mm aluminum wire (99.998%, Morton Thiokol, Inc.) immersed in 52:48 $AlCl_3$:NaCl. The working compartment contained two components. The first was a 0.80 mm diameter

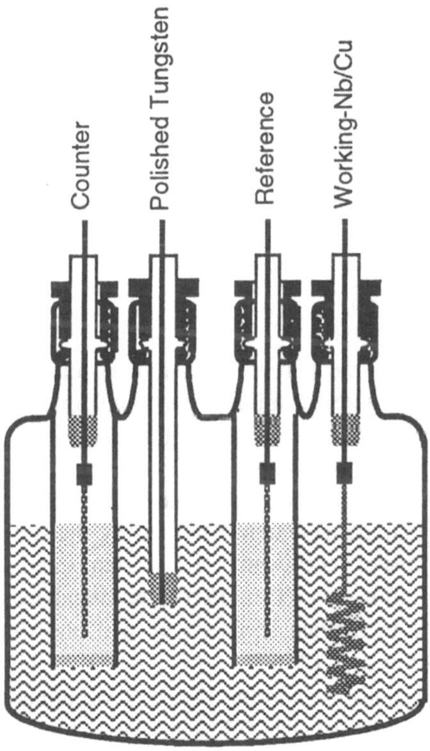

Figure 1. Schematic of electrochemical cell used in niobium dissolution and Al-Nb deposition experiments.

---

* Certain trade names are mentioned for experimental information only; in no case does it imply a recommendation or endorsement by NIST.

tungsten wire sealed in glass and polished (through 0.25 μm diamond) to be used as a working electrode for linear sweep voltammetry, chronoamperometry and chronopotentiometry. The second was either a 1.0 mm niobium wire (99.8%, Alfa) to be electrochemically dissolved, or a 0.80 mm copper wire to serve as the substrate during electrodeposition. The aluminum counter, aluminum reference and niobium or copper working electrodes were connected to tungsten leads by aluminum compression fittings. Each of the tungsten leads then passed through a uranium glass seal prior to exiting the electrode assembly through the top. All four electrode assemblies were held in place by threaded glass compression fittings using Viton o-rings and machined graphite caps.

The electrolyte was a 52:48 mole ratio of $AlCl_3$:NaCl. The $AlCl_3$ (anhydrous, 99.9%) was obtained from Armont Metal Chloride and was used after sublimation at 195 °C in the presence of helium (99.999%, Ultra-Pure, Air Products). The NaCl (Mallinckrodt, Analytical Reagent) was dried for 4 hours at 500 °C. All chemicals were stored in a Vacuum Atmospheres dry box containing ultra pure argon. The powders were thoroughly mixed before being transferred to the cell. The working, counter and reference compartments contained 150, 4 and 4 ml, respectively, of molten electrolyte, initially of identical composition. All experiments were performed with a PAR Model 273 Potentiostat/Galvanostat.

The electrodeposits were examined by scanning electron microscopy (SEM), energy dispersive x-ray spectroscopy (EDS), x-ray diffraction (XRD), transmission electron microscopy (TEM) and x-ray photoelectron spectroscopy (XPS).

4. Results and Discussion

Prior to niobium dissolution, the pure chloroaluminate electrolyte was examined by linear sweep voltammetry. Figure 2 is a linear sweep voltammogram showing the deposition and stripping of aluminum on a tungsten electrode in the 52:48 melt where the electroactive species is $Al_2Cl_7^-$. Although not shown in Figure 2, the magnitude of the deposition wave (I_p) increases linearly with the square root of the sweep rate ($\upsilon^{1/2}$). This is a clear indication that the reduction is limited by the diffusion of $Al_2Cl_7^-$ to the electrode. Assuming a diffusion coefficient of 4×10^{-6} cm^2/s and a reversible deposition process, a $Al_2Cl_7^-$ concentration of 0.495 mol/L can be calculated from the slope of the I_p versus $\upsilon^{1/2}$ plot. This is in reasonable agreement with the solvent equilibria data of Boxall and coworkers for this melt (23).

Several other features can be seen in the Figure 2 voltammogram. The first is the small cathodic current observed between 0.1 and 0.5V. This can be attributed to the underpotential deposition (upd) of aluminum onto polycrystalline tungsten (51). The second is that, other than aluminum upd, no electrochemical activity is observed between the aluminum deposition potential and 1.5 V. Thus, a fairly large electrochemical window exists for the examination of niobium in the 52:48 electrolyte.

When a niobium wire is placed in this melt, a rest potential of 0.85V is measured. Figure 3 is a linear sweep voltammogram of niobium in the 52:48 electrolyte. An anodic current, presumably due to the oxidation of Nb, is observed when the potential

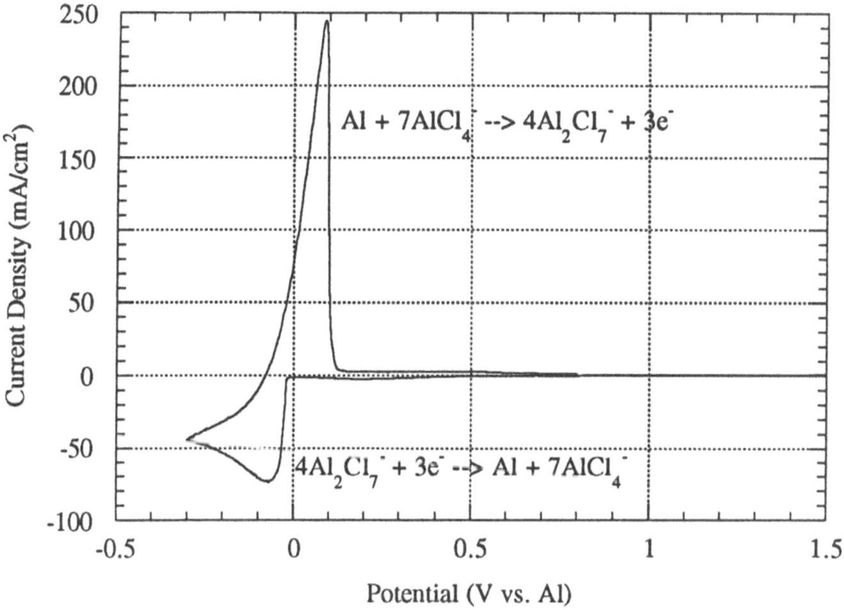

Figure 2. Linear sweep voltammogram in 52:48 $AlCl_3$:NaCl at 190 °C and sweep rate of 100 mV/sec. The working electrode was a 0.005 cm^2 polished tungsten wire.

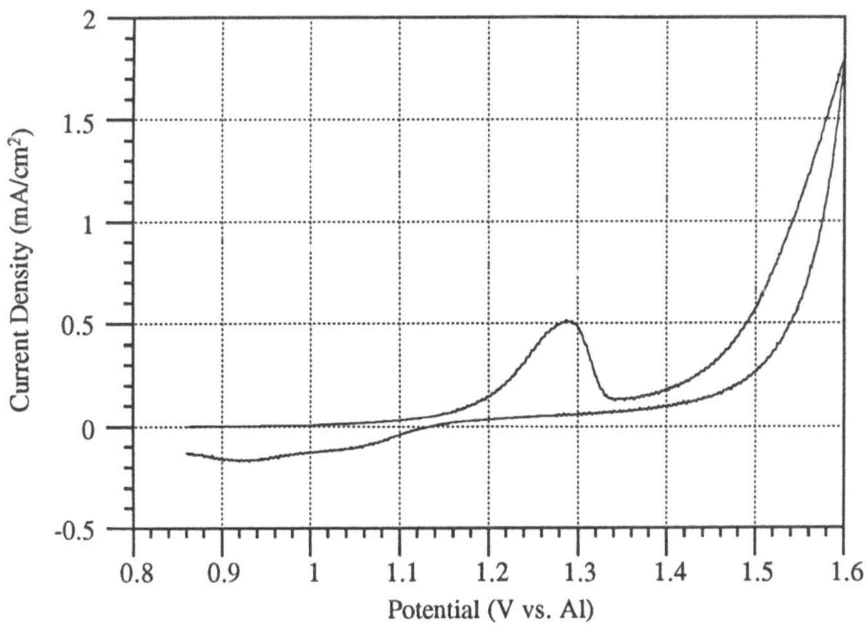

Figure 3. Linear sweep voltammogram in 52:48 $AlCl_3$:NaCl at 190 °C and sweep rate of 100 mV/sec. The working electrode was a 3.13 cm^2 niobium wire.

is increased to about 1.0V. Oxidation continues until about 1.3V where the current is seen to decrease rather dramatically. The shape of the initial oxidation peak suggests that the decrease is due to an inhibition rather than a diffusion limitation and may be an indication that the Nb oxidation product has limited solubility. Oxidation resumes at potentials above 1.4V until the anodic limit of the melt is reached. No discernible transition from niobium oxidation to chlorine evolution is observed when the potential is increased beyond 1.6V. On the return scan, a rather broad reduction peak initiates at about 1.1V.

Information regarding the earliest stages of Nb dissolution can be obtained by examining the current responses to a series of potential steps. Figure 4 shows a series of current responses for potential steps initiating at 1.10V. For low oxidation potentials, Figure 4(a), the current decay is approximately linear with $t^{1/2}$, suggesting the diffusion of soluble species away from the electrode. When the oxidation potential is increased, the shape of the current response deviates from that of simple diffusion. At higher oxidation potentials, the dissolution of Nb becomes inhibited, Figure 4(b), and the rate of current decay is proportional to the magnitude of the potential step; i.e., the dissolution is inhibited faster at the higher oxidation potentials. In Figure 4(b), this is seen as a cross-over in the current transients where the time to cross-over decreases with increasing step potentials. It is interesting to note that although the current drops to very small values, it never actually decreases to zero. Niobium dissolution is inhibited in this range of oxidation potentials, but the electrode never becomes completely passivated.

The dissolution inhibition is easily seen if, at selected times, one generates the current-potential curves from the current transients shown in Figure 4. These selected current voltammograms are shown in Figure 5(a) for times of 1 to 5 seconds. The dissolution inhibition is clearly seen as is the fact that shorter times are required to inhibit the dissolution when larger potential steps are imposed. It is also apparent that when the potential of the niobium wire is stepped to values above 1.4V, inhibition is avoided altogether. It is not clear whether this is due to the oxidation of niobium to a higher, more soluble oxidation state or by some interaction with Cl which may be present on the niobium surface due to the oxidation of $AlCl_4^-$.

The charge associated with the electrochemical oxidation of niobium prior to inhibition can be obtained by integrating the current transients from Figure 4. This is shown in Figure 5(b). In the potential range of 1.2V to 1.4V, about 3.7 mC/cm^2 is passed prior to inhibition. This is approximately equivalent to the formation of a 16/n monolayer film, where n is the number of electrons involved in the oxidation. The amount of charge passed prior to inhibition suggests that either reaction product is diffusing away from the electrode prior to inhibition or that several monolayers of material are formed. In either case, it seems clear that the fixed charge associated with the initial oxidation of niobium is not due to a simple adsorption process.

An alternative approach to characterizing the inhibition process is chronopotentiometry. Figure 6(a) shows a series of chronopotentiograms obtained from the galvanostatic oxidation of a niobium wire. When the anodic current is applied, an oxidation of limited duration occurs in the 1.1V to 1.2V range. The electrode potential then increases dramatically until an oxidation reaction able to support the applied

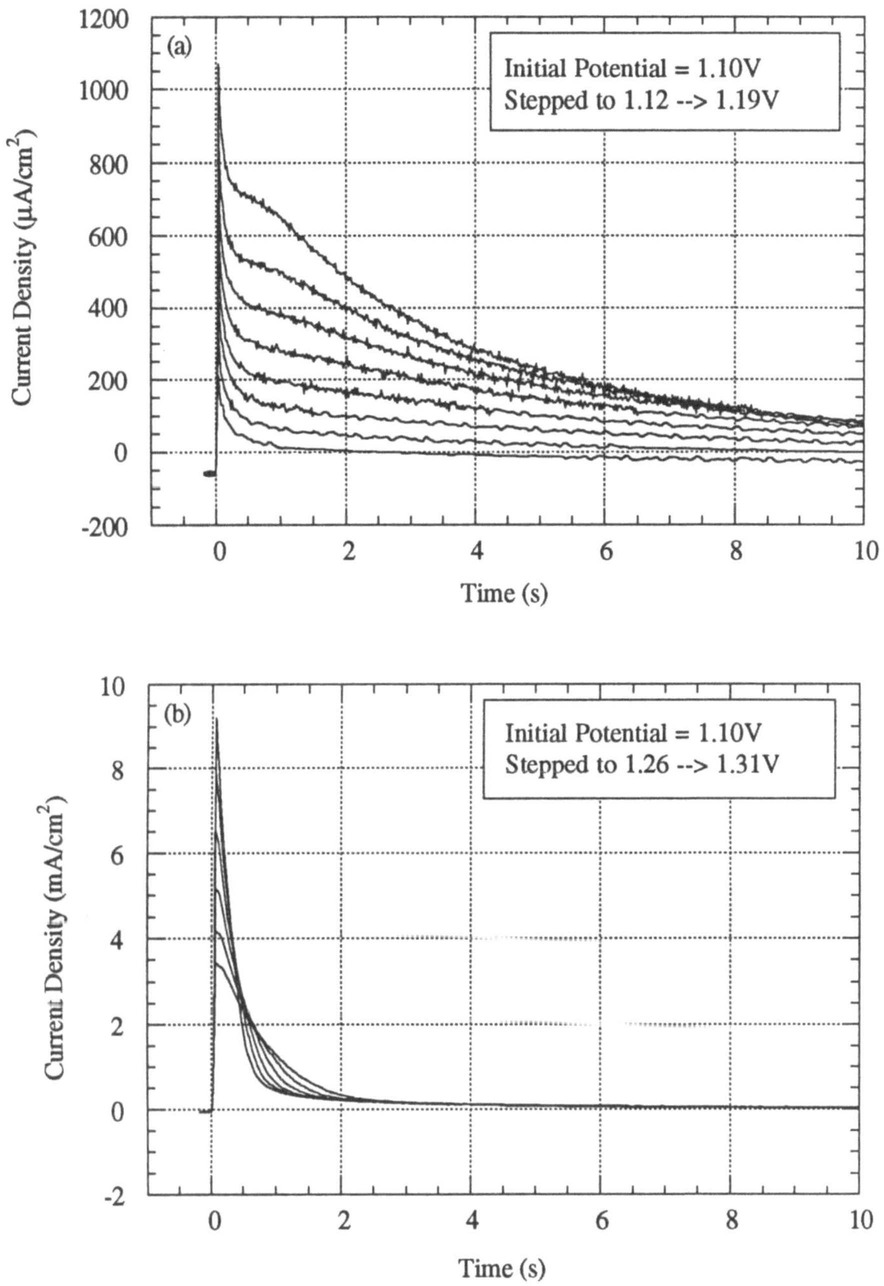

Figure 4. Chronoamperograms for the anodic dissolution of niobium in 52:48 $AlCl_3$:NaCl at 190 °C. (a) Potential was stepped from 1.10V to values ranging from 1.12V to 1.19V in 10mV increments. (b) Potential was stepped from 1.10V to values ranging from 1.26V to 1.31V in 10mV increments.

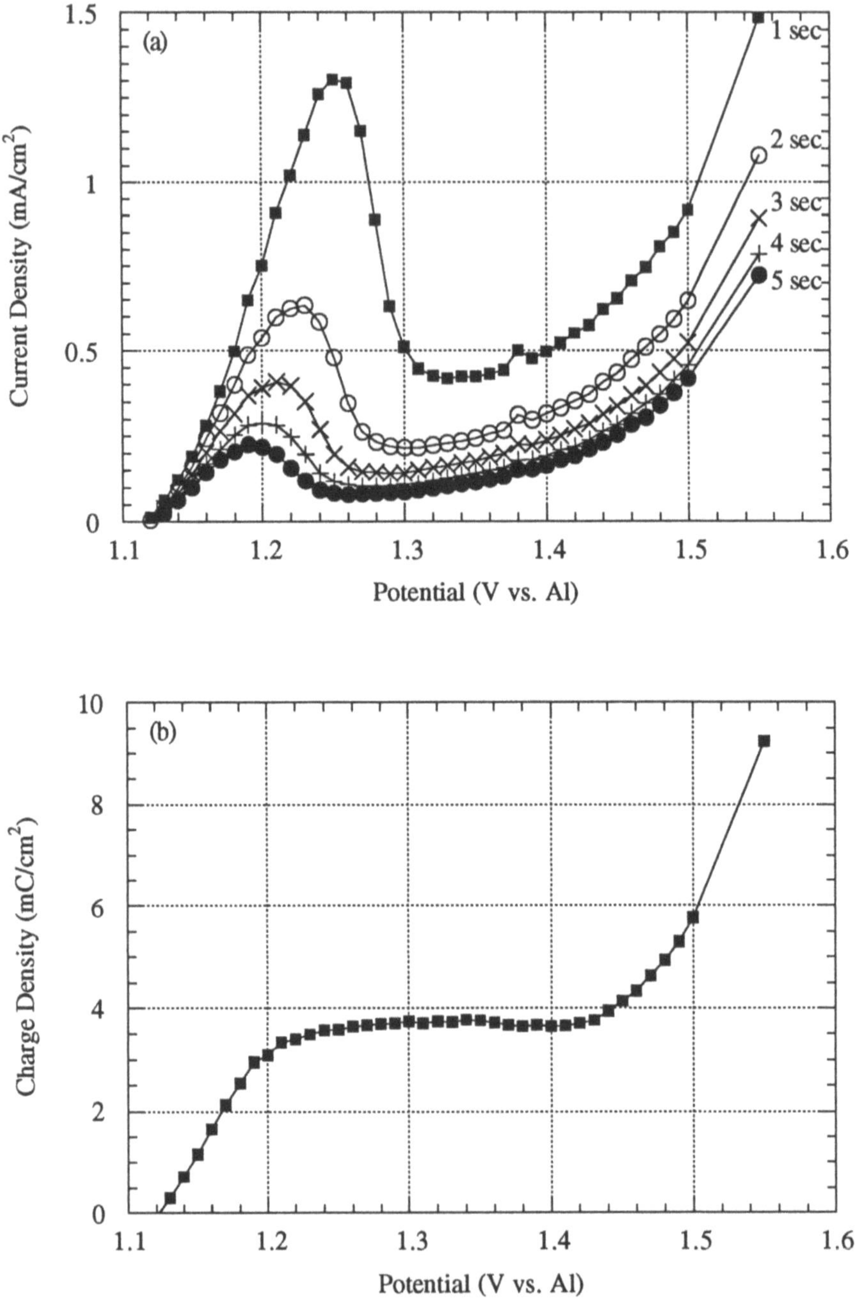

Figure 5. (a) Sampled-current voltammograms for the anodic dissolution of niobium in 52:48 $AlCl_3$:NaCl at 190 °C. The current was measured at 1, 2, 3, 4 and 5 second intervals from the Figure 4 transients. (b) Charge density vs. step potential obtained by integration of Figure 4 chronoamperograms.

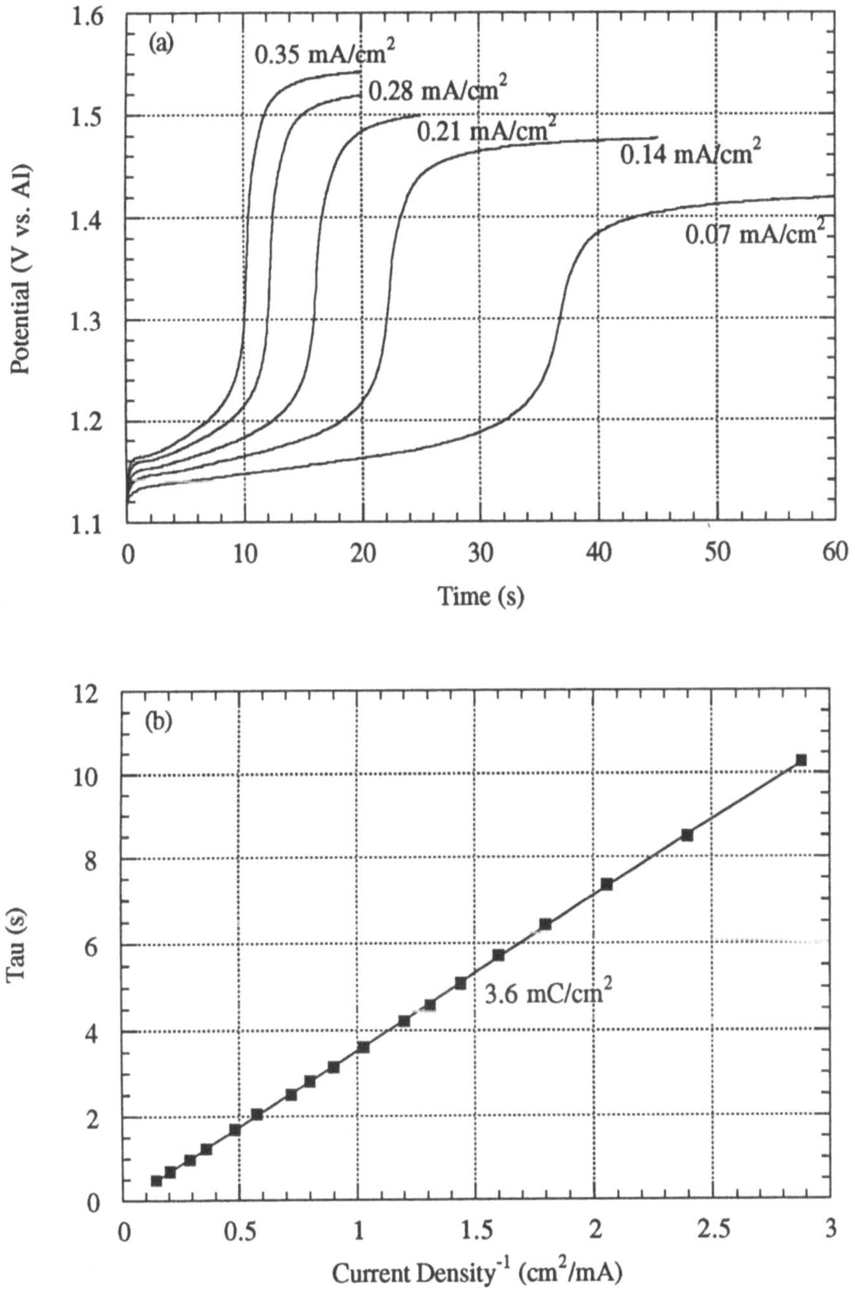

Figure 6. (a) Chronopotentiograms for the anodic dissolution of niobium in 52:48 $AlCl_3$:NaCl at 190 °C. The applied current densities were 0.07, 0.14, 0.21, 0.27 and 0.35 mA/cm^2. (b) Transition times (τ) as a function of current density, measured from the chronopotentiograms shown in (a).

current is achieved. The final potential is a strong function of the applied current density and in all cases exceeds 1.4V. It is clear that the initial oxidation reaction is identical to that which produced the current peaks observed in Figures 3 and 5(a). Transition times for several current densities were measured using the derivative method (52), and the resultant τ vs. i^{-1} plot yielded a straight line with an intercept of zero, Figure 6(b). The linear dependence of τ, rather than $\tau^{1/2}$, on i^{-1} is a clear indication that the initial oxidation is not limited by the diffusion of reactants to or products from the niobium surface. The reaction appears to be confined to the surface, or near surface. The slope of the plot in Figure 6(b) is the charge passed during the initial oxidation. The observed charge density of 3.6 mC/cm^2 is quite close to that obtained from the chronoamperometry experiments and provides evidence that this early stage of oxidation produced the equivalent of multiple monolayers of product prior to its inhibition.

Several steady state, potentiostatic oxidation experiments were performed on the niobium electrode at a potential of 1.5V in an effort to produce a soluble, ionic niobium species. The electrode was removed from the electrolyte on several occasions and the niobium was weighed in order to obtain an average valence for the dissolution product. Dissolution currents ranged from 0.2 to 1.8 mA/cm^2, the larger currents being observed during the latter stages of dissolution after the niobium had roughened considerably. A total of 400 coulombs was passed and the electrode was weighed at 80 coulomb intervals. The average number of electrons passed for the dissolution at 1.5V was 2.08, suggesting that a divalent niobium species is produced.

Linear sweep voltammetry following niobium dissolution shows an electroactive species which can be reduced on a tungsten electrode at about 1.0V. The exact nature of this cathodic reaction is presently unclear. Although experiments to date suggest that the reduction is that of a niobium species in solution, attempts to produce pure niobium deposits at potentials between 0 and 1.0V have been unsuccessful.

Several electrodeposits were formed at potentials ranging from 0 to -0.1V. EDS examination of the as-deposited surface indicated the presence of aluminum and niobium in all of the electrodeposits examined. Chlorine was not detected in any of the samples indicating that the deposits contained no entrained electrolyte. Figure 7 is a plot of alloy composition as a function of deposition potential. The highest niobium concentration detected was 13.5% (atomic fraction). This was observed at a deposition potential of 0.0V. As the deposition potential is made more negative, the niobium concentration is dramatically reduced. This implies that the kinetics for aluminum deposition are much faster than that of niobium, or that the niobium reduction is simply mass transport limited in the potential range examined. The fact that pure niobium deposits are apparently not achievable at potentials more positive of the aluminum deposition potential may be an indication that the codeposition of niobium and aluminum at negative potentials follows an induced codeposition mechanism; i.e., niobium deposition is only possible when aluminum is codeposited.

All of the deposits examined had dense, compact morphologies. The deposits containing low levels of niobium had surface morphologies quite similar to that of pure aluminum. These deposits were nodular and exhibited extensive crystallographic faceting. When the niobium concentration increased, the nodules become quite smooth

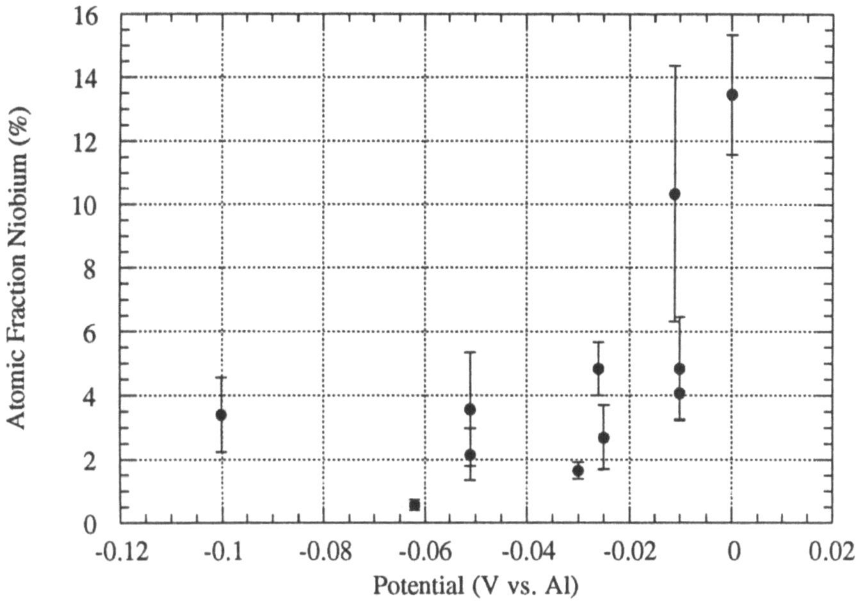

Figure 7. Alloy composition as a function of deposition potential following the anodic dissolution of niobium into 52:48 $AlCl_3$:NaCl at 190 °C.

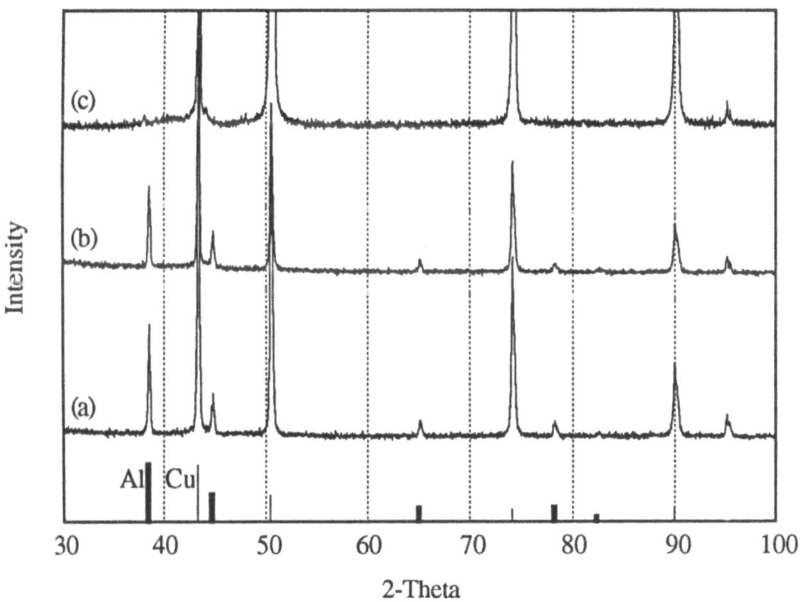

Figure 8. X-ray diffraction patterns from electrodeposits formed following the anodic dissolution of niobium into 52:48 $AlCl_3$:NaCl at 190 °C. The electrodeposits contained (a) 2.16%, (b) 4.84% and (c) 13.46% atomic fraction niobium.

rather than faceted. This may be an indication of reduced grain size or the presence of an amorphous structure.

Deposits containing 0.6, 10.3 and 13.4% atomic fraction niobium were examined by XPS. In all cases the photoelectron lines for aluminum and niobium were present. A detailed examination of the niobium $3d_{5/2}$ and $3d_{3/2}$ lines yielded binding energies which were consistent with that for metallic niobium. Literature values of the binding energies of various niobium oxides and halide salts are 2 to 6 eV higher than that for metallic niobium. These binding energy shifts were not observed in any of the electrodeposits. It can therefore be concluded that the niobium is present in the electrodeposit as a metallic aluminum alloy.

The structures of the electrodeposits were examined by XRD; selected patterns are shown in Figure 8. All of the deposits except that containing 13.5% niobium yielded diffraction patterns consistent with that of face-centered cubic (fcc) aluminum. This is a clear indication that the electrodeposits are either single phase aluminum, supersaturated in niobium or a two phase alloy consisting of saturated fcc aluminum and an amorphous phase which is not detected by x-ray diffraction. The equilibrium solubility of niobium in fcc aluminum is less than 0.3% (53). The existence of an Al-Nb solid solution in a 1.65% Nb deposit was confirmed by TEM. Bright field images of the aluminum grains showed no precipitates while EDS examination of the same single grains yielded clear niobium signatures. The x-ray diffraction pattern of the 13.5% Nb deposit, Figure 8(c), showed only a very broad reflection in the range of 36° to 46° two-theta. No fcc aluminum was detected. The existence of an Al-Nb amorphous phase has yet to be verified by TEM.

5. Conclusions

The electrochemical oxidation of niobium in 52:48 $AlCl_3$:NaCl begins at about 1.10 V vs. Al and appears to be limited to a charge of about 4 mC/cm^2 before the electrode becomes inhibited. The inhibition appears to be due to limited solubility of the niobium oxidation product. The inhibition region extends from about 1.20 V to 1.40 V. At higher potentials, the oxidation of niobium continues. Gravimetric analysis during oxidation indicates that the dissolution of Nb at 1.50 V is a two-electron process. Aluminum-niobium alloys can be electrodeposited from the above electrolyte following Nb dissolution. Alloys containing a Nb atomic fraction of up to 13.5 % have been electrodeposited. X-ray diffraction and transmission electron microscopy examinations of these electrodeposits indicate that alloys containing less than 5% atomic fraction niobium are face centered cubic (fcc) aluminum. As the Nb content increases, an amorphous phase is introduced into the structure.

6. Acknowledgments

The contributions of Frank Gayle, Paul Natishan, Otto Lunder, Tharald Tharaldsen and Maureen Williams, as well as the financial support of SINTEF are gratefully acknowledged.

7. References

1. B.A. Shaw, G.D. Davis, T.L. Fritz, B.J. Rees and W.C. Moshier, J. Electrochem. Soc. 138, 3288 (1991).
2. W. C. Moshier, G.C. Davies, J.S. Ahearn and H.F. Hough, J. Electrochem. Soc. 134, 2677 (1987).
3. W.C. Moshier, G.D. Davies and G.O. Cote, J. Electrochem. Soc. 136, 356 (1989).
4. G.D. Davies, W.C. Moshier, T.L. Fritz and G.O. Cote, J. Electrochem. Soc. 137, 422 (1990).
5. G.S. Frankel, M.A. Russak, C.V. Jahnes, M. Mirzammani, and V.A. Brusic, J. Electrochem. Soc. 136, 1243 (1989).
6. G.S. Frankel, R.C. Newman, C.V. Jahnes and M.A. Russak, J. Electrochem. Soc. 140, 2192 (1993).
7. H. Yoshioka, A. Kawashima, K. Asami and K. Hashimoto, *Corrosion, Electrochemistry, and Catalysis of Metallic Glasses*, pg 242, eds., R.B. Diegle and K. Hashimoto, Proceedings Volume 88-1, The Electrochemical Society, Inc., Pennington, N.J. (1992).
8. P.M. Natishan, E. McCafferty and G.K. Hubler, J. Electrochem. Soc. 135, 321 (1988).
9. M.V. Zeller and J.A. Kargol, Appl. of Sur. Sci. 18, 63 (1984).
10. A.H. Al-Saffar, V. Ashworth, A.K.O. Bairamov, D.J. Chivers, W.A. Grant and R.P.M. Proctor, Corrosion Science 20, 127 (1980).
11. M. Fass, D.Itzhak, D. Eliezer and F.H. Froes, J. Mater. Sci. Lett. 6, 1227 (1987).
12. J. Bouteillon and A. Marguier, Surface Technology 22, 205 (1984).
13. B. S. Del Duca, J. Electrochem. Soc. 118, 405 (1971).
14. B. Nayak and M.M. Misra, J. Appl. Electrochem. 7, 45 (1977).
15. B. Nayak and M.M. Misra, J. Appl. Electrochem. 9, 699 (1979).
16. R.C. Howie and D.W. Macmillan, J. Appl. Electrochem. 2, 217 (1972).
17. P. Rolland and G. Mamantov, J. Electrochem. Soc. 123, 1299 (1976).
18. T. Sato, T. Ishikawa and R. Midorikawa, Denki Kagaku 41, 446 (1973).
19. R. Midorikawa, Denki Kagaku 24, 366 (1956).
20. R. Midorikawa, Denki Kagaku 24, 457 (1956).
21. R. Midorikawa, Denki Kagaku 24, 511 (1956).
22. R. Midorikawa, Denki Kagaku 24, 562 (1956).
23. L.G. Boxall, H.L. Jones and R.A. Osteryoung, J. Electrochem. Soc. 120, 223 (1973).
24. B.J. Welch and R.A. Osteryoung, J. Electroanal. Chem. 118, 455 (1981).

25. A.A. Fannin, Jr., L.A. King and D.W. Seegmiller, J. Electrochem. Soc. 119, 801 (1972).
26. B. Tremillon and G. Letisse, J. Electroanal. Chem. 17, 371 (1968).
27. J. Dartnell, K.E. Johnson and L.L. Shreir, J. Less-Common Metals 6, 85 (1964).
28. Y. Saeki and T. Suzuki, J. Less-Common Metals 9, 362 (1965).
29. S. Senderoff and G.W. Mellors, J. Electrochem. Soc. 113, 66 (1966).
30. L.E. Ivanovskii and M.T. Krasil'nikov, Tr. Inst. Elektrokhim., Akad. Nauk SSSR, Ural'sk. Filial 10, 61 (1967).
31. V.F. Pimenov and Y.V. Baimakov, Elektrokhimiya 4, 1357 (1968).
32. V.I. Konstantinov, E.G. Polyakov and P.T. Stangrit, Electrochimica Acta 26, 445 (1981).
33. S.A. Kuznetsov, A.G. Morachevskii and P.T. Stangrit, Elektrokhimiya 18, 1522 (1982).
34. P. Taxil and J. Mahenc, J. Applied Electrochem. 17, 261 (1987).
35. A. Khalidi, P. Taxil, B. Lafage and A.P. Lamaze, Materials Science Forum 73-75, 421 (1991).
36. A. Barhoun, Y. Berghoute and F. Lantelme, J. Alloys and Compounds 179, 241 (1992).
37. A. Khalidi and J. Bouteillon, J. Applied Electrochem. 23, 801 (1993).
38. F. Lantelme, A. Barhoun and J. Chevalet, J. Electrochem. Soc. 140, 324 (1993).
39. F. Lantelme, and Y. Berghoute, J. Electrochem. Soc. 141, 3306 (1994).
40. F. Lantelme, Y. Berghoute and A. Salmi, J. Applied Electrochem. 24, 361 (1994).
41. C. Rosenkilde and T. Ostvold, Acta Chemica Scandinavica 49, 85 (1995).
42. L. Arurault, J. Bouteillon and J.C. Poignet, J. Electrochem. Soc. 142, 3351 (1995).
43. G. Ting, K.W. Fung and G. Mamantov, J. Electrochem. Soc. 123, 624 (1976).
44. K.D. Sienerth, E.M. Hondrogiannis and G. Mamantov, J. Electrochem. Soc. 141, 1762 (1994).
45. J.H. von Barner, N.J. Bjerrum and G.P. Smith, Acta Chemica Scandinavica A32, 837 (1978).
46. R. Huglen, G. Mamantov, G.P. Smith and G.M. Begun, Journal of Raman Spectroscopy 8, 326 (1979).
47. R. Gut, Helv. Chim. Acta 43, 830 (1960).
48. Y. Sato, K. Iwabuchi, N. Kawaguci, H. Zhu, M. Endo, T. Yamamura and S. Saito, in *Proc. of the Tenth International Symposium on Molten Salts*, PV 96-7, p. 179, The Electrochemical Society, Pennington, NJ (1996).
49. I-Wen Sun and C.L. Hussey, Inorg. Chem. 28, 2731 (1989).
50. R. Quigley, P.A. Barnard, C.L. Hussey and K.R. Seddon, Inorg. Chem. 31, 1255 (1992).
51. G.R. Stafford, J. Electrochem. Soc. 141, 945 (1994).
52. A.J. Bard and L.R. Faulkner, "Electrochemical Methods: Fundamentals and Applications", John Wiley and Sons, New York (1980).
53. Binary Alloy Phase Diagrams, Ed. by T.B. Massalski, ASM (1986).

TECHNOLOGY OF ELECTRODEPOSITION FROM MOLTEN SALTS FOR THE PREPARATION OF ELECTRODE MATERIALS BASED ON REFRACTORY METALS

Pierre TAXIL, B. LAFAGE, O. BOÏKO, P. CHAMELOT and K. SERRANO
Laboratoire de Génie chimique, UMR 5503
Université Paul-Sabatier,
118, route de narbonne, 31062 Toulouse Cedex 04, France

Abstract: This paper reviews all the characteristics of surface coating by refractory metals or alloys including Molten Salts Electrodeposition: Selection of the fluorides as electrolyte; experimental conditions of obtaining massive coatings or alloys; applications of the coatings, focused on the preparation of electrode materials. According to the results in the literature, it seems that the process in some cases can be competitive with other techniques.

Keywords: refractory metals, electrodeposition, electrodeposition of alloys, diffusion coatings, corrosion resistance, electrode materials.

Introduction:

The coating of strategic metals onto more common metals is considered by chemical engeneers as an efficient tool for saving expensive raw material; thus, the processes of coating take an important place in the overall technology of surface treatments.
Refractory metals are of considerable interest when used as a thin coating on more common metals, due to their attractive surface properties, in particular their exceptional corrosion resistance [1, 2]: Tantalum, niobium, titanium and at least some degree the other refractory metals (zirconium, hafnium, tungsten) are named "valve metals", owing to the extreme stability of their oxides in oxidizing conditions, e.g. under high anodic polarisation in concentrated acidic media.

This behavior leads to applications such as the preparation of more stable anode materials . So, for about thirty years, titanium was used as a substrate for the electrocataytic oxide RuO_2 in the composition of DSA (Dimensionnally Stable Anodes) type anodes which is, in this period, well known for its considerable development for the chlorine production by electrolysis.

D.H. Kerridge and E.G. Polyakov (eds.), Refractory Metals in Molten Salts, 131-142.

Nevertheless, the use of Ti substrates in anode, similar to chlorine DSA was shown to be inefficient for anodic reactions involving oxygen production, these reactions requiring higher acidic concentrations and more anodic potential than for chlorine production and should be replaced by Ta which is much more corrosion resistant than Ti [3] but about tenfold more expensive.
The use of tantalum coating for this application, in place of massive tantalum, should be evidently an attractive alternative, as was proposed in a previous paper [4].

Furthermore the suitability of thin coatings of Ta and Nb alloys for electrolytic applications is also expected considering the use of these alloys in their massive form as anodes [6]. Numerous literature data concern the preparation of refractory alloys either by codeposition or by metalliding process.

Nevertheless, the question of the most suitable coating process arises: among all the techniques actually proposed only chemical vapour deposition (CVD) and molten salts electrodeposition (MSE) would be available for large scale production: CVD is characterized by high rates of coating [5], whereas MSE would be a more economic process and it can be improved in terms of quality and rate of coating.

This paper will focus on the optimization of the electrodeposition process with a view of obtaining pure or alloyed refractory metals coatings for industrial applications, including electrocatalysis.
We will report some basic principles concerning (i) the electrolyte (ii) the preparation of these coatings and (iii) their applications with examples from both literature data and our own experimental work.

1- The electrolyte

Among the various molten salts mixtures available in a moderate temperature range (up to 1100°C), only molten alkali chlorides and fluorides are considered suitable as solvent for refractory metals electrodeposition. The properties of these solvents have often been cited [7]: large electroinactivity window, low vapour pressure, high conductivity etc...The specificity of each solvent for electrodeposition process are discussed below:

1-1: ALKALI CHLORIDES OR MIXED CHLORIDES-FLUORIDES

One of the advantages of pure chloride media is their low cost which makes them economically attractive. Unfortunately, in these media, the chemistry of refractory metal ions is very complicated, due to their high degree of oxidation (5 for Ta and Nb, 4 for Hf, 6 for W...); In some cases, each subhalide can intervene in the overall process; for example, Boïko et al have observed on a chronopotentiogram of a solution of niobium ions in molten NaCl-KCl (Fig.1) five steps of reduction of Nb^{V} to Nb^{0} [8]; subsequent to the successive formation of subhalides on the electrode where the coating occurs, chemical reactions such as disproportionation, adsorption or precipitation reactions may occur during the reduction process and act to the formation of powdery coatings.

Fig.1: : Chronopotentiogram of NaCl-KCl-K_2NbF_7 at 750°C on vitreous carbon $[K_2NbF_7] = 1{,}2\ 10^{-1}$mol/L (Ref. [8])

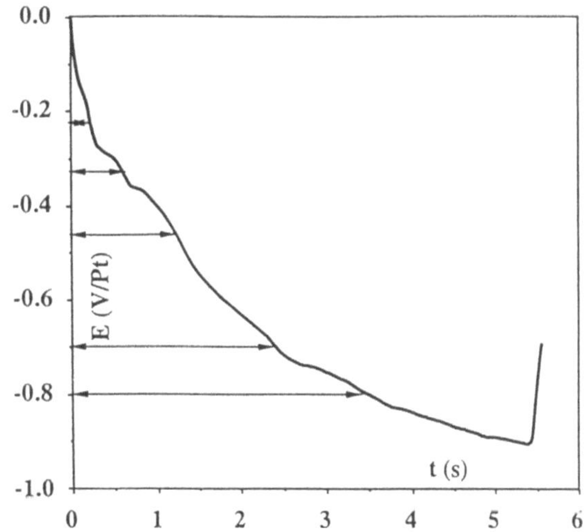

Nevertheless, Khalidi et al [9] and Polyakova et al [10, 11] have reported simpler reduction processes; it is noteworthy that these last authors have used cesium chloride in the solvant composition.

The addition of fluorides in the melt is also considered beneficial to the coating process, due to the complexing properties of fluoride ions, leading to a higher stability of high valencies of refractory metals. As an example, we consider this effect on the reduction of hafnium tetrachloride in molten NaCl-KCl in the square wave voltammogram of Fig 2a and 2-b presented by Serrano et al [12] : in pure chloride melts (Fig. 2a), the reduction path includes two 2 electrons steps from Hf^{IV} to Hf metal. In the presence of fluoride ions (Fig.2b), the decomposition potential of Hf^{IV} is shifted cathodically, so that its reduction takes place in one step with 4 exchanged electrons.

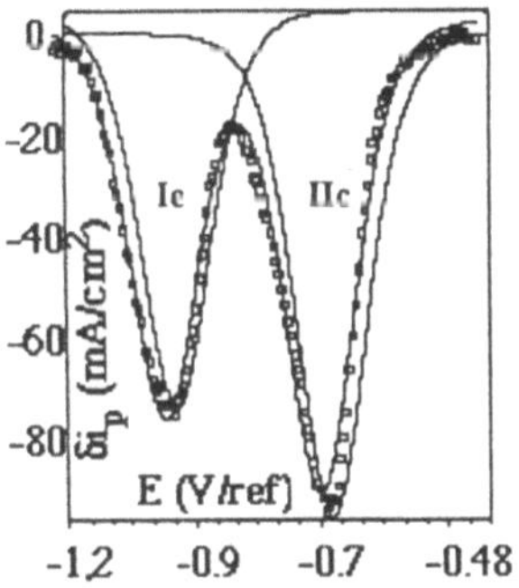

Fig2a: Square Wave Voltammogram of a NaCl-KCl-$HfCl_4$ on a W working electrode at 750°C [12]

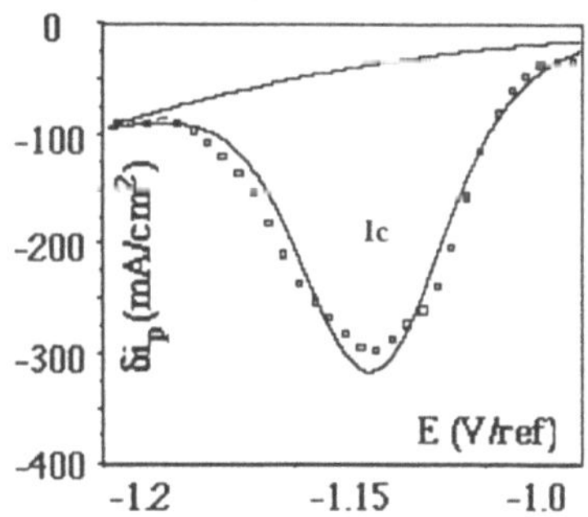

Fig2b: SquareWave Voltammogram of a NaCl-KCl-$HfCl_4$-NaF on a W working electrode at 750°C [12]

Even when fluorides are present in the electrolyte, the refractory metals coatings in chloride media are dendritic or powdery and these media are only used for metals electrowinning in pilot plants, such as for hafnium [13].

1-2 PURE FLUORIDE MEDIA

From the works of Senderoff and Mellors in the sixties, these media are well known to be suitable to obtain coherent, adherent and relatively smooth coatings of refractory metals [14] owing to the beneficial effect of fluoride ions on the coating process, mentionned above.

Reduction paths

According to a recent paper of Girginov which collected literature data on the electroreduction processes of refractory metals in various media [15], the reduction processes of refractory metal ions are simple in all-fluoride melts.

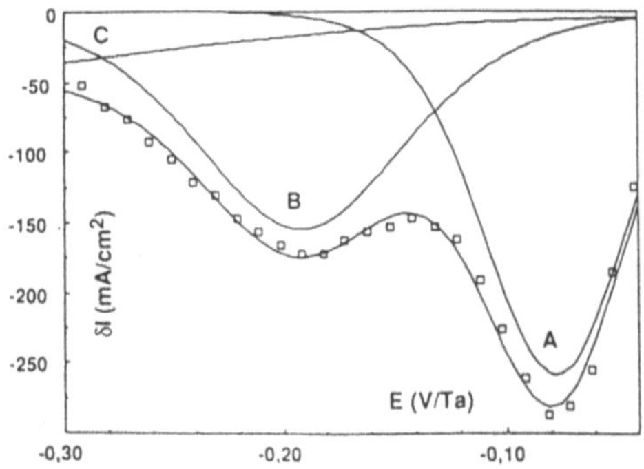

Fig.3: Square Wave Voltammogram of a solution LiF-NaF-K_2TaF_7 (5,0% mass) at 790°C on a Mo working electrode [16].

This is verified in Fig.3 in the case of tantalum, on a square wave voltammogram of a LiF-NaF-K_2TaF_7 mixture, obtained by Chamelot et al [16]: The reduction of TaF_7^{2-} in Ta occurs in a five electron step corresponding to peak A, whereas peak B is attributed to the reduction of Ta oxyfluoride.

Composition of the electrolyte

Fig3 highlights the recurrent question of the effect of oxide ions in molten salts: Oxyfluorides arise from the action of oxide ions which are soluble and reactive with refractory metal ions in fluorides. Fig. 3 shows the possibilities, during the coating process, of interference between the reduction of non oxygenated species, producing the metal and of oxigenated species, which probably give tantalum suboxide, further dissolved in the bath. The same feature was observed for niobium electrodeposition by the same authors [17].

The question of the oxide content in the electrodeposition bath is critical: while it was generally asserted that it should be the lowest possible for the quality of the plates, Christensen et al now claim that the presence of a certain amount of oxide in the melt enhances the crystallinity and the coherence of their deposits [18].

Ref. [17] and [18] emphasize the problem of the bivalence of niobium in molten fluorides and the necessity of a prereduction of Nb^V to Nb^{IV} before the electrolysis in order to avoid a further dissolution of the deposited metal due to its oxidation by pentavalent niobium. The same problem is encountered with tungsten ions existing at

valences VI and IV: Prior to the electrodeposition process, W^{VI} must be reduced on W^{IV}[19].
Obviously, the oxyfluoride content and the valency of the solute must be controlled on line during the electrodeposition; for this, square wave voltammetry is a powerful method, since, one the one hand it allows accurate measurements of these characteristics to be performed, according to ref [16] and [17], and on the other hand, it is relatively easy to be put in place in industrial cells.

2 - Electroplating of massive refractory metals.

2-1- PARAMETERS TO BE CONSIDERED

We have already stated that high compacness, coherence and adherence of plates of tantalum, niobium and tungsten are easy to obtain in an all-fluoride electrolyte. The thickness, the rate of obtention and the surface state of the plate will now be considered. These are important industrially.
An average thickness of 50 µm should be satisfactory, on condition that it is uniform on all the sample, thus the surface state must be smooth; high velocities of electroplating obtained by increasing current densities promote too high a roughness and thus harms the uniformity of the coating.
We have established that in DC current electrolysis, The upper limit is 0,1 mm/h and 0,06 mm/h for the electroplating of Ta and Nb respectively.
We see that the characteristics cited above are correlated with each other; thus a preliminary investigation on the electrocrystallisation of refractory metals is necessary to optimize all the parameters intervening in the growth of the coating.

2 -2 ELECTROCRYSTALLISATION OF REFRACTORY METALS IN MOLTEN FLUORIDES.

As Girginov et al have pointed out [15], very little attention has been paid to the electrocrystallisation kinetics of refractory metals and factors influencing the process, which are fundamental for the monitoring of the process. Nevertheless, we have recently reported chronoamperometric investigations on niobium electrocrystallisation [20] and concluded that the nucleation is instantaneous and further growth of the nuclei is controlled by diffusion. This situation favours an unequal current distibution when high current densities are applied to the cathode and so it is conducive to the inequality of the thickness of the plate.
This conclusion may be extended to other refractory metals, at least to tantalum, knowing the similarity of behavior of Nb and Ta, and may encourage the use of pulsed currents rather than DC current for the electrodeposition.

2-3: THE ROLE OF PULSED CURRENTS

Pulsed electrolysis is beginning to be attempted in molten salt electrodeposition as is reported in ref [21], on titanium electroplating; this method, currently used in aqueous electroplating, allows to increase speed of plating and improve the surface state of the coating. We find the basic principle of this technique, in Ibl's papers, such as [22]:
When an electrode is submitted to high cathodic current densities during a short period (T_c), separated by a rest time (T_r), the growth of the diffusion layer is limited and is small compared to the size of the growing crystals; this "macroprofile" at the interface electrode electrolyte surface favours a secondary current distribution conducive to the

uniformity of crystal growth. An optimized frequency of the pulses and an optimized ratio $\frac{T_r}{Tc}$ may lead to electrodeposition using average electrodeposition currents close to the limit of mass transfer, without deleterious effects on the surface state of the coating. Furthermore, the repetition of pulses and their high current density multiply the number of nucleï, improving the microstructure of the deposited layer and consequently enhancing the compactness of the coating again its corrosion resistance. These favour the possible industrial applications of the process. Good results have been obtained in our work for niobium electrodeposition, in term of electroplating rate (0,3 mm/h) and microstructure [23].

2 -4: APPLICATION OF TANTALUM COATING IN ELECTROCATALYSIS:

As stated in the introduction, Ta coating may be used as substrate in the preparation of DSA oxygen electrodes; it consists of a thin and porous layer of Iridium oxide, which acts as catalyst, obtained by thermal oxidation of an iridium compound on a valve metal. The lifetime of the anode in water electrolysis in extreme conditions of polarization (anodic current = 50 A/m^2), acid concentration (30% m/m) and temperature (T = 80°C) is sensitive to the corrosion resistance of the valve metal; This is shown on table I [24], which standardized life time (lifetime reported for the mass surface density of the catalyst IrO_2) for some varieties of titanium base alloys and a tantalum coating as substrate:

TABLE 1: Standardized life time of DSA O_2 for various substrates

substrate	ASTM gr 4	NFT 40	NFTa6V 4	Cu/Ta (50 μm)
standardized life time h $g^{-1}m^2$	1.0	7.4	6.3	10.1

This table shows that the service life time of DSA Oxygen is significantly increased when the substrate is a 50 μm coating of Ta. Nevertheless, the use of this substrate is hindered in industrial use, because a defect in the coating would have serious consequences for the behavior of the anode. To reduce this risk, we have to condider possible improvements of the quality of the coating. For this, pulse electrolysis seems to be a promising way.

3- Obtention of alloys in molten salts

Electrowinning of alloys, alternatively in a massive form or as a surface coating of metallic substrate is another possibility for obtaining materials with attractive properties by MSE. The literature data gives many examples relevant to this kind of process. Two different paths are to be considered: (i) the simultaneous elecrochemical reduction of the ions of a pair of metals, that is electrodeposition of alloys; (ii): the intermetallic diffusion of a deposited metal in a substrate, that is called metalliding.

3-1: ELECTRODEPOSITION OF ALLOYS:

Thermodynamic features:

All the basic principles of electrodeposition of alloys were presented by Brenner [25]; the alloy is produced by the reduction of the ions of each component of the alloy.

Thermodynamically obtaining the alloy is easier than each component separately, due to the negative free energy of the formation of the alloy. Wendt et al have verified this point in the case of titanium or zirconium diboride electrosynthesis in Flinak [26]: the overall reduction : $Ti^{III} + 2B^{III} + 9e \rightarrow TiB_2$ can be separated into three steps: (1): $Ti^{III} + 3e \rightarrow Ti$; (2): $B^{III} + 3\,e \rightarrow B$; (3): $Ti + 2B \rightarrow TiB_2$. Taking into account that the Gibbs energy of reaction (3) is $\Delta G° = -246$ kJ mol^{-1}, the alloy TiB_2 is more noble than Ti and B and the deposition occurs at a far more positive potential than those of pure titanium or pure boron. The same situation is observed in the case of ZrB_2 ($\Delta G° = -318$ kJ mol^{-1}). This is interesting since it is well known that both pure titanium and zirconium are very difficult to electrodeposit in molten fluorides, due to the too negative reduction potential of their ions, very close to the solvent reduction potential.

Another relevant characteristic of this process was developed by Cohen [27] in the case of niobium-germanium intermetallic compounds; the phase diagram of the system includes the following compounds: $NbGe_2$, Nb_3Ge_2 and Nb_5Ge_3. Cohen produced each of them by varying the composition of the bath in Nb and Ge ions; the dependence of the composition of the alloy and the bath composition is given by the following expression:

$$\frac{a_{Ge}}{a_{Nb}} = \left(\frac{a_{Ge(IV)}}{a_{Nb(IV)}}\right).\ \exp\left[\frac{4F}{RT}(E°_{Ge} + \eta_{Ge} - E°_{Nb} - \eta_{Nb})\right] \qquad (1)$$

($E°_i$: standard potential; η_i : overpotential for each species)

One of the most interesting aspect of MSE, compared to thermic processes, for refractory alloys electrowinning is the moderate temperature of molten salts.
By this way, Stern et al obtained tantalum carbides [28], tantalum silicides [29] and tungsten carbides [30] in molten fluorides at 800°C while the formation of these compounds by thermic processes requires temperature more than 3000°C.

Refractory metal alloys electrowinning and applications

The tantalum carbides and silicides and tungsten carbides cited above are used for their hardness, corrosion resistance in hot environment and their semiconducting properties, while niobium alloys (niobium-germanium, niobium-aluminium...) receive much attention for their properties of superconductivity.
In the electrolysis industry, an important activity now focuses on the use of TiB_2-covered carbon as cathodes in aluminium electrowinning by the Hall-Héroult process, since carbon cathodes have serious shortcomings: carbon is not wetted by aluminium, it reacts with Al to produce Al_4C_3 and its crystal lattice is penetrated by the molten electrolyte during the process; these phenomena are eliminated when carbon is covered by titanium diboride, which has good stability and mechanical resistance in cryolite [26, 31, 32]. Wendt et al have also emphasized the better results of coating of TiB_2 by electrodeposition process than by sintering, sintered TiB_2 cathodes being readily eroded during the Aluminium deposition [33].

3-2 DIFFUSION ELECTROCOATING.

In many cases, intermetallic diffusion can occur between the electrodeposited metal and the cathodic substrate, leading to the formation of a surface alloy. This situation may be expected, taking into account both the temperature of the electrodeposition and the phase diagram of the anodic metal and the cathodic one; the technique, so called "Metalliding", was fully developed by N.C Cook who gives in ref. [34] many examples of such alloys formed in molten fluorides, with attractive, sometimes surprising properties.

Thermodynamic and kinetic aspect:

The Metalliding process is available only if the temperature is sufficient to promote solid state diffusion but also if the cathodic metal is more noble than the cathodic one, so that the alloy can be produced by means of a galvanic cell, schematized as follows:

A/ Electrolyte for A^{z+} ions/A_yB

where A and B are the anodic and cathodic respective metal and y the stoechiometry of the alloy.

When the current passes through the cell, the discharge of A ions promotes the incorporation of the metal in the diffusion layer in the solid state and increases the layer thickness. In some cases where solid state diffusion is rapid, the bulk of metal can be completely alloyed, so that further current brings a change of stoechiometry of the alloy, leading to the obtention of all the binary compounds of the phase diagram: In this way, Weppner and Huggins [35] have performed a thermodynamic investigation of the systems Li/Bi, Li/Sb by measurement of the emf of cells, such as the above, where A = Li and B = Sb or Bi, by correlation with the coulometric titration of lithium transferred; plotting of the emf as a function of the transferred charge exhibited large variations separated by plateaux corresponding to the equilibrium between the intermetallic compounds .

In many cases, solid state diffusion is too slow and the emf plateaux are only observed in a transient period after the opening of the circuit of the cell: So, in the case of the metalliding of nickel with tantalum in molten LiF-NaF, the intermetallic compounds Ta_2Ni, TaNi, $TaNi_2$ and $TaNi_3$, were identified and the corresponding emf of cell such as Ta/ LiF-NaF, K_2NbF_7/ Ta_xNi gave accurate values for the Gibbs Energy of each of them [36] in Table II [36].

Table II: Gibbs Energy of intermetallic Ta-Ni compounds

compound	$TaNi_3$	$TaNi_2$	TaNi	Ta_2Ni
$\Delta_f G°$ kJ mol-1	-117.9	-107.4	-64.4	-36.8

When complete solid state diffusion is achieved, the layer includes one single phase made of the most stable compound $TaNi_3$. The same conclusion was reached in the case of the niobium-nickel system, for which only $NbNi_3$ is present at the surface of the cathode after metalliding of Ni with Nb [37].

From the kinetic viewpoint, modelling of the metalliding process can be based on the control of the process by the intermetallic diffusion; in this field we can cite a basic paper on this subject by Lantelme [38] and the recent work of this author and Salmi who investigated the diffusion of nickel in platinum [39]

In the case of interdiffusion of Ta or Nb in nickel, we have assumed that the external boundary of the diffusion layer does not move (linear semi infinite diffusion), when the expression of the metalliding current is similar to Cottrell's law, leading to a logarithmic growth of the layer during the process; this assumption was verified in the case of $TaNi_3$ [36] and $NbNi_3$ [37]. The linearity of the advance of the diffusion layer was shown by the micrography of a niobium nickel alloy Fig.4.

Fig. 4: micrographic cross setion of a diffusion layer niobium- nickel on nickel
(approximate scale: 1 cm = 10 μm)

Properties of refractory metals surface alloys

Refractory metals give only rise to few intermetallic compounds, due to the very high temperature of the liquidus of their alloy systems, though de Lépinay et al have obtserved the formation of copper-titanium intermetallic compounds during the electrodeposition of titanium [40].
In our own work, we have pointed out the exceptional corrosion resistance of $TaNi_3$ and $NbNi_3$ coatings on nickel, in spite of a very small thickness (10- 20 μm) . This can be explained by the incorporation of the refractory metal into the crystal lattice of nickel, giving the layer ideal characteristic of protection against corrosive reagents. The anodic behavior of these surface alloys (in concentrated acid media at the oxygen evolving potential), exhibited in ref [4], is quite similar to platinum. In fact we have observed no corrosion of these samples during long period at 2.25 V / SHE. In these conditions, the use of these coatings as anodes at the place of the massive alloys cited in ref 6 may now be expectable.

Conclusion

In this paper we have attempted to state the possibilities and limits of the Molten Salts Electrodeposition (MSE) for surface treatment by refractory metals. Industrial applications focuse on the corrosion resistance of the resultant coatings and, as a consequence, their use as electrode materials in aggressive conditions. Tantalum, niobium and tungsten electroplating in molten fluorides were shown to be suitable examples MSE, taking into account of the possible improvements of the method expected with pulsed currents, MSE may be competitive with regard to other techniques such as Chemical or Physical Vapor Deposition. Electrodeposition of alloys including refractory metals offers further possiblilities, in particular the winning of titanium and zirconium alloys which in the new future, may become important in the Hall Héroult process of aluminium electrowinning. Furthermore, the possibility of forming surface alloys by interdiffusion must also considered and more exploited since it can provide new materials with very atttractive properties.

REFERENCES:

1. Burns R.H., Shuker F.S. and Manning P. (1984) Industrial Applications of corrosion resistant Tantalum, Niobium and their alloys in *refractory metals and their industrial applications*, ASTM, Philadelphia, 50-69.

2. Cardarelli F., Taxil P. and Savall A., (1996), Tantalum Protective Coating Techniques for the Chemical Process Industry, *Int. J. of Refractory Metals and Hard Materials*, **14**, 365-381.

3. Vercesi G.P., Rolewicz J. and Comninellis C., Characterisation of DSA type oxygen evolving electrodes. Choice of the base metal, (1991), *Thermochemica Acta,* **171**, 31-47.

4. Taxil P. and Mahenc J. Formation of corrosion-resistant layers by electrodeposition of refractory metals (1987), *J. Appl. Electrochem.,* **17**, 261-269.

5. Spitz J. and Chevalier J., Comparative Study of Tantalum by Chemical Vapor Deposition and Electron Beam Vacuum Evaporation, (1975) *5th international conference on CVD*

6. Trusov G.N., Goliacheva E.P., Goncharenko B.A. and Mikheev V.S., Anodic behavior of niobium- nickel alloys and tantalum-nickel alloys in acid solutions, (1977), *Elektrokhim.*, 1056- 1059

7. Sadoway D.R. Fused salts electrolysis of refractory metals (1988) in *Refractory metals- State of the art- TMS Fall meeting*, Chicago (Illinois), 155-169.

8. Boïko O., Taxil P., Savall A., Lafage B., Chamelot P., The electrodeposition of refractory metals, *Euchem Conference on molten salts,* Smolenice castle, (Slovaquie), 15-20 sept. 1996.

9. Khalidi A., Bouteillon J., Comportement électrochimique de NbCl5 en milieu NaCl-KCl pur et additionné d'ions fluorures, (1993), *J. Appl. Electrochem.*, **23**, 801-807.

10. Polyakova L.P., Kosilo B.I., Polyakov E.G. and Smirnov A.B., Electrochemical behavior of tantalum in CsCl-KCl-NaCl-TaCl5 melt, (1988) *Elektrokhem.*, **24**, 892-97.

11. Elizarova I.R., Polyakov E.G. and Polyakova L.P., Electrochemical behavior of niobium in CsCl-KCl-NaCl-NbCl5, (1991), *Elektrokhem.*, **27**, 640-647.

12. Serrano K., Boïko O., Chamelot P., Lafage B. and Taxil P., Electrochemical reduction of HfCl4 in molten salts, *5th Symp. molten salts chemi.and Techn.* Dechema and V., Dresde, August 1997.

13. Lamaze A.P. and Charquet D., Development of hafnium tetrachloride electrolysis, in *Refractory Metals: Extraction, Processing and Applications*, The Minerals, Metals & Materials Society, 1990.

14.Mellors G.W. and Senderoff S. US patent (1968) 3 444 058

15. Girginov A., Tzvetkov T.Z. and Bojinov M., Electrodeposition of refractory metals (Ti, Zr, Nb, Ta) from molten salt electrolytes (1995), *J. Appl. Electrochem.*, **25**, 993-1003

16. Chamelot P., Taxil P. and Lafage B., Voltammetric studies of tantalum electrodeposition baths, (1994), *Electrochemica Acta,* **39**, 2571-75.

17. Chamelot P., Lafage B and Taxil P., Using square wave voltammetry to monitor molten alkali fluoride baths for electrodeposition of niobium, (1997), *Electrochemica Acta,* **43**, 607-616.

18. Christensen E, Xindong Wang, Von Barner J.H., Ostvold T. and Bjerrum N.J., the influence of oxide on the electrodeposition of niobium from alkali fluoride melts, (1994), *J. Electrochem. Soc.*, **141**, 1212-1220.

19. Broc M,.Dépôts électrolytiques de métaux réfractaires dans les sels fondus, (1978), Rapport CEA R-4921

20. Chamelot P., Lafage B and Taxil P, Studies of niobium electrocristallization phenomena in molten fluorides, (1996), *J. Electrochem. Soc.*, **143**, 1570-76.

21. Wei D., Okido M. and Oki T., Characteristics of titanium deposits by electrolysis in molten chloride-fluoride mixture,(1994), *J. Appl. Electrochem.*, **24**, 923-29.

22. Ibl N., Puippe J.C., Some theoretical aspects of pulsed electrolysis (1980), *Surf. Technology*, **10**, 1081-1104.

23. Chamelot P., Lafage B and Taxil P, Obtention de dépôts à grande vitesse par électrolyse pulsée, (1996), *Récents Progrès en Génie des Procédés*, **9**, 42, 465-70.

24. Cardarelli F.,.Comninellis C., Taxil P., Savall A., Manoli G., and Leclerc O., A new development for the preparation of oxygen evolving electrodes with long service life under extreme conditions, (1996), *4th Europ. Symp. Electrochem. Eng.*, Ed. Institute of Chemical Technology, Pragues, 191-97.

25. Brenner A., *Electrodeposition of alloys, Principles and Practice,* (1963), Academic Press, New york.

26. Wendt H., Reuhl K. and Shwarz V., Cathodic deposition of refractory intermetallic compounds from Flinak melts- I- Voltammetric investigation of Ti, Zr, B, TiB_2 and ZrB_2, (1992), *Electrochemica Acta*, **37**, 237-244.

27. Cohen U., Electrodeposition of Niobium-Germanium alloys from molten fluorides, (1983), *J. Electrochem Soc.*, **130**, 1479-85.

28. Stern K.H. and Gadomski S.T., Electrodeposition of Tantalum Carbide Coatings, (1983), *J. Electrochem. Soc.*, **130**, 1891-95.

29. Stern K.H. and Carlton E.W., Electrodeposition of Tantalum Silicide Coatings from molten salts, (1986),*J. Electrochem. Soc.*, **133**, 2157-60.

30. Stern K.H. and Deanhardt M.L., Electroplating of Tungsten Carbide from molten fluorides, (1985), *J. Electrochem. Soc*, **132**, 1891-95.

31. Wendt H., Reuhl K. and Shwarz V., Cathodic deposition of refractory intermetallic compounds from Flinak melts- II- Preparative cathodic deposition of TiB_2 and ZrB_2 and coatings thereof, (1992), *J. Appl. Electrochem.*, **22**, 161- 165.

32. Makyta M., Danek V., Haarberg G.M. and Thonstad J. Electrodeposition of titanium diboride from fused salts, (1996), *J. Appl. Electrochem.*, **26**, 319-324.

33. Wendt H. and Dermeik S. Erosion of TiB_2- cathodes during cathodic aluminium deposition from $LiCl/AlCl_3$ melts, (1990), *J. Appl. Electrochem.*, **20**, 438-441.

34. Cook N.C. Metalliding, (1969), *Sci. American*, **221**, 2-28

35. Weppner W. and Huggins R.A. thermodynamic properties of the intermetallic systems Lithium-Antimony and Lithium- Bismuth, (1978) *J. Electrochem. Soc.*, **125**, 7-14.

36. Taxil P., Formation d'alliages Tantale-Nickel par voie électrochimique, (1985), *J. Less Comm. Metals*, **113,** 89-101.

37.Taxil P. and Qiao Z.Y., Electrochemical alloying of nickel with niobium in molten fluorides, (1985), *J. Appl. Electrochem.*, **15**, 947-952.

38. Lantelme F. The role of ionic species and of interdiffusion in the solid state in electrodeposition processes from molten salts, (1987), *J. Appl. Electrochem.*, **17**, 243-252.

39. Lantelme F. and Salmi A., Electrochemical Determination of Solid State Interdiffusion in Nickel-Platinum Alloys, (1996), *J. Electrochem. Soc.*, **143**, 3662-3669.

40. De Lépinay J. , Bouteillon J., Traore S., Renaud D. and Barbier M. J., (1987), Electroplating Silicon and Titanium in molten fluoride media, *J. Appl. Electrochem.*, **17,** 294-302.

NEW COMPOUNDS $[Nb_4O_3(O,F)_2].n[K_{1-x}(F,O)]$ (n=1,2,3) OBTAINED BY MOLTEN SALTS ELECTROLYSIS

Preparation and X-Ray diffraction study

A.V.Arakcheeva, V.V.Grinevitch, E.G.Polyakov*, S.A.Kuznetsov*
Institute of Metallurgy RAS, Moscow, Leninsky pr.49, Russia
**Institute of Chemistry KSC RAS, Apatity, Russia*

1. Introduction

The new niobium phases with Nb(II,IV), K and F ions were obtained on cathode by molten salts' electrolysis of the system K_2NbOF_5 +FLINAK. These phases never are produced by electrolysis in the chloride molten system. Niobium phase with similar crystal habit and K ions in composition has been reported by E.Christensen *et al.* [1]. We initiated the present investigation in order to define some individual characteristics of the phases. The aims of our study are following: (a) identification of the niobium phases; (b) determination of their atomic structure; (c) analysis of their transformations under washing out a cathodic product with acid water solution and under the action of the KCl+NaCl melt.

2. Experimental

The electrolysis have been carried out at the temperature 720-780°C, D_c=0.2-0.5 A/cm^2. The NbO-anod was used. An external view of the crystals obtained on cathode is shown in Fig.1. The crystals are formed into a blocks of rectangular tables with maximum size of single crystal 0.5x0.5x0.05 mm. They are dark grey in colour and have a metallic lustre.

Each cathodic product was divided into three parts. From one of them the salt phases entrapped from the electrolysis bath were washed out with acid water solution. Another part was dipped into molten alkaly chlorites , and the obtained product was washed with acid water solution, too. Both initial products and two washed ones were examined by the X-ray diffraction analysis. About 20 cathodic products with new niobium phases have been analyzed.

X-ray diffraction analysis was used for crystal structure determination. Single crystal diffraction experiments were carried out using autodiffractometer Syntex P1

D.H. Kerridge and E.G. Polyakov (eds.), Refractory Metals in Molten Salts, 143-152.

(Mo radiation). Powder diffraction experiments were carried out using DRON equipment (Cu radiation).

Structure determination was performed for two compounds on the base of single crystal data. The theoretical powder diffraction patterns were calculated and compared with experimental ones for identification other new phases.

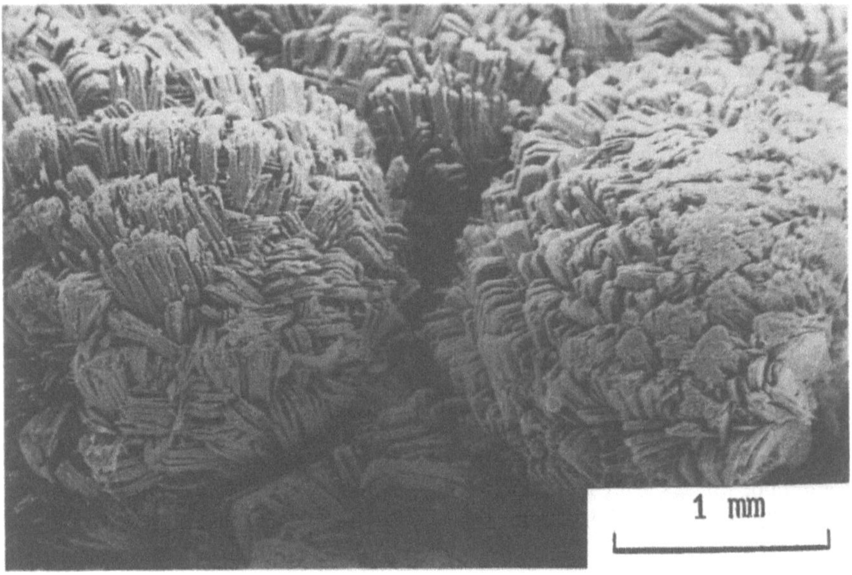

Figure 1. The crystal habit for new phase of Nb(II,IV).

3. The phase identification

Identification of new phases was carried out on the base of X-ray stady in several steps.

3.1. PRELIMINARY POWDER X-RAY DIFFRACTION EXPERIMENTS

According to X-ray powder diffraction analysis there are three main types of experimental diffraction patterns. They are represented in Fig.2. Each pattern is characterized by a strong line on small diffraction angles. But the angles are different for three patterns. In addition there is some angle range for this line in the several patterns of the same type. First pattern (Fig.2,a) contains the line which corresponds to the interplanar distance range from 8.35 to 8.44 A. This type we consider as the pattern of the phase I. The strongest line from the second pattern corresponds to interplanar distance range from 10.7 to 11.0 A. Second type we consider as the diffraction pattern of the new niobium phase II. There are the superposition of lines from second pattern and additional line group in the case of third type of experimental pattern. The additional lines we consider as the pattern of the new niobium phase III. The strongest line from additional group corresponds the interplanar distance range from 13.25 to 13.56 A.

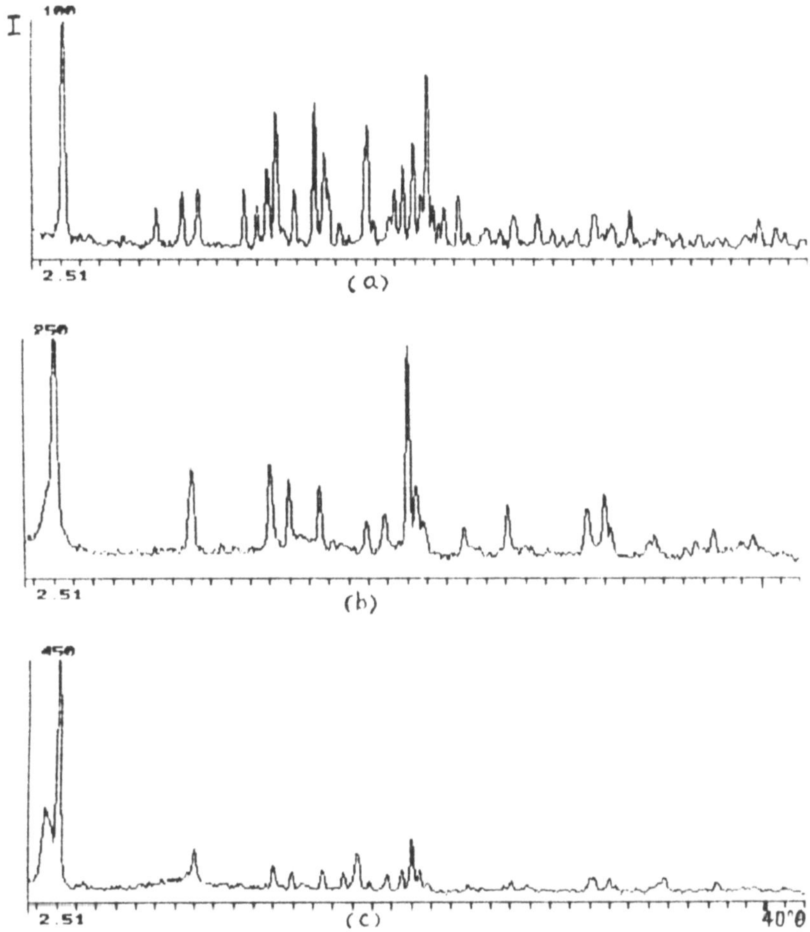

Figure 2. Experimental X-ray diffraction patterns (Cu radiation) for phase I (a), phase II (b) and for phase II + phase III (c).

It should be noted that the first phase is most frequent in our experiments, and the third one is produced only after washing some cathodic products.

The results of the crystal system determination as well as the unit cell parameters are ambiguous on the base of only powder diffraction data. So, the crystal structure determination was necessary to identify these new niobium compounds.

3.2. CRYSTAL STRUCTURE DETERMINATION OF PHASE II

Two single crystals of phase II with different compositions were taken for this purpose. The crystal structure determination [2] was carried out using the single crystal X-ray technique. The main crystallographic and experimental information are given in Table 1.

TABLE 1. The experimental crystallographic data for two compounds of phase II $[Nb_4O_3(O,F)_2].2[K_{1-x}(F,O)]$ (x=0.38 and x=0.16)

	x=0.38	x=0.16
Chemical composition	$K_{1.25}Nb_4O_3(F,O)_3F$	$K_{1.68}Nb_4O_3(F,O)_3F$
Crystal system	Tetragonal	
Space group	P4/mmm (No.123)	
Cell parameters, A	a=4.137(1)	a=4.148(1)
	c=21.860(3)	c=21.582(2)
Total reflections	954	1339
Unique reflections	186	292
Reliability factor of structure determination	0.038	0.055

It can be seen that two crystals vary in composition. They differ in K-content. Both crystals belong to tetragonal space group P4/mmm . The respective unit cell parameters slightly differ in their values for two compounds. There is correlation between the parameter's value and the K content. The more K content, the more 'a' and the less 'c' parameter. Structure parameters, such as atomic coordinates, thermal and occupancy factors, were refined up to reliability factor R=0.038 and R=0.055 for two crystals respectively.

Both crystals of phase II belong to the same crystal type called the 'NbO'.2'KF'. This is the composite structure type because it may be represented as a combination of two parent structure fragments. It is built up by alternation of two layer blocks along the 4 fold axis. One of them is the fragment of NbO structure. Another one is the fragment of KF structure (Fig.3,b). In other hands this structure type may be describe as NbO-blocks separated by 2 KF-atomic nets. The ratio between the number of 'NbO'-blocks and the number of 'KF'-nets is 1 to 2. That is why we use the name 'NbO'.2'KF' for this structure type. The chemical formula of the model differs from the name, because the composition of NbO-block is Nb_4O_5.

Real and schematic structures usually differ in details. This actually is the case. The space group of model is I4/mmm. The space group of real structures is P4/mmm. The 'NbO'-block is essentially the same both in model and in two real structures. The only difference is some substitutions of O atom by F atom in the boundary nets. Real

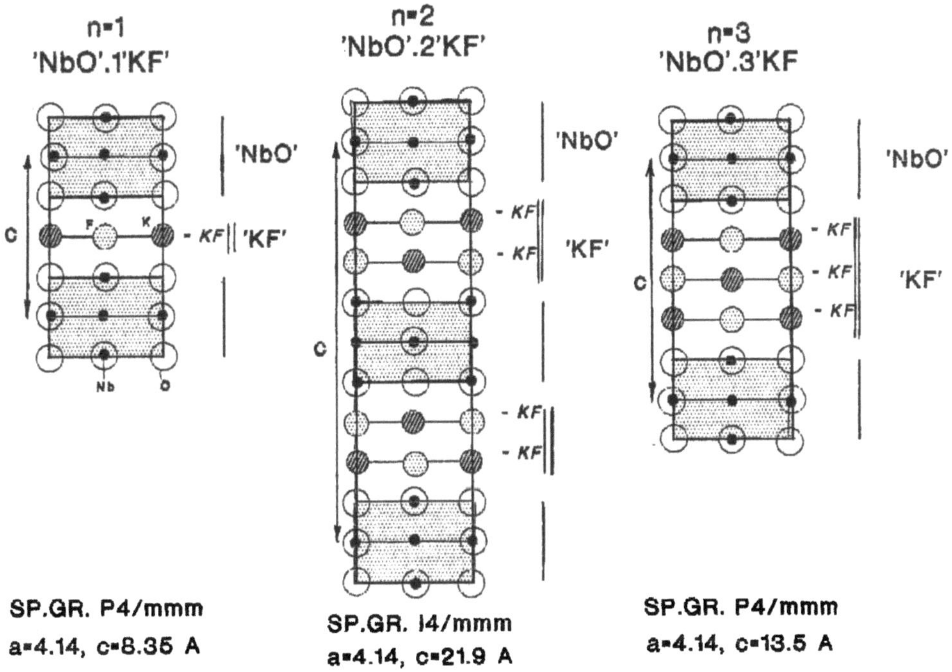

Figure 3. Schematical representation of structure types ((100) projection):
(a) Nb_4O_5.KF ='NbO'.'KF'; (b) Nb_4O_5.2KF ='NbO'.2'KF'; (c) Nb_4O_5.3KF ='NbO'..3'KF'.

and model 'KF'-blocks differ in occupancy of atomic positions and in structure. Moreover, real KF blocks in two different crystals vary in these features. One K position in both real structures is vacant: there are 75% and 32% vacancies for two compounds respectively. Some of F atoms are substituted by O atoms in both cases. The more defects are in K-positions, the more deformations are in respective KF atomic nets and the more is 'c' parameter and the less is 'a' paramcter.

Thus, the structure determination of these two crystals allows us to conclude that the 'KF'-block is a 'weak point' in this structure. It may contain a variable content of K ions resulting in variable unit cell parameters. It is a very important result for later use.

3.3. IDENTIFICATION OF STRUCTURE TYPE FOR PHASES I AND III

As a result of structure determination of phase II we proposed the models for crystal structure of new niobium phases I and III (Fig.3, a and c respectively). They differ in the number of 'KF' atomic nets between "NbO'-blocks. There is only one 'KF' atomic net in the structure of phase I. And there are three 'KF' atomic nets in the structure of phase III. The unit cell parameter 'a' is about the same (4.14 A) in all these structures. The 'c' parameters correlate well with the powder X-ray diffraction data: c_I=8.35 A; c_{II}=13.5 A. The space group is P4/mmm for both I and III. We have calculated the X-ray diffraction patterns for both crystal structure models on a base of atomic positions according to the proposal.

Experimental and calculated patterns for phase I and III are shown on Fig.4 and Fig.5 respectively. In both case two patterns are similar in strong diffraction lines. Some differences between calculated and experimental patterns are due to some differences in schematic and real structures.

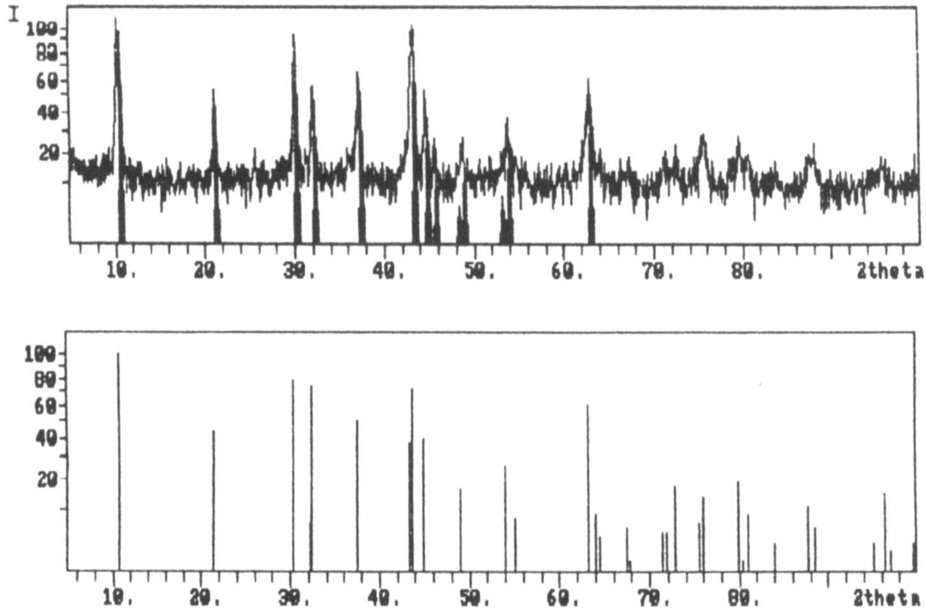

Figure 4. Experimental (high) and calculated (low) X-ray patterns for phase I (Cu radiation)

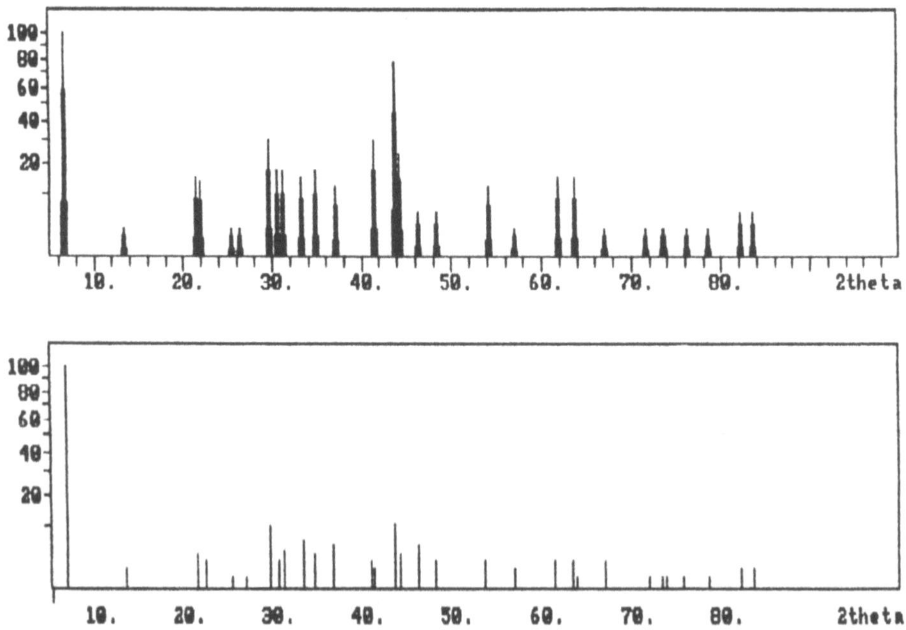

Figure 5. Experimental (high) and calculated (low) X-ray patterns for phase III (Cu radiation)

The strong lines in experimental patterns have been indexed, and the unit cell parameters have been refined after identification of the structure types for some samples of phases I and III. Example of the results for a sample of each phase is represented in Tables 2 and 3.

TABLE 2. Experimental diffraction data, lines' indexes and refined unit cell parameters of phase I

Theta	d, A	I/I_o	d_{cal}, A	h k l	Unit cell parameters, A
5.295	8.3535	98	8.3410	0 0 1	
10.707	4.1493	43	4.1532	1 0 0	
15.196	2.9410	87	2.9429	1 0 2	
16.178	2.7668	46	2.7701	1 1 1	a=4.1532(9)
18.741	2.3994	52	2. 4012	1 1 2	
21.788	2.0769	100	2.0766	2 0 0	c=8.341(4)
22.458	2.0180	37	2.0151	1 1 3	
24.519	1.8576	16	1.8574	2 1 0	V=143.9(1)
27.021	1.6968	25	1.6967	2 1 0	
31.585	1.4718	51	1.4714	2 0 4	

TABLE 3. Experimental diffraction data, lines' indexes and refined unit cell parameters of phase III

Theta	d, A	I/I_o	d_{cal}, A	h k l	Unit cell parameters, A
3.334	13.255	100	13.400	0 0 1	
10.789	4.1182	29	4.1160	1 0 0	
12.716	3.5022	14	3.5071	1 0 2	
13.244	3.3649	14	3.3500	0 0 4	a=4.116(8)
14.868	3.0043	33	3.0268	1 0 3	
15.277	2.9257	20	2.9105	1 1 0	c=13.40(5)
15.653	2.8572	18	2.8440	1 1 1	
16.667	2.6878	16	2.6800	0 0 5	V=227. (1)
18.567	2.4210	10	2.4385	1 1 3	
21.975	2.0601	79	2.0580	2 0 0	

Experimental diffraction pattern for phase III is obtained as a result of subtraction of the phase II pattern from the pattern of sample obtained after washing of cathodic product with acid water solution.

So, the new phase identifications have been carried out. At present time the single crystal X-ray diffraction data have been obtained for phase I. The structural parameters (refined up to R=0.05) confirm the proposal model.

4. Hydration of new niobium phases

As noted above, the initial cathodic products contain some salts entrapped from electrolysis bath and they need in washing out with acid water solution. We obtained the X-ray powder diffraction data for new niobium phases both from initial cathodic product and for ones washed. The comparing of these X-ray data point to changes in the diffraction patterns of the phase II. Thin lines become thicker and the unit cell parameters are changed: 'c' parameter increases from 21.4-21.9 A to 21.9-22.4 A; 'a' parameter decreases from 4.14-4.16 A to 4.12-4.13 A. Three different types of patterns may be for washed products. First type of pattern testifies that the unit cell parameters of initial phase are changed only. Second type of pattern testifies that there are not only some chages in cell parameters, but there occurs the phase III apart from the phase II (Fig.2, c). Third type of pattern is indicative of amorphisation of phases II and III (Fig. 6).

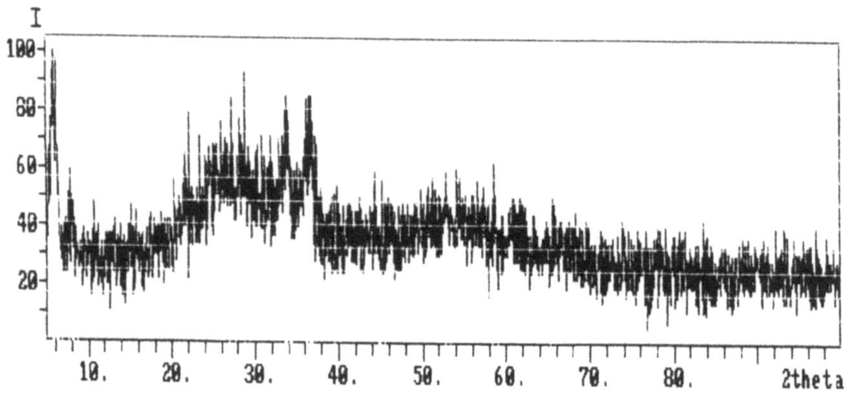

Figure 6. X-Ray diffraction pattern of essentially amorphose sample which washed in acid water solution during 3 mouths.

The change in chemical composition of washed phase II is additional experimental result: the K content changes from 9.3-12.2 to 3-6 wt %; the F content changes from about 7 to 2-6 wt %. So K and F contents essentially decrease in washed phases II.

All these experimental facts are attributable to hydrolysis and hydration of initial niobium phase II. As noted above, there is the 'weak point' in the crystal structure. It is the 'KF'-block with variable K content. K and F ions may be substituted by water according to chemical equation:

$$[KF] + 2H_2O = [H_3O^+ OH^-] + K^+ + F^-.$$

The water may inserts into vacant K-positions according to equation:

$$(K_{1-x}) + xH_2O = (K_{1-x}(H_2O)_x.$$

As the result defects in K positions increase and the unit cell parameters change. In some case, the water inserts in 2'KF' block to form additional net, and 3'KF' blocks are formed. There occurs the phase III. These chemical reactions can work concurrently. If the structure loses K and F ions completely the amorphous phase is formed.

We emphasize that the hydration processes discussed above are characteristic of both II and I phases with one or two 'KF' nets in their structure. The phase III with three such nets is the result of this action.

5. Decomposition of new niobiun phases in KCl+NaCl salt melt

The results of X-ray powder diffraction study for initial phases and for sediments in chloride melt allow us to conclude that the new niobium phases are not stable in chloride melt. It may be the causes of their absence upon the electrolysis in chloride-oxofluoride melts. They decay into various niobium oxides or into combinations of these oxides and the phase I. In any case the structure loses in part the K and F ions in chloride melt. We propose the following schemes of chemical reactions in the melt:

(a) $[Nb^{II-IV}O_3(O,F)_2.n[K_{1-x}(F,O)] \rightarrow NbO+[Nb^{II-IV}O_3(O,F)_2.[K_{1-x}(F,O)] +K^{+}+F^{-}$.

If K ions are lost totally (x=1), the special case will occur: the new niobium oxide Nb_2O_3 or fluori-oxide $Nb_2(O,F)_3$ may be produced. In this case there is no atoms in K positions in 'NbO'.1'KF' structure type;

(b) $[Nb^{II-IV}O_3(O,F)_2.n[K_{1-x}(F,O)] \rightarrow NbO+NbO_2+ K^{+}+F^{-}$;

(c) $[Nb^{II-IV}O_3(O,F)_2.n[K_{1-x}(F,O)] \rightarrow NbO+NbO_2+[Nb^{II-IV}O_3(O,F)_2.[K_{1-x}(F,O)] + K^{+}+F^{-}$;

(d) $[Nb^{II}{}_4O_3F_2].2[KF] \rightarrow NbO+K^{+}+F^{-}$.

The oxidation state of Nb in new phases may be estimated from the oxide compositions.

6. Conclusions

- The new niobium phases with variable composition have been obtained and identified as homologous series with general formula:

  $[Nb_4O_3(O,F)_2]$. $n[K_{1-x}(F,O)]$, where n=1,2,3 and x<1.
- The crystal structure types for three homologous are identified. The structures belong to composite type. They are built up by alternation of two layer blocks along the 4 fold axis. One of them is the fragment of NbO structure. Second block is the fragment of KF structure (rock salt structure type). The structures differ in the number of atomic of atomic nets in 'KF' layer block. The structures have the tetragonal symmetry, space group P4/mmm. Unit cell parameters vary with the chemical composition. The more K content, the more 'a' is and the less 'c' is.
- The key feature of the structures lies with the unstable n'KF' block. The K positions contain vacancies; F atoms may be substituted by O atoms. 'KF' block is the 'weak point' in the structures.
- The new phases are hydrolyzed in acid water solution. Hydrolysis and hydration take place in 'KF'block as in the structure 'weak point'. As a result the K content falls down, the structure becomes more defective, the unit cell parameters change. The phase with 3'KF' nets is formed in the result of hydration of the phase with 1 or 2 'KF' nets.
- The phases under discussion are decomposed in KCl+NaCl melt to form the phase with 1KF net and/or niobium oxides. Products of decomposition depend upon the composition of initial phase and the valence of Nb in it.

References

1. Christensen, E., Xindong Wand, J.H. von Barner, Ostfold, T.and Bjerrum, H.J. (1994) The influence of oxide on the electrodeposition of niobium from alkali fluoride melts, *Journal of the Electrochemical Society* **141,** 5, 1212-1220.
2. Arakcheeva, A.V., Grinevitch, V.V., Lubman, G.U. and Polyakov, E.G. (1998) The crystal structure of new compound of the lowvalence niobium $K_{2-x}Nb_4O_3(F,O)_3F$, *Crystallography Reports* **43,** (in press) .

ELECTROCHEMICAL PROCESSES AS A SUM OF CHEMICAL EQUATIONS

V.G.KREMENETSKIY

Institute of Chemistry Kola Science Centre RAS, 14 Fersman Str., Apatity, 184200, Murmansk region, Russia

To calculate the thermodynamic parameters of model chemical reactions describing, in particular, either the processes of a stepwise discharge of complex species in melts or the processes of forming such particles as a result of anodic dissolution of metals as well as in modelling the above processes' kinetics the electrochemical processes are represented in the form of concrete chemical reactions. But as in many cases the electrochemical stage is accompanied by the preceding and/or subsequent chemical reactions, the necessity arises of considering the electrochemical process as a whole, without limiting oneself by analysis of model reactions describing the electrochemical act of electron transfer proper. Such an approach is particularly essential in analysis of the results of computer modelling of electrochemical reactions in the cluster approximation, when interaction of a limited number of species is considered. The results of such a modelling are unambiguously associated with a definite chemical equation, where to each of the reaction components concrete thermodynamic characteristics can be ascribed. In this case setting up a system of equations describing the electrochemical process reveals the processes which remained outside the computer experiment.

Since until recently there has been no demand for a similar representation of the results of investigations, the methodological basis for setting up such systems of equations is poorly developed. In this work an attempt to partially fill this gap has been made.

As an example let us consider the chemical reaction

D.H. Kerridge and E.G. Polyakov (eds.), Refractory Metals in Molten Salts, 153-158.

$$NbCl_5 = Nb_s + 5/2\ Cl_2\ , \qquad (1)$$

where $NbCl_5$ and Cl_2 are in the gaseous phase and Nb_s is the crystalline phase. The electrochemical equivalent of this reaction is commonly represented as a sum of two electrode semi-reactions:

$$NbCl_5 + 5e = Nb_s + 5Cl^- \qquad (2)$$

$$5Cl^- = 5/2\ Cl_2 + 5e \qquad (3)$$

To calculate the thermodynamic characteristics of these semi-reactions, the Gibbs energy (GE), for instance, the state of the chlorine anions and of the electrons should be specified. Assume that the Cl- anions are in the ideal gas state. The cathode semi-reaction (2) can be represented as a sum of two reactions (to the right of the vertical line the values of standard GE are indicated):

$$5e_c = 5e_g \quad | \quad 5\ A_e(C) \qquad (4)$$

$$NbCl_5 + 5e_g = Nb_s + 5Cl^- \quad | \ \Delta\ G_1{}^o, \qquad (5)$$

where equation (4) corresponds to electron transfer from the cathode material (e_c state) to ideal gas state e_g. Independently of the cathode potential value $\varphi_c{}^i$ the value of the standard GE reaction (4) is equal to A_e (C). This can be shown if two ratios are taken into account [1, p.20-23]: firstly, the equality $\varphi_c{}^i = \Delta\varphi_c{}^s + \varphi_c{}^e$ ($\Delta\varphi_c{}^s$ is the surface jump of potential and $\varphi_c{}^e$ is the external potential of the cathode) and, secondly, an equality determining the notion of the electron work function, $A_e = -\mu_c + F\Delta\varphi_c{}^s$ (μ_c is the chemical potential of electronic gas in the cathode metal). Then we obtain for the equation (4):

$$\Delta G^o = -5F\varphi_c{}^e - 5(\mu_c - F\varphi_c{}^i) = 5(-\mu_c + F\Delta\varphi_c{}^s) = 5A_e$$

The value of standard GE of equation (5) is designated as $\Delta G_1{}^o$. The anodic semi-reaction (3) can be represented as a similar sum:

$$5Cl^- = 5/2Cl_2 + 5e_g \quad | \ \Delta G_2{}^o \qquad (6)$$

$$5e_g = 5e_a \quad | \ -5A_e(A), \qquad (7)$$

where A_e (A) is work function of electron from the anode material. As the result, instead of a conventional notation (equations (2) and (3)) we obtain the following equations of semi-reactions:

$$NbCl_5 + 5e_c = Nb_s + 5Cl^- \quad | \ \Delta G_1{}^o + 5A_e(C) \qquad (8)$$

$$5Cl^- = 5/2\ Cl_2 + 5e_a \quad | \ \Delta G_2{}^o - 5A_e(A) \qquad (9)$$

$$NbCl_5 + 5e_c = Nb_s + 5/2Cl_2 + 5e_a \ | \ \Delta G_1{}^o + \Delta G_2{}^o + 5A_e(C) - 5A_e(A) \qquad (10)$$

The summary equation (10) is an electrochemical equivalent of the chemical reaction (1). However, this system does not take into account the fact that for reaction (9) to occur the Cl- anions obtained in proximity to the cathode surface by reaction (8) should be transferred in the electric field to the anode surface. Having taken into account this circumstance, we obtain a new system:

$NbCl_5 + 5e_c = Nb_s + 5Cl^- (C) \mid \Delta G_1^\circ + 5A_e(C)$ (11)

$5Cl^-(C) = 5Cl^-(A) \mid -5F\Delta\varphi^e = -5F\Delta\varphi^i + 5A_e(A) - 5A_e(C) + 5\mu_a - 5\mu_c$ (12)

$5Cl^-(A) = 5/2Cl_2 + 5e_a \mid \Delta G_2^\circ - 5A_e(A)$ (13)

$NbCl_5 + 5e_c = Nb_s + 5/2\, Cl_2 + 5e_a \mid \Delta G_1^\circ + \Delta G_2^\circ - 5F\Delta\varphi^i + 5(\mu_a - \mu_c)$ (14)

Here $Cl^-(C)$ and $Cl^-(A)$ are chlorine anions localized in the zone of occurrence of reactions (11) and (13) at a distance r_c and r_a from the electrode surface, respectively, whereas $\Delta\varphi^e$ is external potential difference of electrodes ($\Delta\varphi^e = \varphi_a^e - \varphi_c^e$). According to the meaning of reaction (12), the external potentials of electrodes φ_a^e and φ_c^e are determined here, the same as A_e, in points r_a and r_c, and $\varphi^i = \varphi^e + 1/F\,(A_e + \mu)$.

It is commonly supposed that "the external potential is that in a point located at a distance of 10^{-6} m from the metal surface" [1, p.21]. Assuming that the work function of an electron is determined for the same distance, then notwithstanding the fact that the values of standard GE of equations (11)-(13) change, the expression for ΔG° of the summary reaction (14) will remain the same. To make sure let us designate as A_e° the work function of an electron from the electrode to the point removed to a distance R equal to 10^{-6} m from the surface. Then

$$A_e^\circ = A_e + \Delta A,$$

where ΔA is the energy needed to transfer an electron from point r, having the potential φ_r^e, to point R with the potential φ_R^e. As the result we obtain the following expressions for ΔG°:

equation (11): $\Delta G^\circ = \Delta G^\circ_1 + 5[A^\circ_e(C) - \Delta A(C)]$, (11a)

equation (12): $\Delta G^\circ = -5F\,\Delta\varphi^i + 5[A^\circ_e(A) - A^\circ_e(C)] - 5[\Delta A(A) - \Delta A\,(C)] + 5(\mu_a - \mu_c)$, (12a)

equation (13): $\Delta G^\circ = \Delta G_2^\circ - 5[A^\circ_e(A) - \Delta A(A)]$, (13a)

equation (14): $\Delta G^\circ = \Delta G_1^\circ + \Delta G_2^\circ - 5F\Delta\varphi^i + 5(\mu_a - \mu_c)$,

where $\Delta A(C) = F[\varphi^e_R(C) - \varphi^e_r(C)]$ and $\Delta A(A) = F[\varphi^e_R(A) - \varphi^e_r(A)]$.

The $\Delta A(C)$ and $\Delta A(A)$ values can only be calculated in the framework of a concrete model setting the distribution of the difference of potentials $[\varphi^e_r(A) - \varphi^e_r(C)]$ in the space between points r(A) and r(C).

Note that in mixed anion systems, for instance, in chloride-fluoride ones, the cathode reaction may proceed with participation of one anion whereas the anode reaction proceeds with participation of the other anion. In this case the r(A) and r(C) values will already be not equal and the work function $A_e(A)$ and $A_e(C)$ will be set at different distances from the electrode surface.

Setting to zero the expression for GE of the summary reaction (14)

$$\Delta G = \Delta G_1{}^{\circ} + \Delta G_2{}^{\circ} - 5F\Delta\varphi^{i} + 5F\Delta\varphi_{cont} + RT\ln\Pi_o = 0$$

and introducing designations E_o and E° for the equilibrium and standard difference of internal potentials, respectively, we obtain the following expressions for the values indicated:

$$E_0 \equiv \Delta\varphi_0{}^{i} = E^0 + \frac{RT}{5F}\ln \Pi_0;$$

$$E^o = \frac{\Delta G_1{}^o + \Delta G_2{}^0}{5F} + \Delta\varphi_{cont},$$

where $\Delta\varphi_{cont} = \frac{1}{F}(\mu_a - \mu_c)$ - contact difference of electrode potentials,

$$\Pi_0 = (a_{Cl_2})_0^{5/2} \cdot (a_{e_a})_0^{5} / (a_{NbCl_5})_0 \cdot (a_{e_c})_0^{5}$$

The lower zero index is evidence that for all the activities equilibrium values are taken. As follows from the above ratios, the zFE° value is equal (with accuracy of up to $\Delta\varphi_{cont}$) to that of standard GE chemical reaction (1), which is a sum of reactions (5) and (6), but it is not equal to the value of ΔG° for reaction (14), which actually determines the E° value. Note that E° depends on contact difference of potentials notwithstanding the absence of contact between the electrodes (the external circuit is open).

In the case when the external circuit is closed via a resistor (for instance, in measuring E_0), the system of equations (11)-(13) should be added with one more equation describing the electron transfer from one electrode to the other:

$$5e_a = 5e_c \mid 5F(\Delta\varphi^{i} - \Delta\varphi_{cont}), \tag{15}$$

where $\Delta\varphi^{i}$ is still equal to $(\varphi_a{}^{i} - \varphi_c{}^{i})$.

Summing reactions (11)-(13) and (15) we obtain equation (16) equivalent to equation (1):

$$NbCl_5 = Nb_s + 5/2Cl_2 \mid \Delta G_1{}^{\circ} + \Delta G_2{}^{\circ} \tag{16}$$

Thus the equilibrium state of a system with closed circuit is determined by the equilibrium constant of equation (16) and does not depend on the difference of potentials at the electrodes.

If to system (14), which is in equilibrium, an external voltage source with the output voltage $\Delta\varphi_E$ is connected, then the value of standard GE of equation (15) will be:

$$\Delta G^{o} = 5F(E_o - \Delta\varphi_{cont} + \Delta\varphi_E),$$

where E_o is equilibrium difference of potentials of system (14). The expression in equation (12) will change accordingly. So, by varying the $\Delta\varphi_E$ value we thereby change the values of standard GE in reactions (12) and (15) while in the cases when $\Delta A(A)$ and $\Delta A(C)$ are not equal to zero the same is true for reactions (11) and (13) - in keeping with expressions (11a) and (13a).

However, in the cases when the external voltage source is lacking all the chemical equations describing the charge transfer in the electric field have, as is clearly seen, variable ΔG^{o} values of standard Gibbs energy (i.e. depending on $\Delta\varphi^{i}$ values). Since at equilibrium the equality is true,

$$\Delta G^{o} = -RT \ln \Pi_o,$$

then a change in $\Delta\varphi^{i}$ results in the change of the Π_o ratio of equilibrium concentrations, i.e. the disruption of equilibrium and further occurrence of the charge transfer reaction.

In conclusion let us return to equation (12). In reactions similar to the one considered above, this equation appears due to the natural necessity to deliver the product of the cathode reaction to cause the anode reaction (or vice versa). Using this equation provides correct ΔG^{o}, E_0 and E^{0} values of the summary reaction (14). It may seem that in the cases when the reaction is conducted in a medium containing like background ions this equation is no longer necessary. In fact, this is not true.

Let us consider, for instance, the reaction of chloroniobate $MNbCl_6$ reduction occurring in melt of an alkali metal chloride MCl. In the simplest variant the system of model equations for this process is in no way different from the (11) - (14) system:

$$MNbCl_6 + 5e_c = Nb_s + MCl(C) + 5Cl^-(C) \quad (17)$$

$$5Cl^-(C) = 5Cl^-(A) \quad (18)$$

$$5Cl^-(A) = 5/2\ Cl_2 + 5\ e_a \quad (19)$$

$$MNbCl_6 + 5e_c = Nb_s + 5/2\ Cl_2 + MCl(C) + 5\ e_a \quad (20)$$

In this system, equation (18) has no longer the meaning it had in the previous case since here the $Cl^-(A)$-anions for the anode reaction are supplied by the background electrolyte. Nonetheless, the $Cl^-(C)$- anions formed at the cathode, being in the electric field of the interelectrode space, should be redistributed in this field according to Boltzmann distribution. This process leads to the change of the system's GE. The standard GE of equation (18) (see expression (12a)) characterizes the height of the potential barriers of this distribution depending on the $\Delta\varphi^{e}$ value between the surface of the $Cl^-(C)$-anion location at the moment of completion of reaction (17) and the surface of the $Cl^-(A)$-anion location at the initial moment of the anodic reaction (19).

Specifically, the type of $\Delta\varphi^e$ distribution between the aforementioned surfaces depends, as it has already been noted, on the models used.

The essence of the matter will not change if another model of reaction (20), for instance, the following one is accepted:

$$MNbCl_6 + 5e_c = Nb_s + MCl(C) + 5Cl^-(C) \quad (21)$$

$$5M^+(A) + 5Cl^-(C) = 5MCl(A) \quad (22)$$

$$5MCl(A) = 5/2Cl_2 + 5M^+(A) + 5e_a \quad (23)$$

$$MNbCl_6 + 5e_c = Nb_s + 5/2Cl_2 + MCl(C) + 5e_a$$

In this case reaction (18) is incorporated in reaction (22):

$$5Cl^-(C) = 5Cl^-(A)$$

$$5M^+(A) + 5Cl^-(A) = 5MCl(A)$$

$$5M^+(A) + 5Cl^-(C) = 5MCl(A)$$

Naturally, the view of the equations of spatial distribution of ions in the interelectrode space depends on the degree of model complexity. Instead of simple M^+ and Cl^- ions autocomplex species and more complex clusters may be considered, but in all the cases, when among the products of anode or cathode reactions there are charged species, such equations should be taken into account.

Acknowledgment

The work was performed with the support of the *Russian Foundation for Fundamental Investigations (96-03-32175).*

References

1. Damaskin B.B., Petriy O.A. (1983) *Introduction into electrochemical kinetics,* Vysshaya Shkola, Moscow (Rus.).

ROLE OF THE OXOACIDITY AND LIGAND EFFECT IN THE ELECTROWINNING OF TITANIUM IN FUSED SALTS

F. LANTELME, A. BARHOUN [1] and K. KURODA [2]
Laboratoire d'Electrochimie, CNRS, Université P. et M. Curie
4 place Jussieu, 75252 Paris Cedex 05, France.

Abstract - The standard potentials of the reaction steps for the reduction of $TiCl_4$ in alkali chlorides were calculated from transient electrochemical techniques. It is shown that the stability ranges of the various oxidation states of titanium depend on the composition and temperature of the solvent. The presence of strong chloride ion donors, such as cesium chloride, enhances the stability of the high oxidation states. The results are interpreted in terms of formation of chlorocomplexes. The thermodynamic properties of solutes were calculated and the Gibbs energy of dissolution of the titanium salts determined.

1 Introduction

The composition of the solvent has a dramatic influence on the electrodeposition process of refractory metals in fused electrolytes [1]. Also, in contrast to other metals prepared by electrolysis at high temperatures such as alkali metals, magnesium or aluminium, all the refractory metals exist in a variety of oxidation states. The metal chlorides undergo disproportionation reactions leading to formation of seemingly colloidal metal in the melt. The

[1] Present address: Département de Chimie, Université de Tetouan (Maroc)
[2] Present address: CIRSE, Nagoya University, Japan

D.H. Kerridge and E.G. Polyakov (eds.), Refractory Metals in Molten Salts, 159-172.

stability of the various oxidation states depends on the bath composition. The more acidic the cations the more important the effect [2]. Niobium subhalides are stable in acidic melts such as chloroaluminates [3, 4]. In contrast, introducing cesium ions in alkali halide solvents stabilizes the higher oxidation states [5, 6, 7].

The aim of the present paper was to commence a systematic study of the thermodynamics properties of solutions of titanium ions in various fused electrolytes. Experimental determinations were carried out using dynamic electrochemical techniques such as cyclic voltammetry or chronopotentiometry in alkali chloride mixtures.

2 Experimental

Most of the experiments were carried out using the eutectic mixture LiCl(59 mole %)-KCl(41 mole %). In order to examine the influence of the alkali cations of the solvent two other electrolytes were used: the equimolar mixture NaCl-KCl and the eutectic mixture LiCl(59.5 mole %)-CsCl(40.5 mole %). The experiment were carried out over the temperature range 675-995 K in the LiCl-KCl and LiCl-CsCl mixtures and over the range 950-1095 K for the NaCl-KCl mixtures. For experimental details see [8, 9]. In the present paper the potentials are given vs the reference system Cl_2/Cl^- [10].

3 Results

To study the reaction giving rise to metallic titanium, a rod of tungsten, molybdenum or stainless steel (AlSl 304L) was used as a working electrode. The redox reactions occurring at a more anodic potential were studied on glassy carbon (V25, Le Carbone Lorraine) or on tungsten.

3.1 Cyclic voltammetry

In agreement with the literature data [11, 12, 13] the voltammogram in Fig. 1 shows that the following reactions occur in the LiCl-KCl melts

$$Ti^{4+} \overset{e^-}{\leftrightarrow} Ti^{3+} \overset{e^-}{\leftrightarrow} Ti^{2+} \overset{2e^-}{\leftrightarrow} Ti \tag{1}$$

- <u>Reaction $Ti^{4+} + e^- \leftrightarrow Ti^{3+}$</u>

Experiments were carried out in LiCl-KCl mixtures at temperatures ranging from 675 to 1000 K. The voltammograms show that, in addition to the usual increase in the peak current due to the increase in the diffusion coefficient of the electroactive species, there is a positive shift in the peak potentials when the temperature is increased. At high temperatures, the current peaks for the reaction $Ti^{4+} + e^- \rightarrow Ti^{3+}$ are hardly visible since the reaction coincides with the oxidation of chloride ions.

The same trends were obtained for the NaCl-KCl and LiCl-CsCl baths. However, it seems that the oxidation Ti^{3+} occurs at a more negative potential in NaCl-KCl (Fig. 2) and LiCl-CsCl than in LiCl-KCl at the same temperature. In NaCl-KCl the disappearance of the oxidation peak for the reaction Ti^{3+}/Ti^{4+} occurs at high temperature, above 1095 K. In most experiments a change in the scan rate ν did not change the peak potential, indicating that the reaction is reversible. As predicted by the theory, the peak height is proportional to $\nu^{1/2}$. However, a decrease in ν produced a decrease in the ratio of the peak currents and the reverse current scan tends to retrace the forward current scan [9]. This effect is probably due to the low solubility of $TiCl_4$ [14] which evaporates and cannot be reduced during the backward sweep.

- Reactions $Ti^{3+} \overset{e^-}{\leftrightarrow} Ti^{2+} \overset{2e^-}{\leftrightarrow} Ti$

In the LiCl-KCl mixture two associated anodic and cathodic peaks (O_3 and R_3 in Fig. 3) are obtained at -2.05 and -2.3 V on tungsten or molybdenum electrodes. Two bumps, O_2 and R_2 in Fig. 3, are visible on the anodic side of the two peaks. At high sweep rate the anodic peak becomes broader and overlaps the second peak. For high concentrations of titanium salt and low sweep rate, the cathodic peak suffers some distortion due probably to a dendritic growth.

When the temperature is increased there is a positive shift in the potential of the peaks. The shift is more pronounced for the bumps which separate progressively from the main peaks giving rise to a system of two peaks, O_2-R_2, (Fig. 4). In the NaCl-KCl mixture, the separation potential between the peaks, O_2-O_3 and R_2-R_3, is less important. In the LiCl-CsCl eutectic, at 723 K, it is not possible to detect the presence of the O_2-R_2 system. At high temperatures, a small wave appears on the right side of the O_3-R_3 peaks which can be attributed to the O_2-R_2 system.

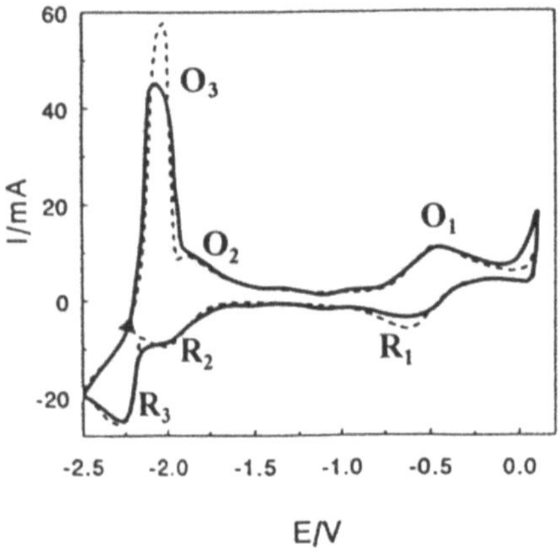

Figure 1. Voltammogram at a tungsten electrode in LiCl-KCl. Area: 0.33 cm^2. Sweep rate: 0.2 V s^{-1}. T = 823 K. Concentration of titanium ions: 0.08 mole dm^{-3}. Reference electrode: Cl_2/Cl^-. Solid line: experimental. Dashed line: calculated.

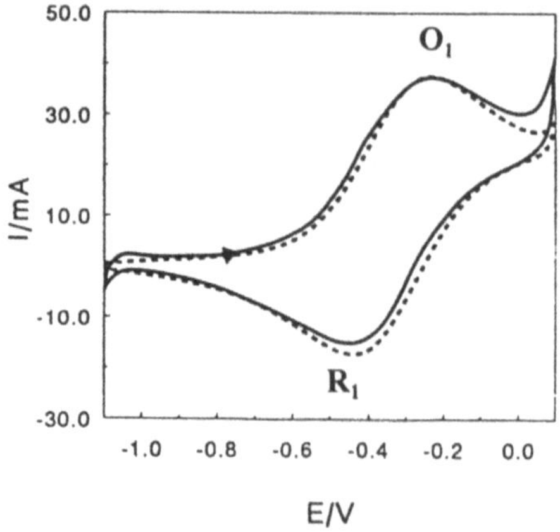

Figure 2. Voltammogram at a glassy carbon electrode in NaCl-KCl. Area: 0.42 cm^2. Sweep rate: 5 V s^{-1}. T : 993 K. Concentration of titanium ions: 0.05 mole dm^{-3}. Reference electrode: Cl_2/Cl^-. Solid line: experimental. Dashed line: calculated.

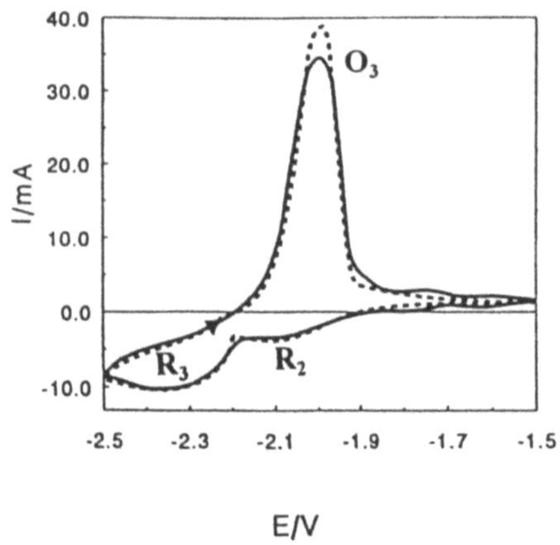

Figure 3. Voltammogram at a tungsten electrode in LiCl-KCl. Area: 0.33 cm^2. Sweep rate: 0.2 V s^{-1}. T = 723 K. Concentration of titanium ions: 0.05 mole dm^{-3}. Solid line: experimental. Dashed line: calculated.

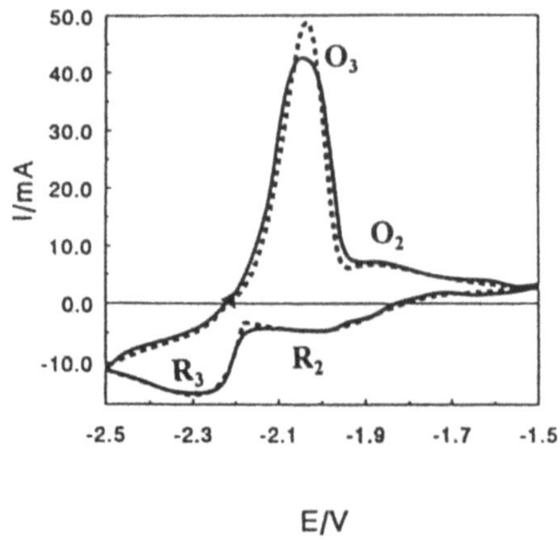

Figure 4. Voltammogram at a tungsten electrode in LiCl-KCl. Area: 0.33 cm^2. Sweep rate: 0.2 V s^{-1}. T = 823 K. Concentration of titanium ions: 0.05 mole dm^{-3}. Solid line: experimental. Dashed line: calculated.

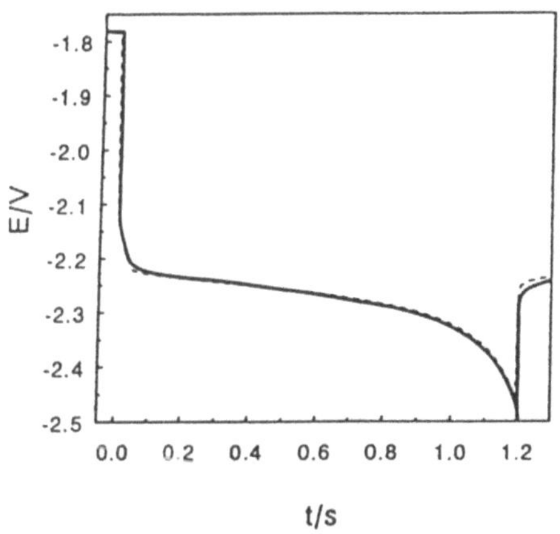

Figure 5. Chronopotentiogram at a tungsten electrode in LiCl-CsCl. Area: 0.33 cm^2. Current intensity: -8.8 mA cm^{-2}. T = 673 K. Concentration of titanium ions: 0.04 mole dm^{-3}. Solid line: experimental. Dashed line: calculated.

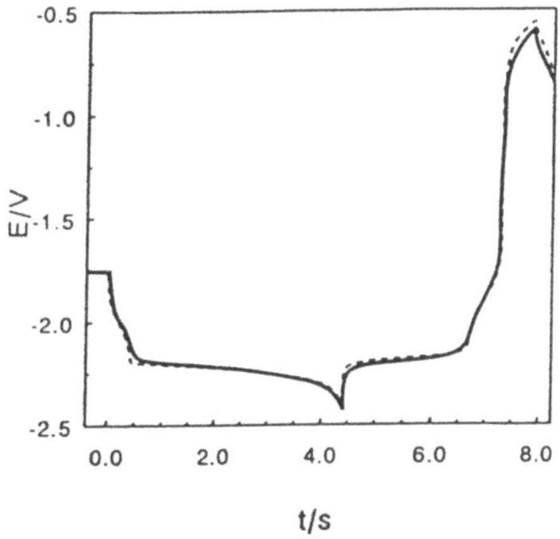

Figure 6. Current reversal chronopotentiogram at a tungsten electrode in LiCl-KCl. Area: 0.33 cm^2. Current intensity: -8.0 mA cm^{-2}. T = 823 K. Concentration of titanium ions: 0.04 mole dm^{-3}. Solid line: experimental. Dashed line: calculated.

3.2 Chronopotentiometry

The analysis by chronopotentiometry of a solution of titanium salt shows the existence of a potential plateau in the potential range -2.2, -2.35 V (Fig. 5). The length of the plateau (transition time, τ) depends on the current intensity and on the concentration of titanium ions. In agreement with the Sand's law, it is shown that τ is proportional to the reverse of the square of the current intensity. The reactions involving metallic titanium were studied by the current reversal technique which is useful for analysing the deposition and dissolution process (Fig. 6). The two bumps at the beginning and at the end of the chronopotentiogram are due to the reaction $Ti^{3+} + e^- \rightleftharpoons Ti^{2+}$. These additional plateaux occur in the same potential region as for the peaks R_2 in cyclic voltammetry. They were not obtained in LiCl-CsCl.

3.3 Modelling

The analysis of electrode reactions composed of diffusion and electrode transfer kinetics is only possible after suitable models have been formulated. Here the scheme (1) is used. The reactions involving the two higher oxidation states are reversible. According to [11] the reaction involving the metallic phase is irreversible. The current intensity of the reaction obeys the current overpotential equation

$$i_1 = i_1^0\{\exp(1.5f\eta) - \frac{c_{TiCl_2}}{c^*_{TiCl_2}}\exp(-0.5f\eta)\} \tag{2}$$

The reactions occuring in the melt are calculated from the numerical simulation of the voltammograms using the program ELEKSIM. The program includes the surface changes due to metal deposition and dissolution [15]. The coverage factor has a great influence on the shape of the dissolution peak. Partial coverage of the electrode induces a shift in the peak potential which must be considered in the calculation [16]. The various parameters involved in the calculations were adjusted to obtain the best representation of experimental curves. Simulated curves are reported in Figures 1-6.

4 Discussion

The experiments were exploited by numerical simulation using the model described in the previous section. In the temperature range of the experiments, the standard potentials of the various reaction steps obey the equations reported in Table I.

When cesium was substituted for potassium in the mixtures, a negative shift in E^0_{43} and E^0_{32} was obtained. The substitution had little effect on the value for the lowest oxidation state, E^0_{20}. In the LiCl-CsCl mixture, the value of E^0_{32} was more negative than E^0_{20}. As a consequence, oxidation state II was hardly visible and Ti^{3+} reduced directly to metal. The emf of the cell corresponding to the formation of solutions of $TiCl_4$, $TiCl_3$ and $TiCl_2$ can be calculated from the values of the standard potentials of the redox couples reported in Table I. The standard Gibbs energy for the formation of these compounds dissolved in the various electrolyte are reported in Table II and compared to the values of the Gibbs energy of formation for the same compounds obtained from literature data [17].

TABLE I. Standard potentials, E^0_{43}, E^0_{32}, E^0_{20} for the redox couples, Ti^{4+}/Ti^{3+}. Ti^{3+}/Ti^{2+}, Ti^{2+}/Ti, respectively. T Kelvin. E^0_{20}, molar scale. Reference system: Cl_2/Cl^-

LiCl-KCl (temperature range 675-995 K)
$E^0_{43} = -0.565 + 1.5 \times 10^{-4}$ (T-723) V
$E^0_{32} = -2.040 + 10 \times 10^{-4}$ (T-723) V
$E^0_{20} = -2.100 + 2.0 \times 10^{-4}$ (T-723) V
LiCl-CsCl (temperature range 675-995 K)
$E^0_{43} = -0.650 + 2.2 \times 10^{-4}$ (T-723) V
$E^0_{32} = -2.115 + 4.8 \times 10^{-4}$ (T-723) V
$E^0_{20} = -2.095 + 2.5 \times 10^{-4}$ (T-723) V
NaCl-KCl (temperature range 950-1095 K)
$E^0_{43} = -0.347 + 2.2 \times 10^{-4}$ (T-993) V
$E^0_{32} = -1.970 + 8.8 \times 10^{-4}$ (T-993) V
$E^0_{20} = -2.098 + 2.5 \times 10^{-4}$ (T-993) V

TABLE II. Standard Gibbs energy of formation, ΔG^0_s, of $TiCl_4$, $TiCl_3$ and $TiCl_2$ dissolved in various electrolytes. Standard Gibbs energy of dissolution, ΔG^0_{diss}. ΔG^0 in kJ mole^{-1} and T Kelvin.

LiCl-KCl (temperature range 675-995 K)	
$\Delta G^0_{(TiCl_4),s}$=-765 + 0.150 T	$\Delta G^0_{(TiCl_4),diss}$=-3 + 0.030 T
$\Delta G^0_{(TiCl_3),s}$=-700 + 0.135 T	$\Delta G^0_{(TiCl_3),diss}$=13 - 0.070 T
$\Delta G^0_{(TiCl_2),s}$=-433 + 0.040 T	$\Delta G^0_{(TiCl_2),diss}$= 77 - 0.115 T
LiCl-CsCl (temperature range 675-995 K)	
$\Delta G^0_{(TiCl_4),s}$=-789 + 0.160 T	$\Delta G^0_{(TiCl_4),diss}$=-27 + 0.039 T
$\Delta G^0_{(TiCl_3),s}$=-710 + 0.135 T	$\Delta G^0_{(TiCl_3),diss}$= 2.5 - 0.070 T
$\Delta G^0_{(TiCl_2)s,}$=-443 + 0.055 T	$\Delta G^0_{(TiCl_2),diss}$= 66 - 0.100 T
NaCl-KCl (temperature range 950-1095 K)	
$\Delta G^0_{(TiCl_4),s}$=-782 + 0.155 T	$\Delta G^0_{(TiCl_4),diss}$=-20 + 0.035 T
$\Delta G^0_{(TiCl_3),s}$=-727 + 0.135 T	$\Delta G^0_{(TiCl_3),diss}$=-14 - 0.075 T
$\Delta G^0_{(TiCl_2),s}$=-453 + 0.050 T	$\Delta G^0_{(TiCl_2),diss}$= 57 - 0.105 T

The Gibbs energy of dissolution of the three compounds in the electrolytes are calculated and reported in Table II. For titanium tetrachloride, it corresponds to the energy of dissolution of one mole of $TiCl_4$ at the gaseous state at one atmosphere. For the other two compounds the energy is relative to the dissolution of one mole of crystallized product. The concentration of dissolved products is one mole per liter (standard state). For dilute solutions, the Gibbs energy of formation of a solution at the concentration, c, is

$$\Delta G_{diss} = \Delta G^0_{diss} + RT \ln c$$

The value $\Delta G_{diss}=0$ is obtained when the pure compound is in equilibrium with the solution at a concentration c_s. For the solid compound it corresponds to the saturation concentration provided that the solution remains ideal up to the concentration c_s.

- $TiCl_4$

The concentration, $c_{(TiCl_4),s}$, is the solubility of $TiCl_4$ at one atmosphere. The values of ln c_s vs T deduced from the electrochemical experiments are reported in Table III. They are in good agreement with literature values [18].

The solubility is a function of the reciprocal radius of the cation of solvent salt. For a mixture, the equivalent radius of the solvent salt is the weighted mean of the radii of the cations of the components. The dissolution of $TiCl_4$ is followed by formation of the anionic complex, $TiCl_6^{2-}$.

TABLE III. Naperian logarithm of the solubility, $c_{(TiCl_4),s}$ of $TiCl_4$ vs absolute temperature T and solubilities at a fixed temperature in various electrolytes. $c_{(TiCl_4),s}$ in mole dm^{-3} .

ln $c_{(TiCl_4),s}$ present work	ln $c_{(TiCl_4),s}$ from ref[18]	$c_{(TiCl_4),s}$ present work	$c_{(TiCl_4),s}$ from ref.[18]
LiCl-KCl (temperature range 675-995 K)		723 K	
-3.67 + 360/T	-3.65 + 941/T	0.04	0.09
LiCl-CsCl (temperature range 675-995 K)		723 K	
-4.63 + 3246/T	-3.44 + 2860/T	0.9	1.6
NaCl-KCl (temperature range 950-1095 K)		993 K	
-4.15 + 2357/T	-3.70 + 3628/T	0.17	0.9

Equations in Table II show that the dissolution energy is made up of two terms $\Delta G_{diss} = \Delta H_{diss} - T\Delta S_{diss}$ [14]. The values of the enthalpies and entropies of dissolution depend on the solvent salt. The change in enthalpy can be explained in terms of complex structure. The electrostatic energy involved in the reaction $TiCl_{4,gas} + 2\ Cl^- \rightarrow TiCl^{2-}_{6,melt}$ has been calculated by Smirnov et al. [18]. The formation of the chlorocomplex during dissolution takes place by break-up of the chloride complex with the alkali metal, M. By assuming that the coordination number of the alkali metal is four, the bond energy of the MCl_4^{3-} complex in molten lithium, sodium, potassium and cesium chlorides are -377, -277, -236 and -184 kJ, respectively [18]. The stronger the bond the less negative the enthalpy of dissolution.

- $\underline{TiCl_3}$

The same model can be used for titanium in oxidation state III. In fused salts $TiCl_3$ exists as two anonic complexes, $TiCl_4^-$ and $TiCl_6^{3-}$. In LiCl-KCl the proportion of hexachlorocomplex decreases when the temperature is increased. This decrease is greatly enhanced in a solvent containing cations strongly bonded to chloride ions [19]. In contrast, in solvents in which the interaction of alkali metal-chloride is weak, such as NaCl-KCl, the formation of $TiCl_6^{3-}$ is greatly enhanced as shown by the spectroscopic investigations of Volkov et al. [20]. This behaviour shows that the value of the Gibbs energy of formation of solution of $TiCl_3$ in a melt becomes more negative when sodium is substituted for lithium, or potassium for cesium. This is in agreement with the results in Table II. Moreover, in agreement with spectroscopic investigations [19], the hexachlorocomplex becomes less abundant as the temperature is increased. It results in a decrease of the Gibbs energy in the melt (Table II).

- $\underline{TiCl_2}$

There is no earlier information concerning the interaction of Ti^{2+} with chloride ions in alkali chlorides melts. However, from the electrochemical investigations it is shown that the interaction should be weak since the standard enthalpy of dissolution of $TiCl_2$ is positive (Table II). Then, the melt composition has a negligible influence on the Gibbs energy of the solute, and the standard potential, E^0_{20}, does not depend on the composition of the solvent. As pointed out in the preceeding paragraph, the formation of the stable hexachlorocomplex, $TiCl_6^{3-}$, is enhanced in basic melts. It results in a negative shift in the standard potential, E^0_{30}. When the solvent is a strong donor of chloride ions the potential for the reduction of Ti^{3+} becomes more negative than the reduction of Ti^{2+}. Then Ti^{3+} reduces directly to the metal and the presence of Ti^{2+} is no longer visible. This is for example the case in LiCl-CsCl. At temperatures lower than 875 K no signal corresponding to the reaction $Ti^{3+} + e^- \rightleftharpoons Ti^{2+}$ is obtained (Fig. 5). This result is in agreement with the work of Chassaing et al. [21].

- Chloroaluminates

In contrast, when acceptors of chloride ions are introduced in the solvent a dramatic positive increase in the Gibbs energy of dissolution $TiCl_3$ is obtained resulting in a positive shift in the value of the standard potential E^0_{30}. In molten $2AlCl_3$-NaCl at 423 K the stability domain of Ti^{2+} becomes very large: from the potential of aluminium deposition (which is taken as the reference point) to the standard potential of the system Ti^{2+}/Ti^{3+} which is 0.43 V vs Al/Al^{3+} [2]. In this system the calculated standard potential E^0_{20} is -0.31 V and the potential difference, E^0_{30} - E^0_{20}, is 0.74 V. In LiCl-KCl at 673 K the difference is nearly zero and becomes negative in LiCl-CsCl which is in agreement with the high stability of the $TiCl_6^{3-}$ complex in chloride melts containing large alkali cations.

- Influence of concentration and temperature

The above results show that the stability domain of the various oxidation states of titanium chlorides varies according to the solvent composition. However, this influence must be studied in connection with the concentration and temperature effect. In the presence of metallic titanium, the ratio, $r = c_{Ti^{3+}}/c_{Ti^{2+}}$, depends dramatically on the concentration and temperature. For a concentration of titanium ions, c_T, the ratio, r, is given by

$$r\sqrt{1+r} = \sqrt{c_T}\exp\left(-\frac{3F}{RT}(E^0_{30} - E^0_{20})\right) \quad (3)$$

The curves in Fig. 7 are calculated from the values of the standard potentials (Table I) for two solvents at two temperatures. It is shown that in the presence of metallic titanium the proportion of Ti^{3+} increases greatly at high concentrations of titanium salts. An increase in concentration has the same effect as the presence of strong donors of chloride ions.

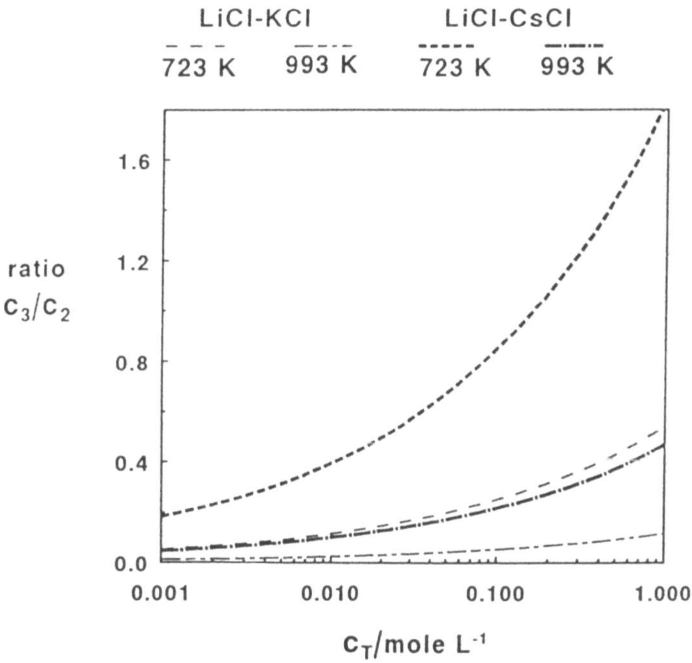

Figure 7. Ratio of the concentrations c_{TiCl_3}/c_{TiCl_2} for a solution of titanium salt at a total concentration c_T in the presence of metallic titanium in two alkali chloride mixtures at two temperatures.

5 Conclusion

Electrochemical techniques have often been used to study the thermodynamics of solutions. The present study shows that the oxidation of titanium ions in fused alkali chlorides depends dramatically on the composition of the melt. Large cations weakly bonded to chloride ions, such as cesium ions, promote the formation of titanium hexachlorocomplexes, $TiCl_6^{2-}$ or $TiCl_6^{3-}$, and results in a large stability range for the high oxidation states. Ti^{3+} reduces directly to metallic titanium. This behaviour is similar to that obtained with stronger ligands such as fluoride ions [9, 22].

In contrast, when small alkali ions such as lithium are introduced in the melt, oxidation state II becomes stable and disproportionation reactions appear giving rise to a colloidal precipitate together with titanium deposition. As for other refractory metals, the stability range of titanium (II) is greatly enlarged in alkali chloroaluminates. This effect is due to the presence of aluminium ions which are strong acceptors of chloride ions.

6 Acknowledgements

The NATO Organizing Committee is acknowledged for its kind invitation. The authors wish to thank Dr. D. H. Kerridge for his help in editing the present manuscript.

References

[1] Girginov, A., Tzvetkoff, T. Z. and Bojinov, M. (1995) Electrodeposition of refractory metals (Ti, Zr, Nb, Ta) from molten salt electrolysis, *J. Appl. Electrochem.* **25**, 993-1003.

[2] Stafford, G. R. and Moffat, T. P. (1995) Electrochemistry of titanium in molten $2AlCl_3$-NaCl, *J. Electrochem. Soc.* **142**, 3288-3296.

[3] Sienerth, K. D., Hondrogiannis, E. M. and Mamantov, G. (1994) A reinvestigation of the electrochemical behavior of Nb(V) in $AlCl_3$-$NaCl_{SAT}$ and related melts, *J. Electrochem. Soc.* **141**, 1762-1769.

[4] Quigley, R. , Barnard, P. A., Hussey, C. L. and Seddon, K. R. (1992) Electrochemical and spectroscopic characterization of $[Nb_6Cl_{12}]^{z+}$ chloride clusters in the aluminum chloride-1-methyl-3-imidazolium chloride molten salt, *Inorg. Chem.* **31**, 1255-1261.

[5] Polyakova, L., Polyakov, E. G., Sorokin, A. I. and Stangrit, P. T. (1992) Secondary processes during tantalum electrodeposition in molten salts, *J. Appl. Electrochem.* **22**, 628-637.

[6] Alimova, Z. A., Elizarova, I. R., Polyakov, E. G. and Polyakova, L. P. (1992) Anodic dissolution of niobium in chloride and mixed chloride-fluoride melts, *Soviet Electrochem.* **28**, 962-966.

[7] Arurault, L., Bouteillon, L. and Poignet, J. C. (1995) Chemical stability of niobium V in molten NaCl-KCl at 750°C, *J. Electrochem. Soc.* **142**, 16-19.

[8] Lantelme, F., Inman D. and Lovering, D. G. (1984) Electrochemistry, in R. J. Gale and D. G. Lovering (eds.), *Molten Salt Techniques*, Plenum Press, New-York, Vol. 2, pp 180-212.

[9] Lantelme, F. and Salmi, A. (1995) Electrochemistry of titanium in NaCl-KCl mixtures and influence of dissolved fluoride ions, *J. Electrochem. Soc.* **142**, 3451-3456.

[10] Lantelme, F., Alexopoulos, H., Devilliers, D. and Chemla, M. (1991) A gas electrode: behavior of the chlorine injection electrode in fused alkali chlorides, *J. Electrochem. Soc.* **138**, 1665-1671.

[11] Popov. B. N., Kimble, M. C., White, R. E. and Wendt, H. (1991) Electrochemical behaviour of titanium(II) and titaniun(III) compounds in molten lithium chloride/potassium chloride eutectic melts, *J. Appl. Electrochem.* **21**, 351-357.

[12] Ferry, D. M., Picard, G. S. and Trémillon, B. L. (1988) Pulse and AC impedance studies of the electrochemical system of titanium in LiCl-KCl eutectic melt at 743 K, *J. Electrochem. Soc.* **135**, 1443-1447.

[13] Haarberg, G. M., Rolland, W., Sterten, Å. and Thonstad, J. (1993) Electrodeposition of titanium from chloride melts, *J. Appl. Electrochem.* **23**, 217-224.

[14] Deviatkin, S. V. and Kaptay, G. (1993) Influence of nature of the alkali metal cation on Ti(III) - e $\rightleftharpoons$ Ti(IV) electrochemical process in chloride melts, *Molten Salt Forum*, **Vol 1-2**, 129-138.

[15] Salmi, A., Berghoute, Y. and Lantelme, F. (1995) Modelling multistep electrochemical reactions in molten salt electrowinninng of refractory metals. *Electrochim. Acta* **40**, 403-411.

[16] Lantelme, F. (1996) Modelling and simulation in fused salt electrowinning of metals, in Y. Zongsen, X. Zeqiang and X. Xishan (eds.), *Modelling and Simulation in Metallurgical Engineering and Materials Science*, Metallurgical Industry Press, Beijing, pp 133-139.

[17] Barin. I. (1993) *Thermochemical Data of Pure Substances*, VCH, Weinheim.

[18] Smirnov, M. V., Maksimov, V. S. and Khaimenov, A. P. (1966) Reaction of gaseous titanium tetrachloride with molten alkali metal chlorides, *Russ. J. Inorg. Chem.* **11**, 945-948.

[19] Gruen. D. M. (1964) Spectroscopy of transition metal ions in fused salts, in B. R. Sundheim (ed.), *Fused Salts*, McGraw-Hill, New-York, pp 301-339.

[20] Volkov. S. V., Shapoval, V. I., Buryak, N. I. and Lutsenko, V. G. (1980) Electronic spectra of titanium(III) complexes in sodium chloride-potassium chloride and sodium chloride-potassium chloride-potassium fluoride melts, *Zhur. Neorg. Khim.* **25** 2993-2997 [*Russ. J. Inorg. Chem.* **25** 1645-1648].

[21] Chassaing, E., Basile, F. and Lorthioir, G. (1979) Electrochemical behaviour of the titanium chlorides in various alkali chloride baths, *J. Less-Common Met.* **68**, 153-158.

[22] Wei, D., Okido, M. and Oki, T. (1994) Characteristics of titanium deposits by electrolysis in molten chloride-fluoride mixture, *J. Appl. Electrochem.* **24**, 923-929.

INFLUENCE OF A GASEOUS ATMOSPHERE ON ANODIC PROCESSES OF GLASSY-CARBON AND GOLD ELECTRODES IN $NaCl$-KCl-Na_2CO_3 MELT

N.M. BARBIN, V.N. NECRASOV, D.I. TERENTIEV, A.P. PEKAR
Institute of High-Temperature Electrochemistry, Ural Division of the Russian Academy of Sciences, 20 Kovalevskaya Str., 620219 Ekaterinburg GSP-146, Russia

1. Introduction

Chloride-carbonate melts are of importance in some metallurgical applications (i.e., production of nonferrous metals). It has been established that the binary chloride salt mixture NaCl-KCl with an addition of Na_2CO_3 is a high-temperature model electrolyte convenient for the study of the regularities involved in anodic oxidation of oxide-containing ions.

The anodic processes taking place on glassy-carbon and gold electrodes at 1053 K in the "(NaCl-KCl) eutectic - X Na_2CO_3 (where X = 0.1, 1 or 5 mole %)" melt mixture in presence of He, CO, CO_2 or $2CO_2$+O_2 atmospheres are discussed in this paper.

2. Experimental Technique

The experiments were performed in a three-electrode electrochemical cell (Fig. 1). The mixture was melted in a beryllium-oxide container (70 mm diameter), which was placed in a quartz tube. The sodium and potassium chlorides were rendered water-free by vacuum with the temperature being elevated gradually. The sodium and potassium chloride melts were treated with dry chlorine, which was subsequently removed by vacuum evaporation. Concentration of oxide impurities in this chloride electrolyte was evaluated as

D.H. Kerridge and E.G. Polyakov (eds.), Refractory Metals in Molten Salts, 173-182.

0.01÷0.05 mole %. The sodium carbonate was calcined in a carbon dioxide atmosphere with the temperature increased stepwise up to 1073 K.

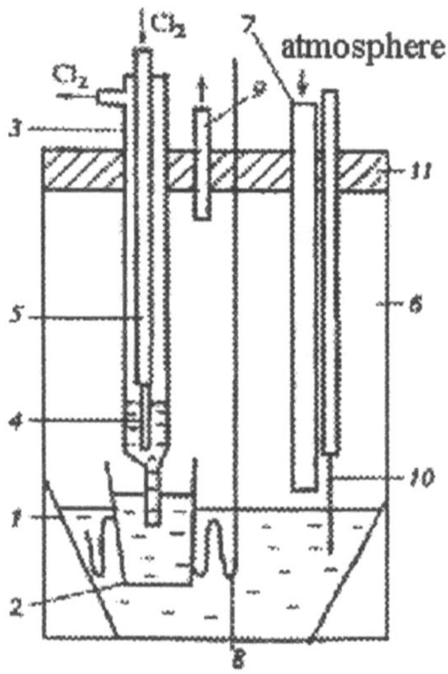

Figure 1. Electrochemical cell
1, 2 - beryllium-oxide crucibles; 3 - quartz casing of the reference chloride electrode; 4 - tip of spectral graphite; 5 - graphite tube; 6 - quartz retort; 7, 9 - gas inlet and outlet pipes; 8 - counter-electrode (platinum mesh); 10 - working electrode; 11 - rubber plug

The working electrode was a 1mm diameter gold rod or 2 mm diameter glassy carbon rod. The electrode-electrolyte contact area was 0.3÷0.5 cm^2. A platinum mesh of 10 cm^2 working surface was used as the counter-electrode.

The reference electrode was a standard chlorine electrode. The quartz casing of a reference electrode was placed in a small beryllium-oxide crucible (~30 mm diameter) with a pure chloride melt. Several recesses were made in this crucible to provide for a better electrolytic contact. This switch had a resistance of about 1 kOhm.

Traces of moisture were removed from the gases used to produce the required atmosphere in the electrochemical cell. The $2CO_2$+O_2 mixture was prepared beforehand from pure components by mixing the components in a separate vessel.

Electrochemical measurements were carried out with a PI-50-1 potentiostatic system controlled by a PR-8 programmer. An ohmic component was compensated. The potential sweep rate was 1×10^{-3} V/s.

3. Results

Some stationary (V = 1×10^{-3} V/s) anodic polarization curves are shown on fig. 2-5. These data were obtained for a large range of potentials – nearly from the initial stationary (quasi-equilibrium) potentials to the potentials, which cause an intensive dissolution of gold electrodes and/or precipitation of chlorine at glassy carbon electrodes.

Initial "micropolarization" E-*i* curves registered in the anodic and cathodic directions up to the current density of 1 to 10 mA/cm^2 are shown on figures 6 and 7.

Stationary polarization curves (Fig. 2) registered at the glassy carbon electrode (in the CO_2 atmosphere) at three selected concentrations of the carbonate are similar: lg*i* increases successively at fixed potentials and increasing carbonate concentration.

Limiting currents depend on the sodium carbonate concentration and the electrode material (Figs 2 and 3). The slopes of the linear sections dE/dlg*i* in the initial and middle parts of the polarization curves are 250÷300 mV for all the electrodes at the sodium carbonate concentration of 0.1 mole %. The same values at the carbonate concentration of 5 mole % are ~200 mV (glassy carbon) and 80÷120 mV (gold).

In the area of a limiting current (Fig. 4) it is difficult to trace a regular change in behavior of the polarization curves at the glassy carbon electrode when one gaseous atmosphere is replaced by another. If the current density is decreased, the polarization curves have breaks ("platforms"), which suggest the trend to the limiting currents from 7 to 10 mA/cm^2. At still lower currents the influence of gaseous phase on the polarization curves becomes discernible (Fig. 6). The CO and CO_2 gases have an influence on the stationary potential, i.e. on the electrode processes. A compromise stationary potential is realized in the helium atmosphere. In this case the character of electrode process may be similar to the previous one. Probably, it corresponds to a certain ratio between CO and CO_2 in the reaction zone. These gases are products of certain interactions that take place in the salt

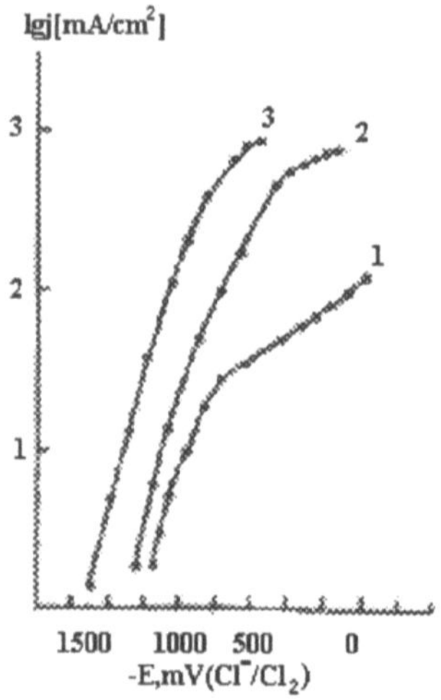

Figure 2. Anodic polarization curves registered at the glassy carbon electrode in the CO2 atmosphere at different contents of Na2CO3, mole %: 1 - 0.1; 2 - 1.0; 3 - 5.0

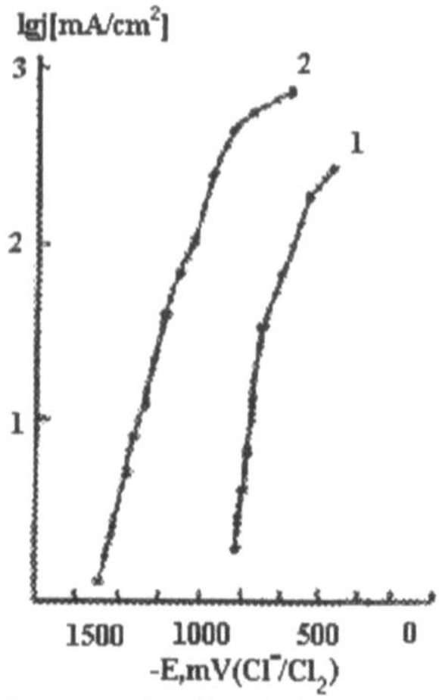

Figure 3. Anodic polarization curves registered at the Na2CO3 content = 5.0 mole %: 1 - gold electrode in the 2CO2+O2 atmosphere; 2 - glassy carbon electrode in the CO2 atmosphere

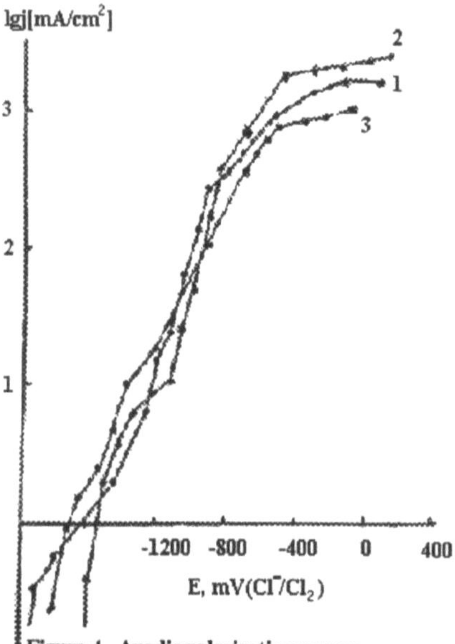

Figure 4. Anodic polarization curves registered at the Na2CO3 content = 1.0 mole % using the glassy carbon electrodes: 1 - CO; 2 - CO2; 3 - He

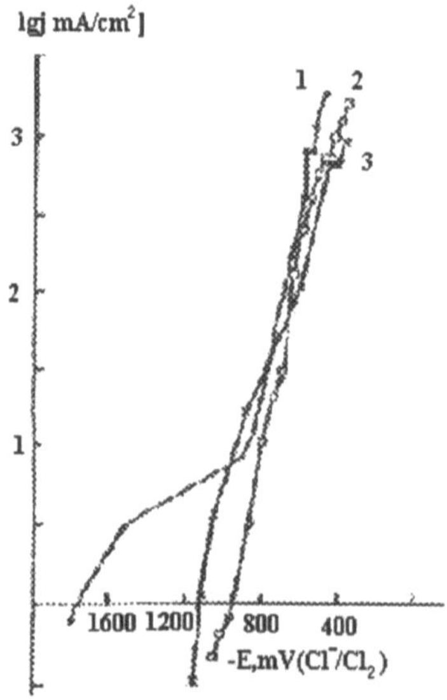

Figure 5. Anodic polarization curves registered at the Na2CO3 content = 1.0 mole % using the gold electrodes: 1 - CO; 2 - CO2; 3 - He

system at hand. When $i > 1$ mA/cm^2 (Fig. 6), in an anodic region the polarization curves registered in various atmospheres approximate the polarization curve registered in CO_2 atmosphere. Probably, starting from this region of potentials and further in the anodic direction, the dominant effect is due to discharge of the carbonate ions followed by the formation of CO_2. In the cathodic region (Fig. 6) the gaseous atmosphere has a considerable effect as before. In the CO_2 atmosphere the polarization is smaller, a fact which suggests that the dissolved CO_2 can be reducted.

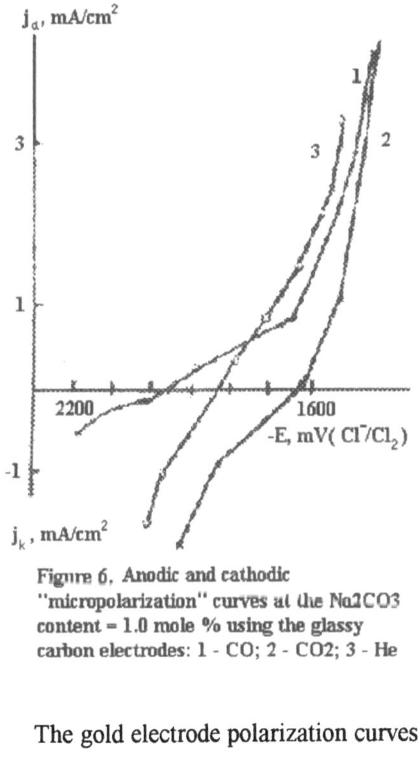

Figure 6. Anodic and cathodic "micropolarization" curves at the Na2CO3 content = 1.0 mole % using the glassy carbon electrodes: 1 - CO; 2 - CO2; 3 - He

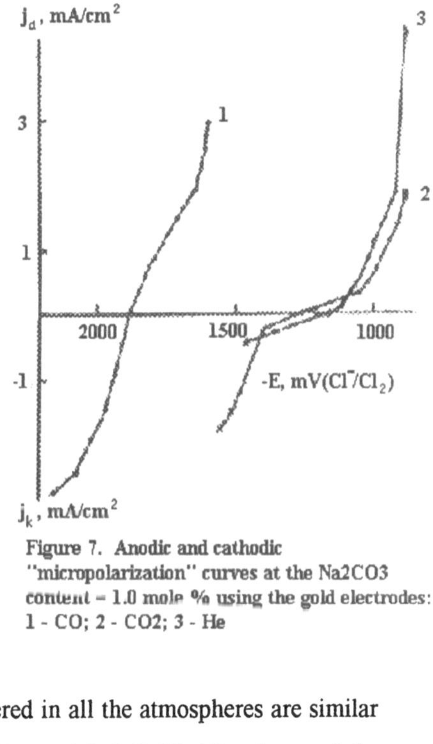

Figure 7. Anodic and cathodic "micropolarization" curves at the Na2CO3 content = 1.0 mole % using the gold electrodes: 1 - CO; 2 - CO2; 3 - He

The gold electrode polarization curves registered in all the atmospheres are similar (Fig. 5) at high current densities ($i > 10$ mA/cm^2, E ~ –(0.9÷0.6) V). The slopes of the Tafel sections are equal to 160 mV in the CO_2 or CO atmosphere and 240 mV in the He atmosphere. If the current is less than 10 mA/cm^2 (Fig. 5), the influence of the gaseous atmosphere is clearly pronounced. In the CO atmosphere the polarization curves have a maximum-current platform extended along the potential axis (up to 0.8÷1.0 V). A small current jog 0.1÷0.15 V long was observed. A smoothly descending polarization curve is obtained in the CO_2 atmosphere. The stationary potentials (Fig. 7) are similar in CO_2 or He atmosphere but differ sharply in CO. In the CO atmosphere the stationary potentials at gold and glassy carbon electrodes (Figs 6 and 7) are similar (within 0.1 V), but in the CO_2 or He atmosphere these values differ by 0.5÷0.7 V. One may suggest that in the CO atmosphere the (quasi-)equilibrium of an electrode process depends mainly on the gas phase and the electrolyte composition. In the CO_2 atmosphere the electrode processes at initial sections of polarization curves on glassy carbon and gold electrodes differ with respect to their character and kinetics.

4. Discussion

The experimental polarization measurements were supplemented by the thermodynamic analysis of the gas-melt systems in question. The method of thermodynamic modeling was used to study the equilibrium in the NaCl-KCl-Na_2CO_3–gas (He, CO_2, CO, $2CO_2$+O_2) systems. The thermodynamic modeling using the program package ASTRA-4, the data base ASTRA.BAS, and the technique described elsewhere [1, 2] was performed. The liquid phases were described in terms of the model of ideal solutions of the interaction products, while the gaseous phases were described as an ideal gas mixture.

Results of modeling of condensed phases in different initial gaseous atmospheres over the temperature interval from 1053 to 1253 K are shown on fig. 8-11. The initial content of Na_2CO_3 was 1 mole %. We shall briefly comment on these results as follows.

In various gaseous atmospheres NaCl and KCl of the chloride electrolyte are highly volatile in the form of monomers and dimers (~10^{-4} atm at 1053 K). Probably, these vapors not have a marked effect on electrochemical transformations of the oxide-containing components.

The main reactions that determine the compositions of the salt and gaseous phases are

$$Na_2CO_3 + 2KCl = K_2CO_3 + 2NaCl \,, \quad (1)$$

$$M_2CO_3 = M_2O + CO_2 \,, \quad (2)$$

$$M_2O + 1/2\ O_2 = M_2O_2 \,, \quad (3)$$

$$M_2O + 3/2\ O_2 = 2MO_2 \,, \quad (4)$$

$$M_2CO_3 + CO = 2Me + 2CO_2 \,, \quad (5)$$

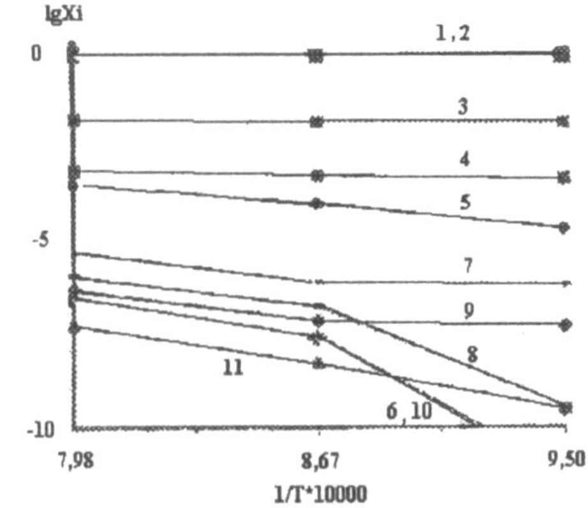

Figure 8. Main components of the melt in NaCl-KCl-(1mole%)Na2CO3-He system.

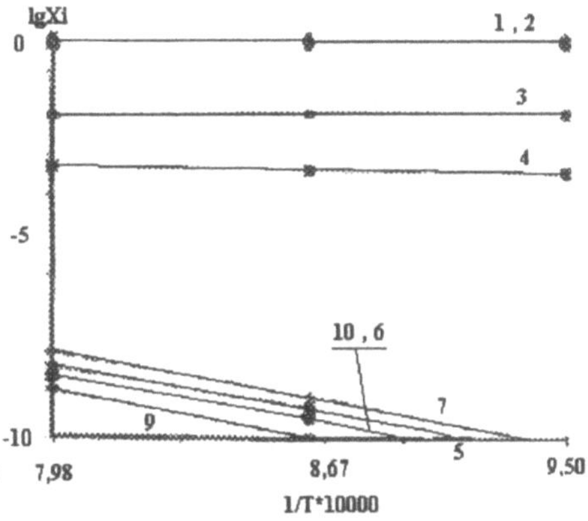

Figure 9. Main components of the melt in NaCl-KCl-(1 mole%)Na2CO3-CO2 system

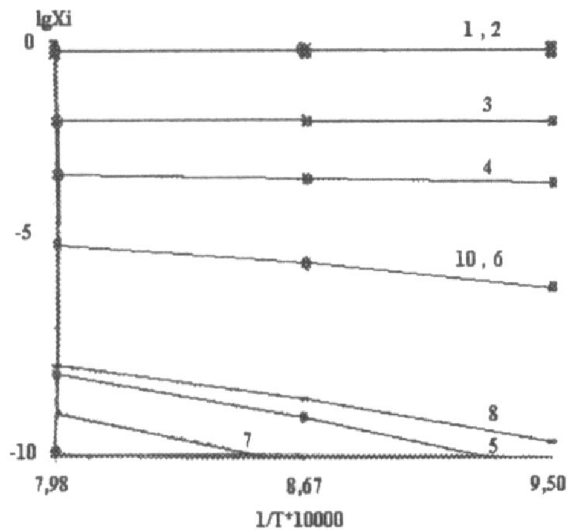

Figure 10. Main components of the melt in NaCl-KCl-(1 mole%)Na2CO3-(2CO2+O2) system.

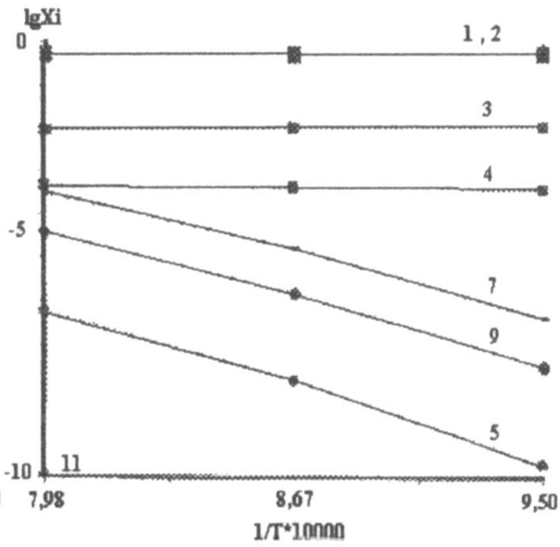

Figure 11. Main components of the melt in NaCl-KCl-(1 mole%)Na2CO3-CO system.

1 - NaCl, 2 - KCl, 3 - Na2CO3, 4 - K2CO3, 5 - Na2O, 6 - NaO2, 7 - Na, 8 - Na2O2, 9 - K, 10 - KO2, 11 - K2O, 12 - K2O2.

$$CO_2 = CO + 1/2\ O_2\ , \tag{6}$$

$$2CO = C + CO_2\ . \tag{7}$$

The carbonate of a chloride melt in various gaseous atmospheres is represented by Na_2CO_3 (actually in the initial concentration) and K_2CO_3. The latter is the result of the exchange reaction (1). At 1053 K $X(K_2CO_3)/X(Na_2CO_3) \cong 0.03$.

In the inert atmosphere the partial pressure of He was 1 atm and remained almost unchanged after equilibrium is established in the system. The gaseous phase also contains CO_2 ($\sim10^{-6}$ atm at 1053 K) and O_2 ($\sim10^{-8}$ atm at 1053 K) being the products of the reactions (2) and (6). The reaction (2) gives Na_2O to the melt. At 1053 K the Na_2O concentration is $\sim10^{-5}$ mole fractions, $X(Na_2O)/X(Na_2CO_3) \cong 10^{-3}$ (Fig. 8). Concentrations of other oxide components (K_2O, K_2O_2, KO_2, Na_2O_2 and NaO_2) as products of the reactions (2-4), are less than 10^{-6} mole fractions at 1053 K.

As compared to the inert atmosphere, the CO_2 atmosphere largely suppresses decomposition of carbonate (Fig. 9). Over interval of 1053 to 1253 K the concentration of all oxides does not exceed 10^{-8} mole fractions. In addition to CO_2, the gaseous phase contains CO and O_2 ($\sim10^{-6}$ atm at 1053 K) formed by the reaction (6).

If oxygen is present in gaseous phase (the $2CO_2$+O_2 atmosphere), peroxide and superoxide particles appear in the melt (Fig. 10). The NaO_2 concentration at 1053 K is $\sim10^{-6}$ mole fractions.

In CO atmosphere the equilibrium gaseous phase will have a considerable amount of CO_2 in accordance with reaction (7) (at 1053 K $P(CO_2) \sim 0.1$ atm). The concentration of the oxide particles in the salt melt will be low (Fig. 11) and comparable with that in the CO_2 atmosphere. The reducting atmosphere causes (reaction 5) a high amount of the alkaline metal (Na, K) appear in the melt ($\sim10^{-6}$ mole fractions at 1053 K) and the gaseous phase ($\sim10^{-6}$ atm at 1053 K).

Simultaneous analysis of results and thermodynamic calculations leads to the following conclusions.

In the $2CO_2$+O_2 atmosphere the stationary potential of gold electrode approximates the equilibrium one for the electrode reaction

$$CO_3^{2-} - 2e^- = CO_2 + 0{,}5\ O_2. \tag{8}$$

The polarization curve 1 in Fig. 3 corresponds to this process in anodic direction.

Probably, an analogous process is observed at gold electrode in the CO_2 atmosphere (Fig. 5, curve 2) at a lower initial partial pressure of O_2.

For the inert atmosphere an alternative electrode process is possible:

$$O^{2-} - 2e^- = 0{,}5\ O_2. \qquad (9)$$

This is attested by results of the thermodynamic modeling (Fig. 8) and a wave in the polarization curve (Fig. 5, curve 3).

A sharp shift of the stationary potential at gold electrode in cathodic direction in the CO atmosphere and a corresponding wave, which is observed in this region of potentials during anodic polarization (Fig. 5, curve 1), suggest an electrode process involving the gaseous (dissolved) CO:

$$CO + CO_3^{2-} - 2e^- = 2\ CO_2. \qquad (10)$$

Generally, anodic processes at the gold electrode in NaCl-KCl-Na_2CO_3 melt will be determined by a sequence of (8) and/or (9), (10) reactions. A specific contribution of these reactions will depend on particular experimental conditions (including a composition of the gaseous phase) and electrode polarization interval. The products of electrode process are CO_2 and O_2. Their ratio depends on the electrode reactions in progress.

The electrode process at glassy carbon electrode in the NaCl-KCl-Na_2CO_3 melt may be described by reaction

$$CO_3^{2-} + (0{,}5X+2Y)C - 2e^- = 1{,}5XCO_2 + 3YCO, \qquad (11)$$

(by combination of reactions $CO_3^{2-} + 2C - 2e^- = 3CO$ and $2CO_3^{2-} + C - 4e^- = 2CO_2$, where x and y are the partial coefficients).

In CO atmosphere the reaction (11) is realized at glassy carbon electrode as a result of the disproportionation reaction ($2CO \rightarrow CO_2 + C$). Here the coefficients x and y are determined by the equilibrium constant of reaction (7).

In CO_2 atmosphere (when the gas flows through the electrochemical cell) the equilibrium by reaction (7) is not achieved because of the carbon oxidation ($C + CO_2 \rightarrow 2CO$). The coefficients x and y of the reaction (11) are not equilibrium ones. As a result, the fraction of CO in the electrode–electrolyte reaction zone is small and the stationary potential is shifted in the anodic direction.

When the potential is close to equilibrium potential, the electrochemical oxidation of carbon by the reaction (11) is delayed. Judging by initial sections of polarization curves (Fig. 6, curve 1), in CO atmosphere the most probable electrode process is the reaction

(10). Under anodic polarization (~0.3÷0.5 V) the electrochemical oxidation of carbon by reaction (11) at a considerable rate (~1 mA/cm^2) is possible. In this case desorption of the $(CO)_x$ compounds from the electrode surface is delayed most. Since under these conditions (i > 1÷10 mA/cm^2) main gaseous product is CO_2, anodic reaction can be written as a stepwise process:

$$CO_3^{2-} + C - 2e^- = CO_{ads} + CO_{2\,gas}, \quad (12)$$

$$CO_{ads} + CO_3^{2-} - 2e^- = 2CO_{2\,gas}. \quad (13)$$

The limiting stage is reaction (13). In this region of potentials the polarization curves registered at the glassy carbon electrode are actually independent of the gaseous atmosphere.

5. References

1. G.B. Sinyarev, N.A. Vatolin, B.G. Trusov, G.K. Moiseyev. Using of Computers for Thermodynamic Calculations of Metallurgical Processes. Moscow: Nauka, 1982, p. 263
2. N.A. Vatolin, G.K. Moiseyev, B.G. Trusov. Thermodynamic Modeling of High-Temperature Inorganic Systems. Moscow: Metallurgiya, 1994, p. 352

MECHANISM OF THE MOLYBDENUM ELECTRODEPOSITION FROM MOLTEN SALTS

A. SILNÝ, V. DANĚK, M. CHRENKOVÁ
Institute of Inorganic Chemistry, Slovak Academy of Sci.
SK - 842 36 Bratislava, Slovak Republic

Abstract

The mechanism of the electrolytic preparation of molybdenum coatings has been investigated in molten KF - K_2MoO_4 - B_2O_3 (SiO_2) system as part of a complex physico-chemical and thermodynamic analysis of the above systems. The influence of the electrolyte composition on the mechanism of the cathodic process in electrodeposition of molybdenum from KF - K_2MoO_4 - B_2O_3 electrolyte was studied by chronopotentiometric and voltammetric methods at 1133 K. It was found, that during dissolution of K_2MoO_4 in the basic KF - B_2O_3 melt, two types of molybdenum species were created, an electrochemically active and an inactive one. Depending on the relative concentration of molybdenum in melt ($x(K_2MoO_4)/x(B_2O_3)$) the reduction process is influenced by the preceding chemical reaction between the molybdenum compounds and B_2O_3 present in the melt. In the KF - K_2MoO_4 - SiO_2 system it was found that SiO_2 has also a beneficial effect and that coherent, smooth and strongly adhesive layers on electrically conductive substrates - nickel, copper, steel and graphite can be obtained. The influence of composition, working temperature, cathodic current density on the quality of the Mo layers was determined. The best results were obtained for an electrolyte with composition 85 mole % of KF, 10 mole% K_2MoO_4 and 5 mole % of SiO_2, at temperatures of 1173 - 1273 K and cathodic current densities ranging from 75 to 400 A/m^2.

1. Introduction

Molybdenum, with respect to its physico-chemical properties and high corrosion resistance, is inevitably used as a construction material for the equipment used in chemical industry. Often only thin layer on the exposed surfaces is preferred. In such cases, electrodeposition from molten salts appears to be a very important method for the formation of compact molybdenum deposits on the electrically conductive substrates. This method may be also considered for the primary production of molybdenum metal.

After considering the literature data on the electrodeposition of molybdenum, several types of molten salts systems were tested as electrolytes. On the basis of the electroactive species used, they can be divided into two main groups:

D.H. Kerridge and E.G. Polyakov (eds.), Refractory Metals in Molten Salts, 183-187.

(i) halide systems containing either K_3MoCl_6 [5-9] or K_3MoF_6 [10] dissolved in alkali metal halides, mainly chlorides and fluorides;

(ii) mixed systems containing oxide compounds of molybdenum, such as molybdenum oxide, MoO_3 [11-13], alkali metal molybdates [1,11,14-16] and $CaMoO_4$ [17].

Supporting electrolytes include: LiCl-KCl mixtures [12], sodium and lithium metaborates [11], KF-$Na_2B_4O_7$ [14], KF-$Li_2B_4O_7$ [1], KF-B_2O_3 [2,15,] and $CaCl_2$-CaO [17].

Comparing the results of molybdenum electrodeposition from several types of supporting electrolytes, it was found, [2] that the process is most successful in electrolytes consisting of a mixture of alkali metal fluorides and boron oxide (or alkali metal borates) to which molybdenum oxide (or alkali metal molybdate) is added as the electrochemically active component.

From the published papers [1-3, 5-17] it seems that the electrodeposition of molybdenum from binary MeF-Me_2MoO_4 mixtures does not occur. However, a small addition (<1 mole %) of boron oxide [1,2,11,14-16] to the above electrolytes facilitates the electrodeposition of molybdenum. From the viscosity [23], as well as the phase equilibrium and density measurements [24 - 25] it is evident, that the system KF-K_2MoO_4-B_2O_3 is a very complex one. Beside chemical reactions, the polymerisation tendency of the melts, especially in the region of higher contents of boron oxide, makes this system difficult to study. From the complex physico-chemical [23 - 25] and thermodynamic analysis [26] of the molten systems KF-K_2MoO_4-B_2O_3 and KF-K_2MoO_4-SiO_2 it follows that the formation of heteropolymolybdates, containing boron, ($[BMo_6O_{24}]^{9-}$), or silicon, ($[SiMo_{12}O_{40}]^{4-}$), as a central atom, most probably facilitate molybdenum deposition.The entrance of fluorine atoms into the coordination sphere of molybdenum in the heteropolyanions lowers the symmetry and thus the electrochemical stability of such electroactive species as well.

The influence of the operational parameters, such as electrolyte composition, current density, temperature and substrate material (iron, copper, nickel, graphite), on the quality of the deposited Mo-layers from KF-K_2MoO_4-SiO_2 electrolyte was determined.

2. System KF - K_2MoO_4 - B_2O_3

The mechanism of the cathode process was studied on platinum working and reference electrode in the system KF - K_2MoO_4 - B_2O_3. In electrolytes consisting of KF - K_2MoO_4, as well as in pure K_2MoO_4, only one process, corresponding to the deposition of potassium, was registered at the potential around -1 V. Small additions of B_2O_3 to the basic electrolyte caused the appearance of a very pronounced second wave at a potential of -0.2 V. Its height corresponded with the B_2O_3 content in the melt. Using chronopotentiometric and voltametric experiments, Makyta et al.[18, 19] found, that the dissolution of K_2MoO_4 in KF - B_2O_3 melts is accompanied by chemical reaction between different boron containing species present in equilibrium in basic melt, such as B_4O^{7-}, BO^{2-}, BF^{4-} and fluoride anions. During these chemical reactions, most probably new electrochemically active molybdenum containing species are formed. Multi-atom

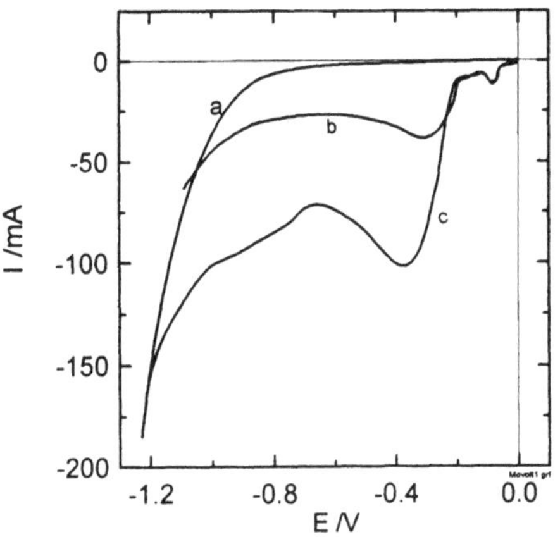

Fig. 1. Voltammetric curves in the system K_2MoO_4 - B_2O_3. a-pure K_2MoO_4, b- 2 mole% B_2O_3, c- 5 mole% B_2O_3.

units consisting of MoO_6 groups with boron as the central atom, have already been described in [20]. According Makyta et al.[18, 19] the reaction of K_2MoO_4 with fluoride anions forms electrochemically inactive, molybdenum containing, structural units. Such units are most probably identical with those present in the KF - K_2MoO_4 mixture. The fact, that only single wave was observed on voltametric curves together with previous results, [2] indicates that the charge transfer reaction Mo^{VI} -> Mo^0 proceeds in one step. Comparison of the cathodic with the anodic area indicates a reversible process. We suppose that this step is preceded by chemical reaction in accordance with a quantitative description of the process [21].

Chronopotenciometric curves recorded in the system KF - K_2MoO_4 , KF - B_2O_3 a KF - K_2MoO_4 - B_2O_3 are shown in Fig. 2 . Curve _1_ indicates a single process only, corresponding to the deposition of potassium. The broad chronopotentiometric wave (curve _2_) corresponds to B^{3+}-> B^0. Addition of K_2MoO_4 to such an electrolyte causes the boron reduction wave vanish while a new wave appears (curve _3_) representing reduction of Mo^{6+} to metallic Mo. Further evaluation of experimental data from chronopotentiometric experiments showed that the reaction kinetics of the system did not correspond with a diffusion controlled charge transfer reaction model. This can be explained as either that the transport of the electrochemically active component towards electrode is not a purely diffusion process, or that the kinetics of the electrode process is influenced by the preceding chemical reaction, as it was indicated later also by voltametric measurements. It was found, that in the concentration region, $0 < (x(K_2MoO_4)/x(B_2O_3)) < 0.05$, the kinetics of the cathode process is not influenced by the preceding chemical reaction. In the concentration region, $0.1 < (x(K_2MoO_4)/x(B_2O_3)) > 0$, the molybdate anion dissolves with formation of an electrochemically active as well as an electrochemically inactive component. In this concentration region boron transforms the electrochemically inactive component into the electrochemically active one. Further increase in the molybdenum concentration in the melt produces the electrochemically inactive component only.

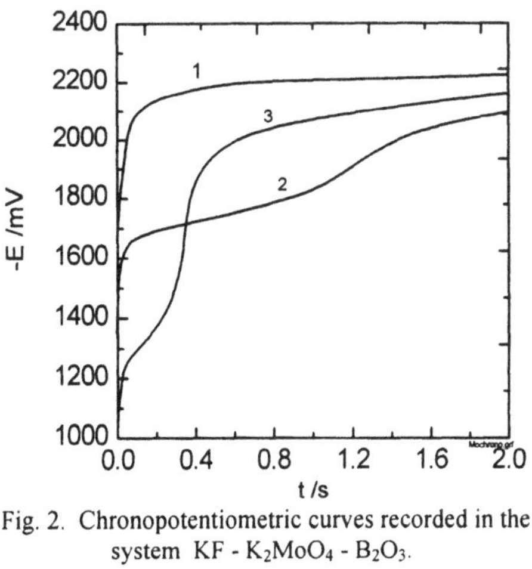

Fig. 2. Chronopotentiometric curves recorded in the system KF - K_2MoO_4 - B_2O_3.

3. System KF - K_2MoO_4 - SiO_2

Further work was intended to find out if the addition of other oxides similar to B_2O_3 (ex. SiO_2) had the same promoting influence on the electrodeposition of molybdenum from the basic electrolyte KF - K_2MoO_4 [22].

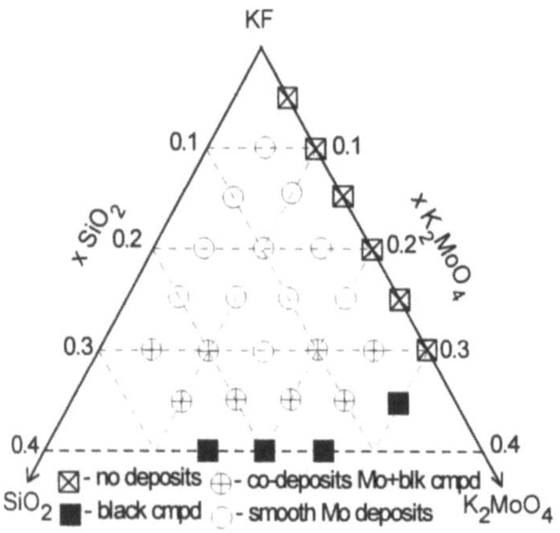

Fig. 3. Influence of the electrolyte composition on the quality of the molybdenum deposits.

In Fig. 3 the influence of the electrolyte composition on quality of the deposited product is shown. As can be seen from the figure, small addition of SiO_2 <(1 mole%) facilitates the deposition of shiny, smooth and strongly adherent Mo layers on different electrically conductive materials.

Addition of SiO_2 to the electrolyte up to 5 mole %, decreased the decomposition voltage, while further increase of SiO_2 concentration did not change it (See Fig. 4). The positive effect of SiO_2 addition can be explained by the structural changes in the ions composition the electrolyte, due to the formation of the hetero-polyanion $[SiMo_{12}O_{40}]^{-4}$ according to the reaction:

$$12\ K_2MoO_4 + 7\ SiO_2 + 36\ KF = K_4[Si(Mo_{12}O_{40})] + 6\ K_2SiF_6 + 22\ K_2O$$

Such heteropolyanions are voluminous and thus much more easily polarisable. In the vicinity of the cathode surface in the electric double layer, these anions are strongly polarized and finally disintegrate to smaller particles from which metallic molybdenum is deposited. By the diffraction analysis of solid deposit on the walls of the reaction chamber it was found, that the deposit was completely composed of K_2SiF_6, which supports the hypothesis of formation of hetero-polyanions according to the above reaction.

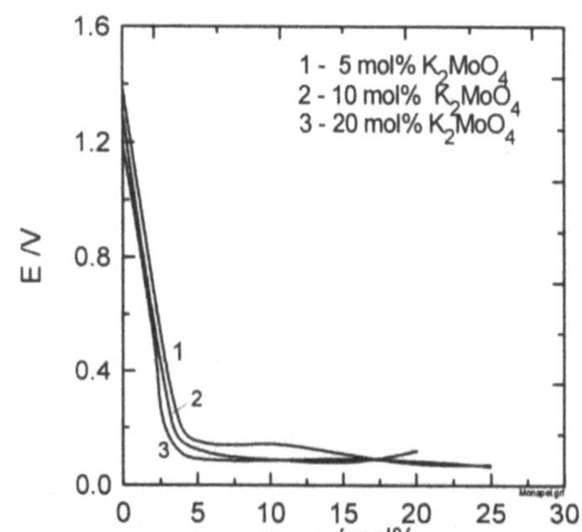

Fig. 4. The dependence of the deposition voltage on the SiO_2 content.

Estimation of the effect of cathodic current density on quality of the deposited layer, electrolyte with composition 85 mole% KF - 10 mole% K_2MoO_4 and 5 mole% SiO_2 was used at 1123 K. Molybdenum was deposited on copper and graphite, whileduration of electrolysis was 2200 s. Excellent deposits were obtained with c.c.d. up to 300 A/m^2. At higher c.c.d. the deposition of MoO_2 and also traces of MoC_2 was observed.

To study influence of temperature, the same electrolyte was used, at c.c.d. 300 A/m^2. Suitable deposits were obtained at temperatures over 1073 K. At lower temperatures the deposition voltage increased and problems connected with electrocrystalisation were also encountered. Current efficiency was over 95%.

Influence of substrate material was investigated using nickel, copper, steel and graphite. For these measurements electrolyte of the above composition, at c.c.d. 300 A/m^2 at 1173 K was used. Under these conditions strongly adhesive and smooth layers without dendrites on the edges were obtained on all the substrates tested. Cathodic current efficiency and deposition voltage were nearly equal for all substrates. The molybdenum layers on steel substrates exhibited cracking at conditions of high mechanical stress.

References

1. Koyama K., Hashimoto Y., Omori S. and Terawaki K. (1987) *J. Less-Common Met.*, **132** 57.
2. Makyta M., Zatko P. and Bezdička P. (1993) *Chem. Papers*, **47** 28.
3. Zatko P., Makyta M., Sýkorová J. and Silný A., (1994) *Chem. Papers*, **48** 10.
4. Stortenbeker W.(1892) *Z. Phys. Chem.*, **10** 183.
5. Senderoff S. and Brenner A.(1954) *J. Electrochem. Soc.* **101** 16.
6. Baraboshkin A. N., Saltykova N. A., Talanova M. I. and Martemyanova Z. S., (1970) *Elektrokhimiya*, **6** 1552.
7. Baraboshkin A. N., Saltykova N. A., Molchanov A. M., Talanova M. I. and Martemyanova Z. S., (1972) *Elektrokhimiya*, **8** 833.
8. Baraboshkin A. N., Saltykova N. A., Talanova M. I. and Molchanov A. M., (1973) *Trudy Inst. Elektrokhim. UFAN, USSR* **20** 65.
9. Suri A. K. and Gupta C. K., (1973) *J. Less-Common Met.* **31** 389.
10. Mellors G. W. and Senderoff S. U.S. Pat. 3 444 058, May 13 1969.
11. McCawley F. X., Wyche Ch. and Schlein D., *J.* (1969) *Electrochem. Soc.* **116** 1028.
12. Popov B. N. and Laitinen H. A. (1973) *J. Electrochem. Soc.*, **120** 1346.
13. Shapoval V. I. and Makogon V. F. (1976) *Elektrokhimiya*, **12** 1723.
14. Koyama K., Hashimoto Y., Omori S. and Teravaki K. (1984) *Trans. Jpn. Inst. Met.* **25** 265.
15. Koyama K., Hashimoto Y., Omori S. and Teravaki K. (1984) *Trans. Jpn. Inst. Met.*, **25** 804.
16. Teravaki K. (1985) *Trans. Jpn. Inst. Met.* **26** 198.
17. Baraboshkin A. N., Shunailov A. F. and Martinov V. A. (1970) *Trudy Inst. Elektrokhim.* Akad. Nauk SSSR, **15** 44.
18. Makyta M., Utigard T. A., Zatko P., Silný A., Daněk V. (1997) *Chem. Papers* **51** 202.
19. Makyta M., Utigard T. A., Zatko P., Silný A., Chrenková M. *Chem. Papers* (1997) **51** 209.
20. Rawson H., Inorganic Glass-Forming Systems., Academic Press London, 1967, pp. 207 -208
21. Delahay P. and Berzins T., J. Amer. Chem. Soc., 75 (1953) 2486, 4205
22. Silný A., Zatko P., Makyta M. and Sýkorová J., Molten Salts Forum, Vol. 1-2, Ed. C.A.C. Sequeira and G.S. Picard, Trans Tech. Publications, Switzerland, 1993, pp. 155-160.
23. Silný A., Daněk V., Chrenková M. (1997) *Ber. Bunsenges. Phys. Chem.* **99** 74.
24. Patarák O. (1995) PhD Thesis, Institute of Inorganic Chemistry, Slovak Academy of Sciences, Bratislava.
25. Chrenková M., Daněk V., Silný A. (1994) *Z. anorg. allg. Chem.*, **620** 385.
26. Kosa L., Nerád I., Patarák O., Proks I., Strečko J. and Adamkovičová K.(1994) *Themochim. Acta* **244** 69.

CHEMICAL REACTIONS ACCOMPANYING THE PROCESSES OF REFRACTORY METAL ELECTRODEPOSITION

S.A. KUZNETSOV
Institute of Chemistry, Kola Science Centre,
14 Fersman St., Apatity, 184200, Russia

1. Introduction

For developing the processes of refractory metal electrodeposition and the electrochemical synthesis of metal-like refractory compounds one must have the information on not only the electrochemical behaviour of the complex in molten salts but on the chemical reactions accompanying these processes as well.

2. Reactions of disproportionation

In the electrolytic refining and preparation of coatings of refractory metals in molten salts, there is often observed on the electrolyte surface the formation of a metallic film, which, by growing, can short - circuit the cathode to the anode and disturb the normal course of electrolysis. The reason of formation of metallic films is reactions of disproportionation [1].

In another investigation for the conditions of stabilization of highest oxidation states of refractory metals against the background of NaCl-KCl melt have been determined[2]. Changes of the oxidation states of d - elements in molten NaCl-KCl salt from pure chloride to oxochloride, to chloride-fluoride and oxofluoride complexes are given in the Table 1. Stabilization of higher oxidation states is often observed when passing from pure chloride melts to oxochloride, chloride-fluoride and oxofluoride melts. The highest oxidation state +7 was obtained for rhenium in the oxofluoride melt [3].

Stabilization of the higher oxidation state of metals is accompanied by disproportionation if the anion composition of the melt is changed. It was noted [4] that the disproportionation reactions:

$$4TiCl_2 \Leftrightarrow Ti + 3TiCl_4 \quad (1)$$

$$2TiCl_2 \Leftrightarrow Ti + TiCl_4 \quad (2)$$

$$3TiCl_2 \Leftrightarrow Ti + 2TiCl_4 \quad (3)$$

are accelerated in the presence of solid particles of active magnesium oxide, but the role of magnesium oxide remains unclear if it is considered only as a surface on which reactions (1-3) occur. In this case, the equilibrium constants of reactions (1-3) depend only on the temperature, but the role of magnesium oxide becomes quite understandable as a substance participating in the reaction:

$$3TiCl_2 + 2MgO \Leftrightarrow 2TiOCl \downarrow + 2MgCl_2 + Ti \quad (4)$$

D.H. Kerridge and E.G. Polyakov (eds.), Refractory Metals in Molten Salts, 189-196.

TABLE. Oxidation states of refractory metals in NaCl-KCl molten salt

3d *)	Composition of complexes				4d *)	Composition of complexes				5d *)	Composition of complexes			
	Cl	O-Cl	Cl-F	O-F		Cl	O-Cl	Cl-F	O-F		Cl	O-Cl	Cl-F	O-F
Ti	II-IV	III, IV	III, IV	IV	Zr	II, IV	IV	IV	IV	Hf	II, IV	IV	IV	IV
Cr	II,III	III,V, VI	II, III	VI	Nb	IV-V	V	IV,V	V	Ta	IV-V	V	V	V
										Re	III, IV	V-VII	III, IV	VI, VII
										Ir	III, IV	V, VI	IV	VI

*) - elements; Cl-chloride; O-Cl - oxochloride; Cl-F - chloride-fluoride; O-F-oxofluoride

The reaction between titanium chlorides and oxide refractory materials with the formation of lower titanium oxides is mentioned in [5]. The following schemes of the processes are proposed:

$$MeO + TiCl_2 \Leftrightarrow MeCl_2 + TiO \quad (5)$$

$$3MeO + 2TiCl_2 \Leftrightarrow 3MeCl_2 + Ti_2O_3 \quad (6)$$

However, in the electrolysis of melts containing $TiCl_2$, a metallic coating is formed on the walls of the container (SiO_2), and the cathodic metal has high purity with respect to silicon.This metal formation is stated [5] be due to disproportionation:

$$2TiCl_2 \Leftrightarrow TiCl_4 + Ti \quad (7)$$

however the constant of equilibrium (7) in NaCl-KCl melt at 1000K has the value 10^{-5} and even establishment of nonequilibrium conditions does not lead to significant metallization of crucibles by equilibrium (7). These contradictions can be removed if we assume that coating formation on oxide materials in the case of titanium (zirconium) occurs mainly because of reactions of the type:

$$3TiCl_2 + MeO_2 \Leftrightarrow 2TiOCl\downarrow + MeCl_4 + Ti \quad (8)$$

This viewpoint is also confirmed to some degree by the fact that in the electrolysis of melts containing $TiCl_3$ in SiO_2 containers it is not possible to obtain a metal that is pure with respect to silicon [3] (although the reaction $4TiCl_3 \Leftrightarrow Ti + 3TiCl_4$ is possible) because the reaction with the oxide occurs without an increase of the valence of titanium and the formation of metallic titanium:

$$2TiCl_3 + MeO_2 \Leftrightarrow 2TiOCl\downarrow + MeCl_4 \quad (9)$$

Thus reactions (4) and (8) are due to a change of the anionic composition of the melt - the appearance of oxygen anions from material of containers.

Another reason for the formation of a metallic film may be insufficient purity of the inert gases. Thus we have noted that an oxygen content of 0.5-0.7 w/o in helium or argon leads to disproportionation reactions accompanied by the formation of metallic titanium (hafnium) film on the surface of a NaCl-KCl melt containing lower titanium (hafnium) chlorides.

We also observed the formation of a metallic film in the electrolysis of a NaCl-KCl-K_2HfF_6 melt. Hafnium is reduced at the cathode in one irreversible four-electron stage [6]:

$$HfF_6^{2-} + 4e \rightarrow Hf + 6F^- \qquad (10)$$

A study of the interaction of hafnium with the NaCl-KCl-K_2HfF_6 melt indicates the absence of metal -salt reaction with the formation of lower oxidation states [7]. It has been shown [8] that at low anodic current densities (up to about 0.02 A/cm^2) the mean valence of hafnium is close to four in the melt NaCl-KCl-K_2HfF_6 (10 w/o). This is due to the fact that this current density is not above the limiting diffusion current density for hafnium dissolution according to reaction:

$$Hf - 4e + 6F^- \rightarrow HfF_6^{2-} \qquad (11)$$

At current densities which are higher than the limiting diffusion current of process (11), hafnium dissolves also in the form of $HfCl_4^{2-}$ complexes [8]. We have pointed out repeatedly that the formation of complexes of different oxidation states during anodic dissolution is an undesirable effect in the melts of a heteroligand composition, because a disproportionation reaction will occur in the bulk and at the melt surface on account of the convection of fluoride ions liberated during cathodic discharge [1,9]:

$$2HfCl_4^{2-} + 6F^- \rightarrow HfF_6^{2-} + Hf + 8Cl^- \qquad (12)$$

and a film will form and grow on the melt surface during electrolysis. Film formation reduces the cathodic current efficiency of metal in the form of coherent coatings. When the melt NaCl-KCl-NaF (5 w/o)-K_2HfF_6 (10 w/o) is used at least up to an anodic current density of 0.5 A/cm^2 the hafnium will dissolve in the melt, only in the form of HfF_6^{2-} complexes, which prevents reaction (12) and leads to higher cathodic current efficiency efficiency in forming metal coatings [9].

Thus, the presence of a disproportionation reaction in the melts, containing refractory metals may be due to a number of reasons and to a particular reason in each individual case. Apparently, firstly, an important role is played by the organization of the anodic and cathodic processes. Secondly, thorough removal of solid particles from the melt, removal of traces of moisture and oxygen anions, selection of an inert container and high purity of the inert gas under the melt are necessary for prevention of disproportionation reaction and film formation.

3. Metal-salt reactions

Study of metal-salt reactions in molten salts is of great practical and scientific value in the electrolysis of refractory metals and in the study of the electrochemical properties of compounds of low oxidation states which are formed in contact with metal. Metal-salt reaction has a great influence on current efficiency and the granulometric composition of niobium powders during electrolysis of the KCl-KF- K_2NbF_7 melt with graphite anode [10]. Thus a considerable drop of current efficiency and reduction of size of niobium powders when concentration of K_2NbF_7 increased from 8-10 w/o to 14 -18 w/o are connected with an ever-increasing influence on these parameters of reaction:

$$4Nb(V) + Nb \Leftrightarrow 5Nb(IV) \qquad (13)$$

At the same time the presence of Hf(IV) complexes in the chloride melts and their reaction with the coating [11]:

$$Hf(IV) + Hf \Leftrightarrow 2Hf(II) \qquad (14)$$

causes smoothing of the cathodic deposit and has the controlling influence on the retention of planar crystallization front.
Elsewhere interaction of niobium with its monooxofluoride complexes in the NaCl-KCl melt has been studied[12]. Such interaction results in «deoxidation» of the melt, the appearance in the melt of Nb(IV) fluoride complexes and formation of oxide phases on the metal surface. This interaction plays an extremely important role in synthesis of oxygen-containing niobium compounds.

4. Reactions of currentless transfer

An equimolar mixture of sodium and potassium chloride containing 10-30 w/o potassium heptafluoroniobate (K_2NbF_7) was used as the melt. When the metal is immersed in the NaCl-KCl-K_2NbF_7 melt, the metal-salt reaction spontaneously takes place with formation of the reduced form of niobium:

$$4NbF_7^{2-} + Nb \rightarrow 5NbF_5^- + 3F^- \qquad (15)$$

The metal-salt reaction takes place against the background of the NaCl-KCl melt, and for this reason formation of NbF_5Cl^{2-} complexes is favoured. The formation of NbF_6^{2-} and NbF_6Cl^{3-} complexes is also possible. Nb(IV) complexes diffuse to the steel substrate and then disproportionate on its surface with the formation of niobium carbide:

$$5NbF_5^- + 3F + C \text{ (surface of the steel)} \rightarrow NbC + 4NbF_7^{2-} \qquad (16)$$

The energy of carbide formation ΔG_{NbC} is the driving force of reaction (16). The Nb(V) cations in the NbF_7^{2-} complexes liberated according to reaction (16) diffuse to the metallic niobium and participate with it in reaction (15), forming ions of Nb(IV). The process of transfer of niobium to the surface of the St.3 sample is thus closed in a cycle, and the resulting reaction with consideration of reactions (15) and (16) can be written as:

$$Nb + C \rightarrow NbC \qquad (17)$$

The scanning electron photograph of the surface of the St.3 substrate after holding in NaCl-KCl-K_2NbF_7 melt (10 w/o) for 3 h at a temperature of 1073 K shows spheroidal crystals of niobium carbide NbC on the substrate surface according to the X-ray data. The rate of formation of niobium carbide is low in the case of currentless transfer, but it increases with both increase in the temperature of the melt, and in the concentration of potassium fluoroniobate, and when metallic niobium powder is used as the metal. The experiments showed that for a 30 w/o concentration of K_2NbF_7, the use of metallic niobium powder at 1073 K for 6 h produced obtaining a coherent, poreless coating of 1.5-2 μm thick on the surface of St.3 steel. The thickness of the coating is a function of the concentration of carbon in the steel; for brand U10 steel (1.0 w/o C), the thickness of the coatings can thus be 5-7 μm [13]. Using the electrolyte $NaCl$-KCl-K_2HfF_6 (10 w/o) - NaF(10 w/o) and different substrates Ni, Co or Fe intermetallic compounds $HfNi_5$, $HfCo_6$, Fe_2Hf. were obtained. The formation, for example, of intermetallic compound $HfNi_5$ can be written as:

$$Hf^{4+} + 4Na^{o} + 5Ni \Leftrightarrow HfNi_5 + 4Na^{+} \qquad (18)$$

The transfer of electrons in this case being by cations of alkali metals, which dissolved in the melt.

The phenomenon of catalytic dissolution of copper in the melt, containing complexes of refractory metals occurs by the reactions:

$$\rightarrow 4Cu(I) + NbF_6^{2-} \Leftrightarrow Nb + 4\,Cu\,(II) + 6F \qquad (19)$$

$$Cu\,(II) + Cu \Leftrightarrow Cu(I) \qquad (20)$$

with formation of a porous niobium coating with a bad adhesion to substrate[14]. The formation of coatings by the phenomenon of currentless transfer prior to electrodeposition offer influence the quality of the latter deposits, obtained by electrolysis of molten salts.

5. Reactions of cluster formation

Electroreduction of chloride complexes of niobium at a temperature below 903-923 K proceeds by a three-step mechanism [15]:

$$\text{I}\ Nb(V) + e \rightarrow Nb(IV) \qquad \text{(reversible)} \qquad (21)$$

$$\text{II}\ Nb(IV) + 2e \rightarrow Nb(II) \qquad \text{(irreversible)} \qquad (22)$$

$$\text{III}\ Nb(II) + 2e \rightarrow Nb \qquad \text{(reversible)} \qquad (23)$$

The second step of electroreduction is complicated by the formation of insoluble cluster compounds owing to the reaction of Nb(IV) and Nb(II) complexes. At the second step, cluster compounds $NbCl_{2.33}$ and a large assortment of nonstoichiometric compounds in the composition range $NbCl_{2.67}$ - $NbCl_{3.13}$ have been isolated by potentiostatic electrolysis. Electrolysis at more negative potentials leads to an increased amount of niobium in the cathodic deposit. At temperatures higher than 923 K, the second and third waves merge into a single four-electron wave, and the formation of cluster compounds of the $NbCl_x$ type becomes thermodynamically unfeasible. The voltammograms display only two reduction waves:

$$Nb(V) + e \rightarrow Nb(IV) \qquad \text{(reversible)} \qquad (24)$$

$$Nb(IV) + 4e \rightarrow Nb \qquad \text{(reversible)} \qquad (25)$$

Addition of fluoride ions at a temperature below 903 K causes dissolution of cluster compounds $NbCl_x$ and formation of NbF_6^{2-} complexes and metallic niobium in the melt. In this case we have a disproportionation reaction [2].

6. Reactions of polymerization

During the electrodeposition of rhenium coatings from the NaCl-KCl melt at a rhenium concentration of 5.0 w/o and a current density of higher than 0.05 to 0.07 A/cm^2, rhenium metal powder and the salt are deposited at the cathode. Voltammograms for the NaCl-KCl electrolyte containing 5.0 w/o of rhenium show an extra wave at more negative potentials. Potentiostatic electrolysis at the potentials of the first wave results in the formation of a compact phase of rhenium metal. Electrolysis at the potentials of the second wave yields rhenium powder or rhenium sponge. In addition to the metal, a darkgreen salt was discovered on the cathode. Diffractometric and chemical analyses identify this salt as $K_2Re_2Cl_8$ whose formation can be explained only by the participation in the second discharge process, of the polynuclear rhenium complexes that are likely to appear in melts of high rhenium concentration [3]. Strubinger et.al. [16,17] reported the following polynuclear complexes of rhenium: $[Re_2Cl_8]^{2-}$, $[Re_2Cl_8]^{3-}$, $[Re_3Cl_{12}]^{3-}$ and $[Re_3Cl_{11}]^{3-}$. The formation of $K_2Re_2Cl_8$ at the cathode indicates that trinuclear rhenium complexes must be participating in the discharge process along with the formation of binuclear complexes. The reduction of trinuclear rhenium complexes may be assumed to proceed according to:

$$4[Re_3Cl_{12}]^{3-} + 2K^+ + 5e \rightarrow K_2Re_2Cl_8 + 5[Re_2Cl_8]^{3-} \quad (26)$$

7. Reactions of competitive complex formation

Additional purification of the NaCl-KCl-K_2NbF_7 melt from oxofluoride complexes may be achieved if an acceptor is introduced into the melt with higher affinity that niobium for oxygen, and such that its own reduction potential and the reduction potential of its reaction products are more negative than the reduction potential of niobium. We found that the complex fluorides: K_2TaF_7, K_2TiF_6 , K_2ZrF_6 , K_2HfF_6, K_3AlF_6 may be used as such additives [18]. In the case of K_2TaF_7 redistribution of oxygen between complexes occurs :

$$NbOF_6^{3-} + TaF_7^{2-} \Leftrightarrow NbF_7^{2-} + TaOF_6^{3-} \quad (27)$$

A NaCl-KCl-K_2NbF_7 melt with addition of oxygen acceptors was used for production of nonporous coatings having good mechanical properties[18].

8. Acidic - basic reactions

A NaCl-KCl equimolar mixture doped with 5-10 w/o NaF and containing K_2NbF_7 (K_2TaF_7) and K_2CO_3 was used for electrochemical synthesis of niobium and tantalum carbides. The electrochemical behaviour of K_2NbF_7 and K_2TaF_7 in chloride-fluoride melts was considered by us earlier. The study of sodium and potassium carbonate electrochemical behaviour in NaCl-KCl melt showed no waves of carbonate-ion discharge until the background electrolyte started decomposing. However, with niobium or tantalum electrodes a carbonate-ion discharge wave was recorded due to depolarization caused by the formation of tantalum and niobium carbides.

The composition of products formed at the cathode was determined by acidic-basic reactions:

$$K_2MeF_7 + K_2CO_3 \rightarrow K_3MeOF_6 + KF + CO_2 \quad (28)$$
$$K_2MeF_7 + 2K_2CO_3 \rightarrow K_3MeO_2F_4 + 3KF + 2CO_2 \quad (29)$$
$$K_2MeF_7 + 3K_2CO_3 \rightarrow K_2MeO_3F + 6KF + 3CO_2 \quad (30)$$

where Me - Nb, Ta.
Time intervals for reactions (28)-(30) have been determined. Thus, for instance, at the ratio $K_2TaF_7/K_2CO_3 \geq 1:1$, , TaF_7^{2-}, $TaOF_6^{3-}$ complexes and CO_3^{2-} ions co-exist in the melt and on their discharge electrochemical synthesis of tantalum carbides is possible. After a certain period of time reaction (28), only TaF_7^{2-} and $TaOF_6^{3-}$ complexes remain in the melt and metallic tantalum is formed at the cathode, because no carbonate is present (in the case of a 1:1 ratio the only complex is $TaOF_6^{3-}$) .Thus, the composition of cathodic deposits is different with the same electrolysis parameters, depending on how far reactions (28)-(30) have proceeded.

9. Reactions of heteronuclear complex formation

Introduction of $AlCl_3$ into the $NaCl$-KCl-$CrCl_2$-$CrCl_3$ melt was shown to result in the formation of chromium and aluminium heteronuclear complexes which are characterized by their own electroreduction wave[19]. The multielectron discharge of Cr(III) and Al(III) heteronuclear complexes was shown to occur in a narrow interval of potentials with the formation of a chromium-aluminium alloy at the cathode. The formation of heteronuclear complexes is accompanied by Cr(III) stabilization, presumably due to the following reaction:

$$2xAl^{3+} + 3CrCl_4^{2-} + (2y-4)Cl^- \Leftrightarrow 2[Al_xCrCl_{4+y}]^{(3x-1-y)} + Cr \quad (31)$$

Cr(III) stabilization is simultaneously accompanied by dispersing of the fine powder of chromium in the bulk of the melt and formation of a coating on the container walls.

10. Conclusions

It should be noted that, as a rule chemical reactions complicate the processes of electrodeposition of refractory metals. At the same time we successfully used disproportionation reactions for producing fine metal powder and surface coatings.
Our findings have shown that linear sweep voltammetry is an effective method for studying chemical reactions in molten salts.

11. Acknowledgments

The author is grateful to the Russian Foundation for Basic Research (grant 96-03-32175) for the financial support of this study.

12. References

1. Glagolevskaya,A.L., Kuznetsov, S.A. and Stangrit, P.T. (1988) Reasons for the formation of surface metallic films on the electrolysis of melts, *Zh. Prikl. Khim* **61,** 1036-1040.
2. Kuznetsov, S.A. and Stangrit, P.T. (1993) Complex formation and stabilization of highest oxidation states of d-elements in molten salts, *Zh. Prikl. Khim* **66,** 2702-2709.
3. Kuznetsov, S.A (1994) Electroreduction of rhenium complexes in molten halides and oxohalides, *Russ. J. Electrochemistry* **30,** 1337-1344.
4. Gopienko, V.G. (1965) Development of methods for preparation of melts containing lower titanium chlorides, *Thesis,* Leningrad Polytechn. Inst., Leningrad.
5. Sukhodskii, V.A., and Gopienko, V.G. (1960) Electrolytic refining of titanium in melts, *Trudy VAMI* **46,** 146-158.
6. Kuznetsov, S.A. Polyakov, E.G., Kuznetsova, S.V., and Stangrit, P.T. (1987) Cathodic process during the electrolysis of potassium hexafluorohafnate in molten NaCl-KCl mixture, *Chemical Techn. Investigation of Raw Minerals of the Kola Peninsula,* Nauka, Leningrad.
7.Glagolevskaya,A.L., Kuznetsova, S.V and Kuznetsov, S.A. (1989) Corrosion of hafnium in chloride and mixed chloride-fluoride melts, *Zh. Prikl. Khim* **63,** 2673-2677.
8. Kuznetsova, S.V, Glagolevskaya,A.L., and Kuznetsov, S.A. (1993) Anodic processes occurring during hafnium dissolution in chloride-fluoride melts, *Soviet Electrochemistry* **28,** 558-563.
9. Kuznetsova, S.V, Kuznetsov, S.A, Polyakov, E.G., and Stangrit, P.T. (1989) Cathodic current efficiency during the electrodeposition of hafnium coherent coatings, *Rasplavy* **3,** 118-120.
10. Kuznetsov, S.A. Polyakov, E.G., and Stangrit, P.T. (1978) Current yield in the electrolysis of chloride-fluoroniobate melts, *Zh. Prikl. Khim* **52,** 1308-1312.
11. Kuznetsova, S.V, Glagolevskaya,A.L., and Kuznetsov, S.A. (1992) Electrodeposition of hafnium coatings from molten NaCl-KCl-$HfCl_4$ mixtures using direct and reversible currents, *Rasplavy* **6,** 29-35.
12. Kuznetsov, S.A., and Grinevich, V.V. (1994) Niobium interaction with its chloride, fluoride and oxofluoride complexes in melts of alkali chlorides, *Zh. Prikl. Khim* **67,** 1423-1430.
13. Kuznetsov, S.A, Glagolevskaya,A.L., and Kuznetsova, S.V. (1992) Preparation of high-melting metal carbide coatings in molten salts using disproportionation reactions, *Zh. Prikl. Khim* **63,** 2078-2080.
14. Kuznetsov, S.A, Kuznetsova, S.V., and Glagolevskaya,A.L., (1994) Catalytic dissolution of copper in equimolar mixture of sodium chloride - potassium chloride, containing complexes of rare refractory metals, *Rasplavy* **8,** 38-47.
15. Kuznetsov, S.A, Morachevskii, A.G., and Stangrit P.T. (1982) Electrochemical behaviour of niobium (V) in chloride melts, *Soviet Electrochemistry* **18,** 1522-1525.
16. Strubinger, S.K.D., Sun, I-W, Cleland, W.E., and Hussey, C.L. (1990) Electrochemistry and spectroelectrochemistry complexes and their one-electron reduction products in a basic room-temperature chloroaluminate molten salts, *Inorg. Chem.* **29,** 993-999.
17. Strubinger, S.K.D., Sun, I-W, Cleland, W.E., and Hussey, C.L. (1990) Electrochemical and spectroscopic studies of rhenium (IV) monomeric and dimeric chloride complexes in the basic aluminium chloride-1-methyl-3-ethylimidazolium chloride room- temperature molten salts, *Inorg. Chem.* **29,** 4246-4252.
18. Kuznetsov, S.A. Polyakov, E.G., and Stangrit, P.T. (1983) Electrolytic deposition of niobium coatings from melts on a long conductors, *Zh. Prikl. Khim* **56,** 427-429.
19. Kuznetsov, S.A., Glagolevskaya,A.L. (1995) Stabilization of Cr(III) in the formation of heteronuclear complexes of chromium and aluminium., *Russ, J. Electrochemistry* **31,** 1285-1288

PRODUCTION OF TANTALUM BY ELECTROLYSIS OF OXOHALIDE MELT WITH OXIDE ANODES

Y.P.Zaikov, A.P.Khramov, V.P.Batukhtin,
N.N.Chikhaleva, A.V.Frolov, L.E.Ivanovskiy,
N.G.Molchanova and N.I.Moskalenko

Institute of High Temperature Electrochemistry of the Ural Branch of the Russian Academy of Sciences,
S.Kovalevskaja str. 20,
620219, Ekaterinburg, Russia.

1. Introduction

Earlier we reported [1] that tantalum could not be produced by electrolysis of a melt of the composition (KCl-KF-K_2TaF_7-Ta_2O_5 [2]) using an anode composed of such oxides as NiO, SnO_2, Cr_2O_3, ZnO, ZrO_2, CoO, CuO, Fe_2O_3 because of their high solubility. This conclusion was confirmed, in particular, by tests of a NiO anode during electrolysis of this melt.

Electrolysis of this melt using an anode of TiO_2 failed to give positive results: since there was no tantalum metal in the cathode product, but solid intermediate products of Ta_2O_5 and TiO_2 reduction. In addition the anode area failed and was slurried with solid products of the melt interaction with oxygen evolving on the anode. Measurements showed the high solubility of TiO_2 in this melt, Table 1.

TABLE 1. TiO_2 solubility in KCl (58,3 w/o) - KF (29,2 w/o) -K_2TaF_7 (12,5 w/o) melt at 800°C.

Additives of Ta_2O_5 to a melt, % of the salt mass	Solubility of TiO_2 (concentration [Ti]) w/o
0	1,5
2	1,35
5	0,27

It was therefore necessary to find another salt system, the main

D.H. Kerridge and E.G. Polyakov (eds.), Refractory Metals in Molten Salts, 197-203.

requirements of such a melt being a high solubility for Ta_2O_5 and a low solubility of the anode material. Test electrolysis with different melts established that none of the salt systems tested was suitable for process realization due to low solubility of Ta_2O_5. Thus we have developed a metal-thermal reduction process for tantalum from Ta_2O_5 using an oxide oxygen-generating anode.

2. Metal-thermal reduction of tantalum from Ta_2O_5 during electrolysis of $CaCl_2$-CaO melt with a ceramic anode of TiO_2

$CaCl_2$ was chosen as a melt for the metal-thermal reduction process, as much CaO can be dissolved in it , which is required for an anode process with oxygen evolution. In addition electrolysis of $CaCl_2$ can be conducted at high temperatures without danger of considerable evaporation and high temperatures are necessary for operation of a TiO_2 anode. The possibility of electrolysis of $CaCl_2$-CaO melt with a ceramic anode was shown earlier [3].

A preliminary experiment, without electrolysis, showed the possibility of the process: calcium metal was put into a reaction cell together with $CaCl_2$ salt and the Ta_2O_5 powder and was heated up to the $CaCl_2$ melting point, resulting in Ta_2O_5 reduction to tantalum metal.

The electrolysis experiments were carried out with TiO_2 anodes, as preliminary experiments had showed that such anodes are rather stable on electrolysis in $CaCl_2$ melt. Measurements of TiO_2 solubility in different melts [4] are given in Table 2.

TABLE 2. Solubility of TiO_2 in different melts [4].

Solvent, mol %	T, °C	[TiO_2] wt%
$CaCl_2$ - CaF_2 (10)	720	0,003
$CaCl_2$ - CaF_2 (24)	720	0,005
	800	0,01
	900	0,02
LiF(53)-NaF(36)-CaF_2(11)	750	0,03
	850	0,07
	950	0,15
LiF - BaF_2 (18)	800	0,01
$BaCl_2$ - BaF_2 (14,5)	850	0,06

Titanium oxide was obtained from $TiCl_4$ (reaction with water and subsequent addition of ammonia solution). Ta_2O_5 or Nb_2O_5 [5] were used as a doping additive obtained from the corresponding chlorides. Oxides were mixed in a ball mill,then pressed and sintered into slabs. The open porosity, water absorption and apparent density of the oxide samples of TiO_2-Ta_2O_5(5 mol.%), prepared under different conditions were measured by hydrostatic weighing (Table 3).

TABLE 3. Ceramic features of TiO_2-Ta_2O_5 oxides (5 mol%)

Pressing pressure kgf/cm^2	Sintering time, h	Sintering T°C	Open porosity, %	Water absorption%	Apparent density, g/cm^3	Note
3200	8	800	37,2	13,6	2,7	
3800			37,7	13,1	2,9	
4500			32,5	11,1	2,9	
3200	18	1000	2,7	0,7	4,0	
3800			7,9	2,0	3,9	
4500			11,1	2,8	4,1	
2600	72	1100	0	0	4,4	
3200			5,26	1,25	4,2	
3800			3,45	0,81	4,2	
4500	48	1200	0,83	0,2	4,1	
			~4,5	~1,1	~4,2	*
			~5,7	~1,5	~3,9	

* The samples cracked.

This table shows the open porosity decreases with increases in temperature and sintering time. The relation of the open porosity to the pressing pressure is more complex and careful selection of this parameter is required for the production of rather dense samples.

Next attempts were made to find a TiO_2 anode composition and method of construction so as to improve the cell design as well as the conditions for the Ta_2O_5 reduction process by calcium deposition on the cathode during electrolysis of $CaCl_2$-CaO melt. Some results are given in Table 4.

The electrolysis was carried out in an alumina crucible in the air atmosphere in most cases and the mass of salt was 300-500 g. Several experiments were conducted with a diaphragm, or a baffle limiting the cathode area. In the case of a diaphragm, the cathode area was limited by the inner crucible made from BeO, or alumina in which holes were made a little above the bottom. These holes were covered inside by a layer of Ta_2O_5 powder which was reduced during the process to the metal. In the case of a baffle there were no holes in the inner crucible and the level of the melt in the outer crucible was higher than the walls of the inner one.

TABLE 4. Some experimental results on the optimization of electrolysis operating conditions and construction of a cell for metal-thermal reduction of Ta from Ta_2O_5 at the presence of an oxide anode made from TiO_2-Nb_2O_5 (15 wt%) [1)]

1	Melt wt%	T °C	Current À	Voltage V	Electrical power À_h	Cathode current density, À/cm²	Current efficiency on Ta, %: with a diaphragm or a baffle	Current efficiency on Ta, %: Without a diaphragm	Notes
2	$BaCl_2$-BaO (~5)	1000	8-10	12-14	~34	~8		~0,5	
		1000	6-9	9-20	~24	~8	33		
3	$CaCl_2$-CaO (~5)	750-800	~5	13-24	~23	~4,5	~3		
4		770	3	6-8	~3,8	1,5	~15		2)
5		850		6-14	~3,8		~23		2)
6		950		5-9	4		~65		2)
7		950	5-6	14-17	~11	~5	36		
8		950	~4	11-20	~16	~4	21		
9		940	5	7-12	~12	0,2		65	
10		950	4	4-11	~120	1		~25	2)
11		950	4,25	5-15	~100	~1		~20	2)
12		900	0,9	2-3	3,3	0,44	~72		2)
13			2	4-6	3,5	1	~72		2)
14			4	5-10	3-8	2	40-75		2) 3)
15			4,5	7-15	~21	2,1		~11	2)
16			5,3	6-13	~27	0,77		~14	2)
17		920	~5,5	5-6	~18	0,2-0,3		~40	2)

1) Anode current density 0,2-0,3 À/cm².
2) Anode with an inner current lead.
3) Results of seven experiments.

Several experiments were conducted with the $BaCl_2$-BaO melt, because the process of Ta_2O_5 reduction by barium deposition on the cathode will occur more efficiently than that by calcium, as barium possesses higher solubility. Such experiments were carried out both with a vibrating cathode (for efficient agitation of Ta_2O_5 powder) and with a stable one. The experiment with a vibrating cathode in a cell without a diaphragm gave a very low current efficiency for tantalum (<0,5%). Those with a stable cathode in a cell with a diaphragm gave a considerably better result ($\sim$33%)A similar result was obtained for electrolysis of $CaCl_2$-CaO melt with a vibrating cathode the output amounted to several per cent, the metal being spread over the bath as a fine powder.

In several runs, e.g. run 17 , Table 4, stable operation of the anode was obtained at a rather low voltage. This was due to the effect of the anode construction (the inner current lead had been pressed into a hole in the anode body) and also that the temperature in this run was high, which resulted first in a relatively low electrical resistance of the ceramic anode and secondly in a more complete and faster metal-thermal reduction reaction. As a result tantalum powder was not spread over the container volume, but was concentrated on the cathode rod as a compact conglomerate which could be removed from the melt together with the cathode. In addition there was practically no Ta_2O_5 powder in thc container and after the experiment it was not difficult to separate the tantalum powder from that of Ta_2O_5. The low current efficiency in this run is due to the non-optimized operating conditions of electrolysis; in particular that an excess amount of electricity had been passed.

The cathode construction is of considerable importance in this process. Cathodes were tested made of a steel (nikel) rod as well as a rod to the lower end of which a perforated steel disc of 30 mm in diameter was fastened (the disc surface was perpendicular to the rod axis and immersed in the layer of Ta_2O_5 powder on the container bottom.

From Table 4 it is seen that current efficiency is considerably higher for a cell with a diaphragm (or a baffle) than one without it. The current efficiency increased with increase in temperature and decreased with increase of the cathode current density. At the optinum selection of electrolysis parameters and cell construction a current efficiency in the range of 70-75% was obtained.

In nearly in all the experiments a film formed on the anode after a rather long electrolysis. In the course of the time it cracked and separated from the base and was perhaps the cause of voltage jumps in the bath which were observed from time to time. The composition of the film could not be determined with x-ray phase analysis, but probably consisted of tantalum and calcium oxides.

TABLE 5. Impurity analysis of tantalum powder obtained from the electrolysis of oxide-halide melts with different anodes, weight %.

Impurity	Melt (anode)		
	$BaCl_2$-CaO (TiO_2-Nb_2O_5)	$CaCl_2$-CaO (TiO_2-Nb_2O_5)	KCl-KF-K_2TaF_7-Ta_2O_5 (Graphite) [6]
Nb	3.10^{-2}	1.10^{-2}	2.10^{-2} ±3,3
Fe	$2{,}7.10^{-2}$	$4{,}5.10^{-3}$	$(6{,}5 \pm 2) .10^{-2}$
Ni	0,25	2,06	$(100 \pm <4) .10^{-4}$
Ti	2.10^{-2}	2.10^{-2}	$<6.10^{-4}$
Ca	$7{,}3.10^{-3}$	1,6	$<6.10^{-3}$
Ba	8.10^{-2}	$<4.10^{-3}$	
W	$\leq 4.10^{-3}$	$\leq 8.10^{-3}$	$<1.10^{-2}$
Mo	$\leq 3{,}7.10^{-3}$	$<4.10^{-3}$	$(2{,}7 \pm <1) .10^{-3}$
Zr	$1{,}4.10^{-3}$	1.10^{-3}	5.10^{-4}
Mg	3.10^{-2}	$7{,}3.10^{-3}$	$<5.10^{-4}$
Al	5.10^{-3}	$\leq 1.10^{-3}$	$(7 \pm 2) .10^{-3}$
Si	6.10^{-2}	0,11	$1{,}1.10^{-2}$
Na	$1{,}3.10^{-3}$	6.10^{-3}	$<2.10^{-2}$
Cu	$1{,}7.10^{-3}$	$2{,}8.10^{-2}$	$<6.10^{-4}$
Sn			$<5\_10^{-4}$
Cr	$\leq 5.10^{-4}$	$\leq 8.10^{-4}$	$(5 \pm 20) .10^{-3}$
Co	$\leq 1.10^{-3}$	$\leq 1.10^{-3}$	1.10^{-4}
Mn	$\leq 6.10^{-4}$	3.10^{-3}	$5{,}5.10^{-4}$
C		5.10^{-2}	$0{,}39 \pm 0{,}19$
O		$9{,}1.10^{-2}$	$(25 \pm 2{,}6) .10^{-2}$

The results of tantalum metal analysis obtained from the electrolysis of two melts are given in Table 5, together with results from an electrolyzer with a carbon anode (i.e. without the process of metal-thermal reduction) [6]. The ranges of impurity concentrations in the tantalum powders are given for those cases where these concentrations changed at the transition from one electrolysis cycle to another.

The high Ni impurity content is significant and results from the dissolution of the inner current lead made from Ni which diffused through pores of the ceramic anode body. The increased content of Ca in tantalum produced in electrolysis of $CaCl_2$-CaO melt is connected with the entrainment of complex oxides containing CaO. A great advantage of the electrolysis process with a ceramic anode over that with a carbon anode is the production of tantalum with a relatively low carbon content.

Thus, it may be concluded that reduction of tantalum oxide occurs in the electrolysis of an alkaline earth chloride melt containing the corresponding dissolved oxide with the use of a ceramic anode at which oxygen evolves. Such a process needs a rather high temperature ($\geq$900°C in the case of calcium chloride) and it is desirable to use a baffle, or a diaphragm, limiting the cathode area. Work on optimising the material for the ceramic anode continues.

Acknowledgment

The authors wish to acknowledge their thanks to the Russian Foundation for Basic Research for the financial support of this work.

References

1. Zaikov, Y.P., Khramov, À.P., Batukhtin, V.P., Ivanovskiy, L.E., Frolov, À.V. (1996) The perspectives of the development of a nonconsumable anode in the production of Ta by electrolysis of oxide-halide melts, *Khimija tverdogo tela i novye materialy* **2**, Collection of reports of the All-Russia conference, Ekaterinburg, 14-18 Oct. The institute of solid body chemistry of the Ural Branch of the Russian Academy of Sciences: Ekaterinburg Ur RAN, 176-177.
2. Konstantinov, V.I. (1977) Electrolytic production of tantalum niobium and their alloys, Metallurgy, Moscow.
3. Zaikov, Y.P., Dubovtsev, À.B., Batukhtin, V.P., Khramov, À.P., Ivanovskiy, L.E. (1995) Electrolysis with oxide anodes in a chloride-oxide melt, *J. Rasplavy* **2** , 41-46.
4. Zaikov, Y.P., Batukhtin, V.P, Khramov, À.P. (1996) Solubility of TiO_2 in some melt systems, *Construction and featuris of metal and slag melts* **2,** Collection of reports of the VIIIth All-Russian conference, Ekaterinburg, September. The Institute of Metallurgy of the Ural Branch of the Russian Academy of Sciences: Ekaterinburg, Ur RAN, 111.
5. Perfiljev, Ì.V., Demin, À.Ê., Êuzin, B.L., Lipilin, À.S. (1988) *High-temperature electrolysis of gases*, Nauka, Moscow.
6. Zaikov, Y.P., Khramov, À.P., Batukhtin, V.P., Ivanovskiy, L.E., Pazdnikov, I.P., Spitsin, P.Ê. (1995) Electrolytic production of tantalum from Ta_2O_5 dissolved in chloride-fluoride melt, *Rare Earth Metals: raw material processing, technology of compounds and realated products*, International conference, Abstracts, Krasnojarsk, 157-160.

TANTALUM CURRENTLESS DEPOSITION ON NICKEL FROM MOLTEN SALTS

E.MATYCHENKO, V.NOVICHKOV
Institute of Chemistry Kola Science Ctr. Russian Acad. Sci., 14 Fersman Str., Apatity, RUSSIA

Nickel is widely applied as structural material in molten salts owing to its high corrosion resistance, including a use in sodium-thermic production of tantalum powders. In this connection it is of interest to consider its behaviour in melts containing tantalum ions. For the experiments KCl-NaCl(1:1 mole), (Cs-K-Na)Cl(eutectic), KCl-KF(3:2 mole), (Li-Na-K)F(eutectic) melts containing 2.5% mole of potassium heptafluorotantalate K_2TaF_7 each were selected.

The studies were carried out in a hermetic stainless steel vessel in helium at 750°C. As the melt container a nickel crucible was used. Alkali metal halides were carefully dehydrated by stepwise heating up to ~650°C under vacuum followed by melting in an inert atmosphere. Potassium fluorotantalate was obtained from tantalum metal. Prior to conducting the experiments the melt was subjected to electrolysis with tantalum electrodes. The working specimens cut from nickel rolled stock (~10 or 5 x 50 x 0.4 mm) were treated in the (HNO_3 +H_2SO_4+H_3PO_4) mixture. The reference electrode was a chemically polished tantalum rod. In the experiments PI-50-1.1 potentiostat and PDA-1A recorder were used.

During the experiment we checked the specimen weight prior and following the tests, the change in the nickel electrode potential during holding, the presence of nickel in the melt, the current by the connection of the external circuit of the nickel and tantalum electrodes as well as the surface layer composition by X-ray diffraction (CuK_α-radiation).

The results of the experiments on determining the relationship between the specific change in the nickel specimen weight and duration of their stay in the melt are presented in Fig.1. A stable increase of weight is apparent in all the melts studied. No traces of nickel in the electrolyte were found by AES, *i.e.* its content was less than 10^{-4} % weight. At the same time, the X-ray diffraction data point to the formation of predominantly Ni_3Ta on the nickel surface in practically all the cases. So, tantalum is currentlessly transferred to the nickel substrate. Deposition of a more electronegative metal (Ta) onto the less negative one (Ni) is possible owing to the gain in power during the alloy formation.

D.H. Kerridge and E.G. Polyakov (eds.), Refractory Metals in Molten Salts, 205-210.

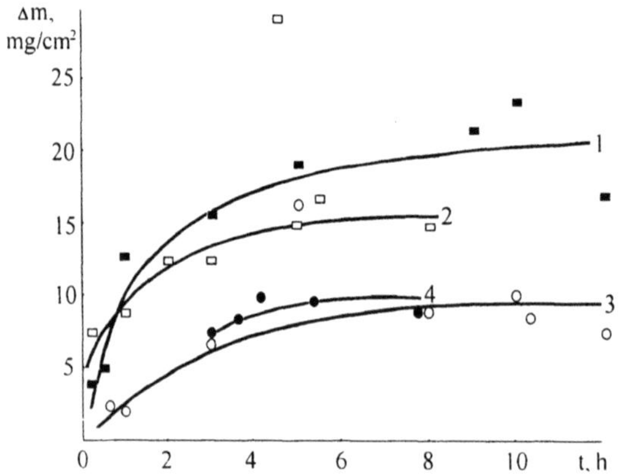

Figure 1. The time dependence of the specific change in the nichel speciemens weight. Electrolyte on the base: 1 - KCl-NaCl; 2 - KCl-KF; 3 - (Li-Na-K)F; 4 - (Cs-K-Na)Cl

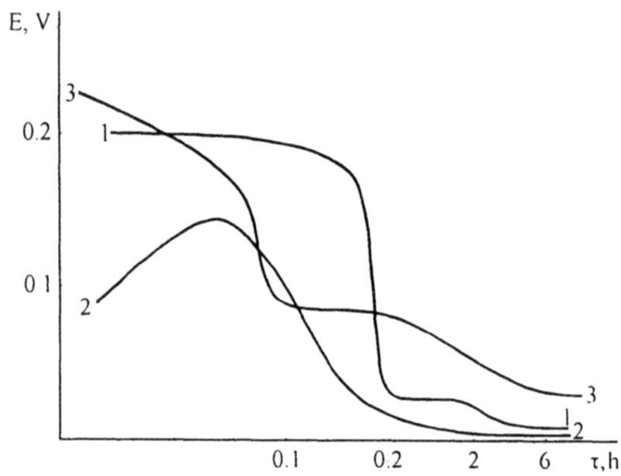

Figure 2. Zero-current chronopotentiograms for the nickel electrodes. Electrolyte on the base: 1 - KCl-NaCl; 2 - KCl-KF; 3 - (Li-Na-K)F;

Zero-current chronopotentiograms of the nickel electrodes are shown in Fig.2. At the moment of touching the melt the sample's potential relative tantalum in chloride melts is ~0.200 V. This agrees with the difference of equilibrium potentials of Ni^{2+}/Ni and Ta^{3+}/Ta in the KCl-NaCl equimolar mixture at 727°C[1]. For the fluoride melt (FLINAK) this value is somewhat higher at ~0.230 V and close to the value of 0.245 V, which in P.Taxil's opinion [2] corresponds to the formation of the Ni_3Ta compound during the tantalum deposition onto nickel at 950°C. In a mixed chloride-fluoride electrolyte the potential at the moment of touching is ~0.090 V, it grows fast up to ~0.150 V and then began to decrease. These differences are, apparently, caused by anionic composition of the electrolyte, *i.e.* the first coordination sphere. But may be influenced of microimpurities as well. As is seen in Fig.2, this influence spreads to the character of subsequent change of the potential. Currentless transport is commonly due to the presence in the melt of two different valencies of the metal deposited [3]. According to the voltammetric studies of various authors [2,4-8], in melts of different compositions Ta(V) ions are discharged in one 5-electron stage, *i.e.* only the pentavalent form is stable. However there is both indirect [7] and direct [9,10] evidence of the presence of tantalum ions in molten alkali metal halides in a lower oxidation state. Our findings also prove the presence of low-valent tantalum ions in the melts. The different stability of low-valent tantalum complexes in melts of varying compositions results in different concentrations of tantalum ions of the lower oxidation state. The fluoride ion stabilizes the ions in the higher oxidation state, consequently, in melts containing great quantities of F^--ions the content of ions in lower oxidation state should be smaller. This in its turn should result in lower rate of the tantalum currentless deposition. The slower tantalum deposition from fluoride melts in comparison with chloride ones is consistent with this viewpoint. At first glance the values of increases of weight for the melt of ternary eutectic (Cs-K-Na)Cl depart from the general regularity. However, in this case two factors exert an influence in the same direction: 1) the impact of the second coordination sphere, *i.e.* Cs^+ cation, which is similar but weaker than that of the fluoride ion; 2) as it follows from Table 1, the diffusion coefficients of Ta(V) ions in this melts are essentially smaller as compared with other melts, therefore one should expect lower diffusion coefficients for the low-valent form as well. Commonly the diffusion of the higher oxidation state ions in the salt phase does not limit the rate of the currentless transfer process.

TABLE 1. Diffusion coefficients of Ta(V) ions

	T, ^{0}C	$D_{Ta(V)}$, cm^2/s	method	reference
KCl-NaCl(1:1)	720	$1.35\ 10^{-5}$	LVA	[6]
	720	$1.65\ 10^{-5}$	CPM	[6]
	750	$3.38\ 10^{-5}$	LVA	[7]
	750	$1.35\ 10^{-5}$	LVA	[8]
(Cs-K-Na)Cl eutectic	749	$3.57\ 10^{-6}$	LVA	[4]
(Li-Na-K)F eutectic	750	$1.74\ 10^{-5}$	LVA	[2]
	750	$2.83\ 10^{-5}$	LVA	[5]
	750	$1.5\ 10^{-5}$	CPM	[9]

LVA - linear sweep voltammetry; CPM - chronopotentiometry

Furthermore, the specific increases of weight do not depend on the distance between the nickel and tantalum electrodes. On the other hand, this agrees with the results of experiments, which connected the external circuit of electrodes via small resistance. Though the specific increments of weight obtained are little different from those in the case of an open circuit, still at R=1Ω then are higher than in the case of R=10 Ω (KCl-NaCl+K_2TaF_7, T=750^{0}C, τ=1 h):

R, Ω	1	10	1	10
Δm, mg/cm^2	16.7	11.4	12.9	8.1

After 1 h the current density measured in the external circuit is ~2÷3 mA/cm^2 and changes very insignificantly (Fig.3). Assuming that this value corresponds to the stationary growth of the Ni_3Ta layer, *i.e.* to equality of the flow of Ta(V) ions through to the nickel substrate surface to the flow of atoms diffusing through the intermetallic compound formed, it is possible to determine the interdiffusion coefficient D from the ratio:

$$J = i/nF = -D(\Delta C/\Delta x),$$

where i is current density, mA/cm^2; ΔC is difference of concentrations on the diffusion layer boundaries, mole/cm^3; Δx - thickness of the diffusion layer, cm.

The Δx value determined from the metallographic cross polished section is ~10 μm. Calculation of the Ni_3Ta density by the measured thickness layer Δx and the specific change mass Δm of the sample results in a value ρ~12.5 g/cm³. It agrees well with the additive value ρ =12.6 g/cm³ at 750°C. The error in this case is ~10%, which is much lower than that of experimental determination of metallic interdiffusion coefficients. The result of calculation $D \sim 3 \cdot 10^{-12}$ cm²/s agrees satisfactorily with the data [11] (D=2.5·10^{-12} cm²/s) and [2] (D=2.6·10^{-11} cm²/s), revised to 750°C, which points to the validity of the assumptions and the possibility of evaluating the interdiffsion coefficients by the above method.

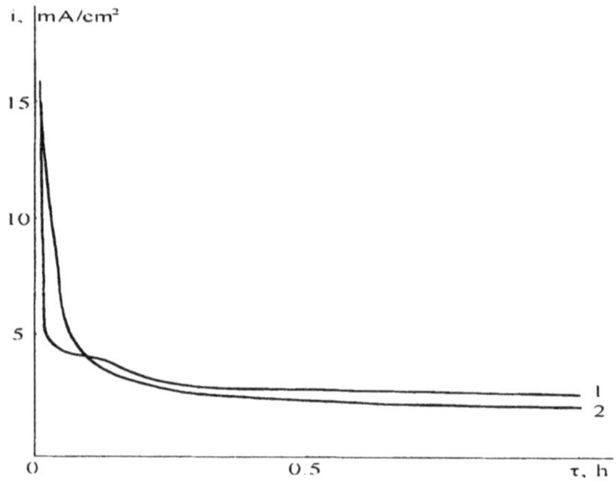

Figure 3. Chronoamperograms for the connected nickel and tantalum electrodes. The resistance of the external circuit: 1 - 1 Ω; 2 - 10 Ω

Consider the possible causes of predominant formation of the Ni_3Ta compound from five possible componds according to the phase diagram (Fig.4) [12]. Firstly, this compound is the strongest, as is seen from Table 2. Enthalpy of its formation measured by the direct calorimetric method is -81.6 kJ/mole at 1200°C [15] whereas according to [13] ΔH^0_f =-116.8 kJ/mole at 900÷1050°C. At the same time the thermodynamic characteristics of the Ni_8Ta compound were not determined.

Secondly, the formation of Ni_3Ta is preferable for kinetic reasons as well since the constant of the given phase growth rate at 750°C is maximal in comparison with other phases (Table 2).

TABLE 2. Change in the free Gibbs' energy during the formation of compounds and constants of the phase growth in the nickel-tantalum system

compound	-ΔG, kJ/mole			k, cm²/s
	[2]	[13]	[14]	[11]
Ni_8Ta	-	-	-	3.6 10^{-18}
Ni_3Ta	118.0	112.2±28.5	132.7	1.3 10^{-12}
Ni_2Ta	107.6	95.9±26.8	130.0	1.1 10^{-16}
NiTa	64.5	-	-	1.5 10^{-17}
$NiTa_2$	36.8	-	-	1.7 10^{-16}

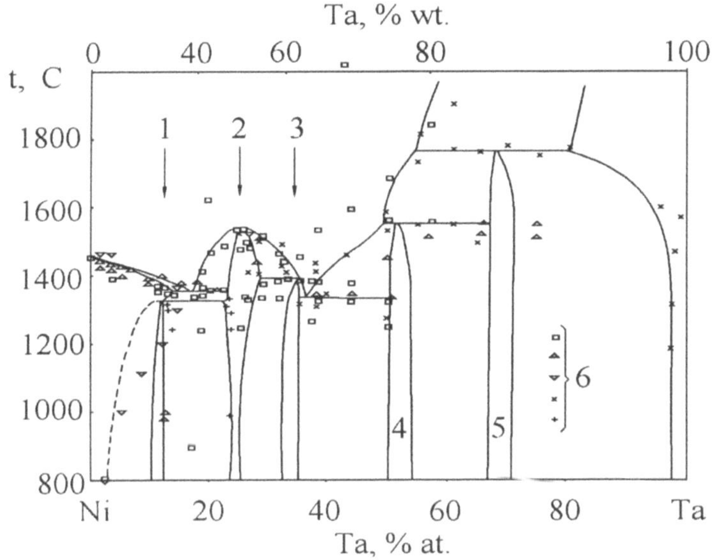

Figure 4. Phase diagram of the nickel - tantalum system [12]. 1 - Ni_8Ta; 2 - Ni_3Ta; 3 - Ni_2Ta; 4 - NiTa; 5 - $NiTa_2$; 6 - literature experimental data.

Thus we are fully justified in observing that spontaneous interaction between nickel and tantalum containing chloride, chloride-fluoride, and fluoride melts results in the predominant formation of Ni_3Ta intermetallic compound on the surface of the nickel samples. This corresponds with the published thermodynamics and kinetics data for the nickel-tantalum system.

References

1. Baraboshkin, A.N. (1976) *Electrocrystallisation of Metals from Molten Salts,* Nauka, Moscow (in Russian).
2. Taxil, P. (1985) Formation d'alliages tantale-nickel par voie electrochimique, *J. Less-Comm. Metals* **113**, 89-101.
3. Ilyshenko, N.G., Anfinogenov A.I., Shurov N.I. (1991) *Interaction of Metals in Ion Melts,* Nauka, Moscow (in Russian).
4. Polyakova, L.P., Kosilo, B.I., Polyakov, E.G., Smirnov, A.B. (1988) Electrochemical Behaviour of Tantalum in the Melt CsCl-KCl-NaCl-K_2TaF_7, *Rasplavy* **2**, 83-87 (in Russian).
5. Polyakova, L.P., Polyakov, E.G., Matthiesen, F., Christensen, E., Bjerrum, N.J. (1994) Electrochemical Study of Tantalum in Fluoride and Oxofluoride Melts, *J. Electrochem. Soc.* **141**, 2982-2988.
6. Lantelme, F., Barhoun, A., Li, G., Besse, J.-P. (1992) Electrodeposition of Tantalum in NaCl-KCl-K_2TaF_7 Melts, *J. Electrochem. Soc.* **139**, 1249-1255.
7. Kuznetsov, S.A., Glagolevskaya, A.L. (1996) Cathodic Processes Involved in the Electrosynthesis of Tantalum Borides, *Elektrokhimiya* **32**, 344-351 *(Russ.J. Electrochem.* **32**, №3).
8. Polyakova, L.P., Kononova, Z.A., Kremenetsky, V.G., Polyakov, E.G. (1996) The Influence of Caesium Chloride on the Diffusion Coefficients of Tantalum Complexes in KCl-NaCl-K_2TaF_7 Melt, *Zhurnal Prikladnoi Khimii (J. Applied Chemistry)* **69**, 1307-1313 (in Russian).

9. Senderoff, S., Mellors, G.M., Reinhart, W.J. (1965) The Electrodeposition of Coherent Deposits of Refractory Metals. II.The Electrode Reactions in the Deposition of Tantalum, *J. Electrochem. Soc.* **112**, 840-845.

10. Alimova, Z.A., Polyakov, E.G., Polyakova, L.P. (1988) Anodic Dissolution of Tantalum in Chloride and Chloride-Fluoride Melts, *Elektrokhimiya.* **24**, 1205-1211(in Russian).

11. Pimenov, V.N., Ugaste, Yu.E., Akkushkarova, K.A.(1977) The Study of the Interdiffusion and phase equilibrium diagram in the System Ni-Ta, *Izvestiya AN SSSR. Metally,* **№ 1**, 184-188 (in Russian).

12. Nash, A., Nash, P. (1984) The Ni-Ta (Nickel-Tantalum) System, *Bull. Alloy Phase Diagr.* **5**,259-271.

13. Liakishev, N.P., Snitko, Iu.P., Alexeev, V.I., Levshin, G.A. (1981) Thermodynamic of the Reaction of Ni_3Ta and Ni_2Ta Intermetallide Formation, *Doklady AN SSSR (Transactions of the Acad. of Sci. USSR)* **258**, 1404-1406 (in Russian).

14. Nesterenko, S.N., Meshkov, L.L., Sokolovskaya, E.M. (1981) Thermodynamic Properties of Alloys of Nickel-Tantalum System, *Vestnik Mosk. Univ. Ser.2. Khimiya (Herald of the Moscow Univ. Ser 2 Chemistry).* **22**, 305-307 (in Russian).

15. Guo, Q., Kleppa, O.J. (1994) Standard Enthalpies of Formation of Ni_3Ta, Pd_3Ta, Pt_3Nb, Pt_2V and Pt_3V by High-Temperature Direct Synthesis Calorimetry, *J. Alloys and Compounds* **205**, 63-67.

ELECTROCHEMICAL PRODUCTION OF HAFNIUM-BASED COMPOSITE MATERIALS

S.A.Kuznetsov, S.V.Kuznetsova,
E.G.Polyakov, P.T.Stangrit
Institute of Chemistry, KSC RAS.
14 Fersman Str., Apatity, 184200, Russia

1. Introduction

The fields of hafnium use have been expanding in the recent years. Most of them pertain to the latest achievements in engineering and technology.

2. Electrolytic modification of niobium surface by hafnium in molten salts

In the present part of the work we studied alloy formation upon the deposition of hafnium on a niobium cathode from a molten equimolar mixture of NaCl and KCl containing 10 w/o K_2HfF_6 at a temperature of 1023K. The investigation was carried out by a method involving the recording of chronopotentiometric curves. Potassium hexafluorohafnate was obtained by dissolving type GFI-1 metallic hafnium or "pure-grade" hafnium tetrachloride in a solution of hydrofluoric acid followed by precipitation by potassium fluoride. After being washed with ethanol and dried in a vacuum, the precipitate was identified with the aid for X-ray powder diffraction, IR-spectroscopic, and optical crystallography analyses. The electrolysis of the melt, which was placed in a container made from type SU-2000 glassy carbon, was carried out with the aid of a P-5848 potentiostat in an electrochemical cell [1]. The recording of the potential of the

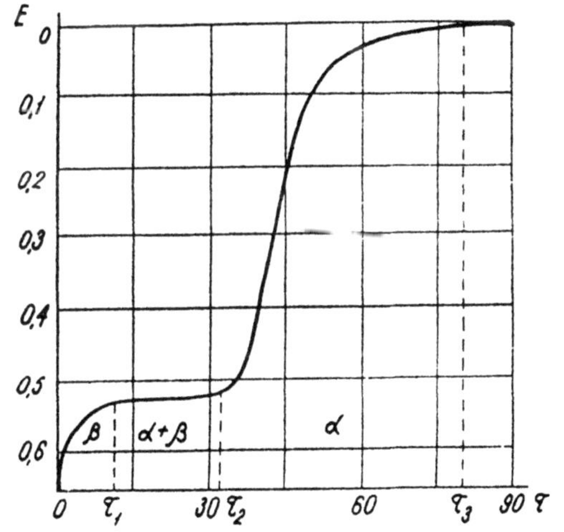

Figure 1. Dependence of the potential E(V) formed on the surface of the Nb-Hf cathode on the electrolysis time τ (s). The cathodic current density $8.3 \cdot 10^{-3}$ A/cm^2; T = 750°C

D.H. Kerridge and E.G. Polyakov (eds.), Refractory Metals in Molten Salts, 211-218.

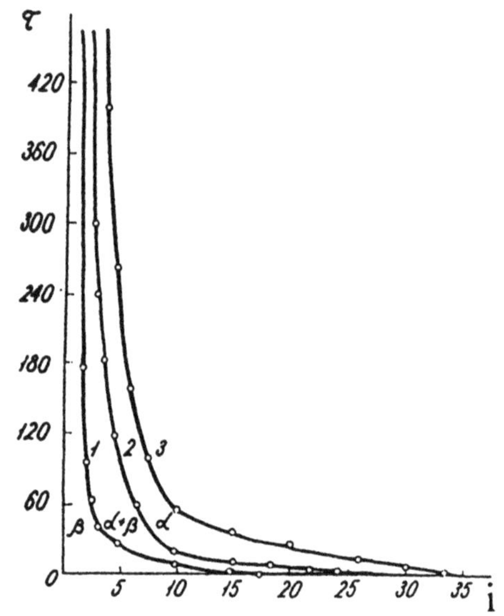

Figure 2. Dependence of the time τ(s) of the achievement of the limiting concentration of hafnium on the surface of a niobium cathode on the current density i (A/cm^2). Concentration of Hf (at.%) on sections: 1) C(0 to τ_1)=5.9, β-solid solution; 2) C(τ_1 to τ_2)=95.9, (α+β)-solid solution; 3) C (τ_1 to τ_3)=100, α-solid solution.

niobium cathode (a rod with a diameter of 2 mm, which was obtained from an ingot by electron-beam melting) relative to a hafnium reference electrode (a 5-cm^2 plate) was carried out by a KSP-4 potentiometer and a V7-21 digital voltmeter.

On the charging curves (the cathodic chronopotentiograms) in potential-time coordinates (Fig.1), which allowed for the ohmic component, there are four sections with different types of variation of the potential, in agreement with the niobium-hafnium equilibrium diagram [2]. The formation of a β-solid solution of hafnium in niobium takes place on the section from 0 to τ_1. The concentration of hafnium in it increases. The formation of an (α+β)-solid solution takes place on the section from τ_2 to τ_3. After the limiting concentration of hafnium in the α-solid solution is achieved (at τ_3), a phase of hafnium forms, and pure hafnium grows.

The absence of a dependence of the potential on the composition (the electrolysis time) is observed in the two-phase α+β-region owing to the constant activity of hafnium within the two-phase region. Abrupt variation of the potential was discovered within the α-phase i.e., the activity of hafnium in the α- phase is greatly dependent on the concentration.

Charging curves were recorded during the deposition of hafnium on niobium at current densities from $4 \cdot 10^{-4}$ to $5 \cdot 10^{-2}$ A/cm^2. Plots of the dependence of the limiting concentration of hafnium in the β, α+β and α-phases (Fig.2) were obtained on the basis of these curves. The curves obtained made it possible to establish the variation of the phase composition during the electrodeposition of hafnium on niobium and to determine the time intervals for the formation and growth of the phases on the surface of the cathode at different current densities and a temperature of 750°C [3].

It follows from Fig.2 that in order to obtain a β-solid solution of hafnium in niobium, the electrolysis process must be carried out at a very low current density, i.e., at a current density not exceeding $1 \cdot 10^{-3}$ to $2 \cdot 10^{-3}$ A/cm^2. Hafnium coatings can be obtained on a niobium substrate with intermediate layers of solid solutions of hafnium in niobium at current densities rangind from $4 \cdot 10^{-3}$ to $3 \cdot 10^{-2}$ A/cm^2. The formation of intermediate layers of solid solutions guarantees good adherence of the hafnium coating to the niobium substrate. The use of a current density higher than $3\cdot10^{-2}$ A/cm^2 results in the formation of only a phase of metallic hafnium on the niobium substrate.

The deposits with the solid solution $\alpha+\beta$ - layer have the greatest thermal resistance and best mechanical characteristics [4].

3. Electrodeposition of hafnium articles by the method of galvanoplactics

Coherent coatings of hafnium were obtained in [5] by electrolysis of molten NaCl-KCl-$HfCl_4$ mixture. The structure of hafnium coatings deposited from a molten chloride have already been investigated [6].

The construction of our electrolytic cell for deposition of hafnium coatings has been described [7]. Distilled water was used to wash the cathodic deposits. Niobium and steel (St.3) substrates 1 mm thick were used for electrodeposition of coherent hafnium coatings. These substrates had a mean deviation in the profile of R_a=0,11-0,13 μm.

We have shown in [8] that electroreduction of hafnium tetrachloride to the metal in molten NaCl-KCl mixtures takes place in two successive stages. The Hf(II) complexes are irreversibly discharged to the metal [8,9], whereas the standard rate constant of the discharging process Hf(II) + 2e $\rightarrow$ Hf is equal to $2{,}2\cdot10^{-4}$ cm/s at 750°C.

As a result of the reaction between the hafnium anode and the molten NaCl-KCl-$HfCl_4$ electrolyte, Hf(II) complexes are formed in the melt. At low concentrations of the tetrachloride, all of the tetravalent hafnium is transformed to the metal in the divalent state [9]. Wave I (Fig.3), which corresponds to the charge transfer Hf(IV) + 2e $\rightarrow$ Hf(II) and transforms to a wave corresponding to the oxidation process: Hf(II) - 2e $\rightarrow$ Hf(IV) whereas, in the cathodic region, wave II remains due to electroreduction of Hf(II) to the metal. In the case of the melt with 10 w/o $HfCl_4$, even upon prolonged contact with the metal, Hf(IV) complexes remain in the electrolyte since the voltammograms show that wave I is of a cathodic-anodic nature.

Coexistence of the Hf(II) complexes in the melt and a small fraction of Hf(IV) can be positive factors since, in the presence of cations of different degrees of oxidation, the stability of the plane crystallization front increases [10]. However, a shortcoming of this electrolyte is that it tends towards disproportionation. For example, at Hf(II) concentrations greater than $6\cdot10^{-5}$ mol/cm^3, the following reaction was observed to occur [8]:

$$2Hf(II) \leftrightarrow Hf(IV) + Hf \quad (1)$$

It can cause formation of a metallic film on the surface of the electrolyte.

The first experiments on the electrodeposition of hafnium-containing coatings from the molten NaCl-KCl-$HfCl_4$ (10 w/o) mixture supported our assumptions about the formation of a metallic film on the surface of the melt.

However, the metallic film did not prevent the growth of coherent hafnium deposits, at least up to 200 μm in thickness.

The cathodic current efficiency of hafnium in the coherent coating in the range of current densities $5\cdot10^{-3}$-$5\cdot10^{-2}$ A/cm^2 varied significantly and was equal to 1.68-1.72 g/(A·h) whereas, at current densities of $1\cdot10^{-2}$ and $2\cdot10^{-2}$ A/cm^2, it was

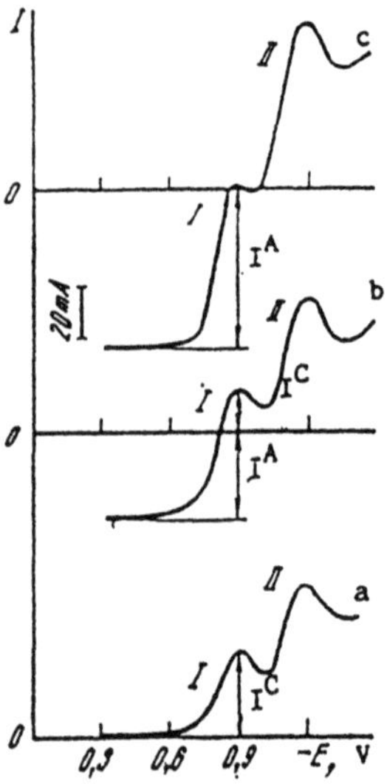

Figure 3. Transformation of the voltammograms during reaction between hafnium and molten NaCl-KCl-$HfCl_4$ mixture: C_{HfCl_4}=1,98·10^{-5} mol/cm^3; V=1,0 V/s; T=800°C;
a - before introducing the metal into the melt;
b, c - time of reaction 2h(b) and 8h(c)

observed to decrease to 1.2-1.3 g /(A·h). To a large extent, this decrease is caused by the fact that the fraction of the material going into forming the film and not the coherent coating increases with increasing current density. To a small extent, it is caused by the formation of dendrites poorly adherent to the substrates, which are removed during the washing.

We need to note that, upon electrolyzing the NaCl-KCl-K_2HfF_6 (10 w/o), and NaCl-KCl-K_2HfF_6 (10 w/o)-NaF(5 w/o) molten salts, it is possible to prepare coatings at least 60-70 μm thick by using diverse current conditions beyond a single cycle of electrolysis on a St.3 substrate. At the above values of the thickness l, the stability of the planar crystallization front is observed to be lost. Application of a second cycle (before immersing into the melt, the dendrites forming on the substrate are mechanically removed) enabled us to increase l up to 100-110 μm. Meanwhile, by electrolyzing the pure molten chloride for more than one cycle, it was possible to prepare significantly thicker coatings.

Let us analyze the possible reasons for such differences. A great influence is exerted on the growth of the coherent deposits by the initial stage of the crystallization, which is related to the number of the nuclei formed. The greater the number of particles formed per unit surface area in the initial period of the electrocrystallization, the more rapidly and at smaller thicknesses does the cathode surface get filled up with the deposit and the grains merge together to form a pore-free coherent layer with a minimum roughness. It is well known that, compared to pure chloride melts, application of chloride-fluoride electrolyte mixtures decreases the mobility of the discharging complexes and increases the polarizability of the cathode, which increases the number of the nuclei in the crystals generated at the cathode [10]. Apparently, the number of crystals is most important for their merger to form a coherent layer but other factors have a more substantial effect on their growth in the coherent deposit and the stability of the planar crystallization front in this case. The distinctive feature of a chloride melt compared to a mixed chloride-fluoride melt is the possibility of the coexistence of complexes of different valences Hf(II) and Hf(IV). It is precisely the presence of these Hf(IV) complexes in the molten chloride and their reaction with the coating by the scheme:

$$Hf(IV) + Hf \leftrightarrow 2Hf(II) \quad (2)$$

that exerts the controlling influence on the retention of the planar crystallization front. The cause for the loss of stability of the planar front in the molten chloridefluorohafnate mixture is the absence of any reaction between the metal and HfF_6^{2-} complexes [11-12].

Upon electrodepositing coatings from the molten NaCl-KCl-$HfCl_4$ mixture with thicknesses greater than 100 μm, they were found to have a rather high level of roughness, which was primarily determined by the presence of dendrites and defects of the "core" type on the surface of the coating. In order to decrease the coating and prevent the formation of dendrites, we carried out electrodeposition with reverse of current.

For the reverse current during the electrodeposition of coating from molten salts diverse parameters have been presented in [13-14], whereas the conditions for the application of currents with variable polarity giving rise to a smoothing effect have been formulated [10]. The most commonly used conditions involve the application of a cathodic current density much smaller than the diffusion limit but, in the anodic period, of a high current density leading to a salt passivation of the object [14]. Chemical dissolution of the salt occurs during the cathodic period and this partially compensates the depletion of the metal ion in the layer near the cathode, which is equivalent to some decrease in the thickness of the diffusional layer and improvement in microdistribution of the current.

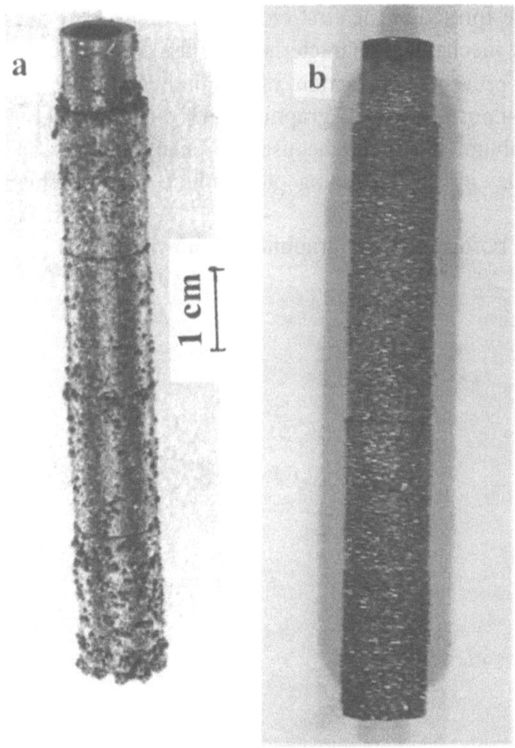

Figure 4. Hafnium rods, obtained by electrolysis of melt NaCl-KCl-$HfCl_4$(10 w/o) at 750°C.
a - cathodic current density 0,03 A/cm^2; time of electrolysis 6h; thickness 228 μm; b - cathodic current density 0,03 A/cm^2; time of cathodic period 100 s; anodic current density 0,4 A/cm^2; time of anodic period 7 s; total time of electrolysis 6h; thickness 193 μm

At first, we used the conditions suggested above. Electrodeposition was carried out on plates of St.3. These conditions did not produce a smooth deposit and dendrites remained on the surface.

In another investigation [15] reverse pulse current parameters were determined experimentally : the cathodic current density 0,05 A/cm^2, time of cathodic period 115 s, the anodic current density 0,4 A/cm^2, time of anodic period 6 s. These parameters gave coatings without dendrites. Similar conditions were used for obtaining hafnium articles-cathodes for electronic lamps by the method of galvanoplastics. Hafnium rods formed using direct and reverse currents are illustrated in Fig.4, which shows that using the reverse pulse current reduced the coating roughness.

4. Heat resistant composites on the basis of carbon and hafnium

Articles based on graphite are widely used in present day, technology because of the unique properties of graphite. However, a drawback of graphite-based composites is a low heat resistance in oxidizing atmosphere. To increase the heat resistance of articles on the basis of graphite, siliconizing and borosiliconizing are currently used. However, articles of borosiliconated graphite are operable in a limited period of time at the temperatures up to 1200°C only.

In order to prevent graphite oxidation at the temperatures of 1500-1700°C and higher, protective coatings of hafnium and hafnium-niobium alloy with a mixed (α+β)-structure can be used. One of the most promising methods of coherent hafnium layers and hafnium-niobium alloy deposition on graphite (borosiliconated graphite), articles of complex shape is molten salt electrolysis, which gives coherent, poreless coatings of desired compositions, a factor important for the Nb-Hf system since the most heat resistant are (α+β)-alloys and of them the alloy with 20 at.% of niobium is the best. Another merit of the method is a unifom distribution of one alloy component in the other. A thin layer of hafnium carbide formed at the first electrolysis stage [16] does not influence the physico-chemical and mechanical characteristics of the composite due to its small thickness (<1 μm), but prevents interaction of hafnium and hafnium-niobium alloys with the graphite. Composites with a graphite matrix and protective coatings of hafnium and hafnium-niobium alloy can be used for making units and mechanisms operated for some time in an oxidizing atmosphere at superhigh temperatures.

Coatings obtained on samples from borosiliconized graphite are shown in Fig.5.

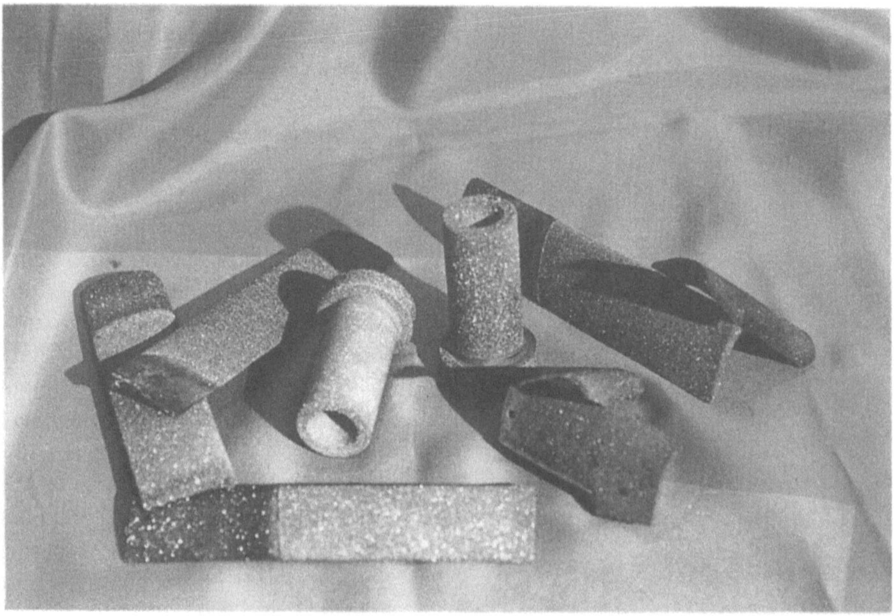

Figure 5. Hafnium deposits on borosiliconized graphite obtained from the melt NaCl-KCl-K_2HfF_6(10w/o)-NaF(5w/o); Current density 0,01 A/cm^2; T=750^{0}C; time of electrodeposition 4-6h; thickness of coatings 40-60 μm

5. Electrochemical production of copper-hafnium compositions

In another investigation [17] it was shown by metallography, X-ray microprobe analysis and scanning electron microscopy that the cathodic incorporation of hafnium into copper primarily yields a $HfCu_4$ compound at the electrode surface. Cu-Hf alloy formation is possible, by electrochemical deposition and also by interdiffusion of the metals.

When examining the bimetallic composites produced at a constant current density 0,015 A/cm^2 (for a total time of 11h) using an MS-46 Cameca X-ray microprobe three intermetallic compounds were seen between the copper and the hafnium (Fig.6a). They are also clearly visible in a cross section of the bimetallic composition (Fig.6b) with a REM-200 scanning electron microscope. Layers 1 and 2 have thicknesses of 24 and 8 μm, respectively, while the third layer is very thin (2-3 μm). Local microanalysis of layer 1 yields copper and hafnium concentrations close to those expected for the $HfCu_4$ compound, while the analysis of layer 2 yields values corresponding to $HfCu_3$. The determination of the copper and hafnium concentration in layer 3 is not possible because its thickness is in the range of the probe diameter.

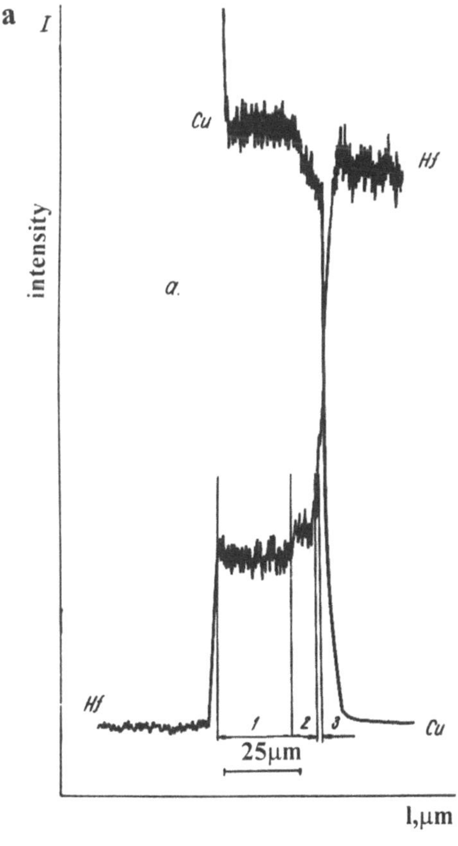

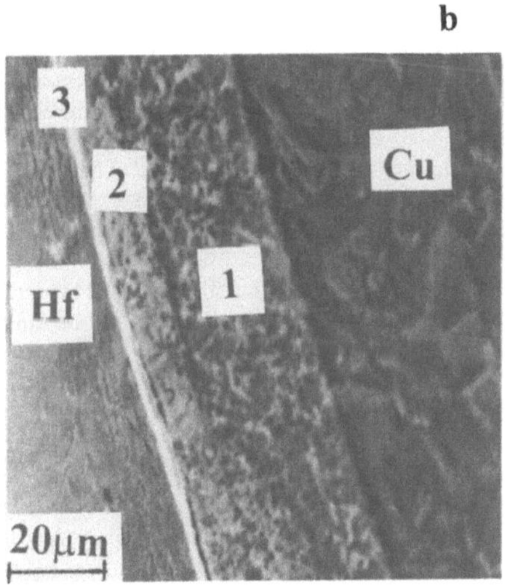

Figure 6. Distribution curves for the element concentrations (a); scanning electron micrograph of the Cu-Hf composite (b)

At certain parameters of electrolysis [17] it is possible to obtain five intermetallic compounds $HfCu_4$, $HfCu_3$, Hf_2Cu_5, Hf_2Cu_3, Hf_2Cu between the copper substrate and the hafnium coatings.

Laminated metallic copper-hafnium compositions were used as high-temperature solders for diffusional soldering of refractory metals.

6. Conclusion

The diverse problems of creating hafnium-based composite materials can be solved by hafnium electrodeposition from molten salts.

7. Acknowledgment

The authors are grateful to the Russian Foundation for Basic Research (grant 96-03-32175) for the financial support of this study.

8. References

1. Kuznetsov, S.A., Polyakov, E.G., Kuznetsova, S.V., and Stangrit, P.T. (1987). Cathodic process during the electrolysis of potassium hexafluorohafnate in molten NaCl-KCl mixture. *Chemical Tech. Investigations of Raw Minerals of the Kola Peninsula, Nauka, Leningrad,* 107-114.
2. Tylkina, M.A. and Tsygonova, I.A. (1964) System hafnium-niobium, *Zh.Neorg.Khim.* **9**, 1650-1652.
3. Kuznetsov, S.A., Polyakov, E.G., Kuznetsova, S.V., and Stangrit, P.T. (1988). Alloy formation upon the deposition of hafnium on a niobium cathode, *Zh.Prikl.Khim.* **61**, 16-161.
4. Ruda, G.I., Samgina O.Yu., and Smirnov, V.P. (1983). Phase composition and thermal resistance of Hf-Nb alloys, *Zaschita Metallov* **19**, 984-987.
5. Ivanovskii, L.E., Chashchikhin, V.A., and Batukhtin V.P. (1970). Refinement of hafnium and zirconium in molten chlorides, *Tr.Inst.Electrokhim., Ural.Nauchn.Tsentr., Akad.Nauk. SSSR* **15**, 18-23.
6. Martem'yanova, Z.S., Petenev, O.S., Baraboshkin, A.N., and Ivanovskii, L.E.(1974). Structure of coherent hafnium deposits prepared by the electrolysis of a molten chloride, *Ibid* **21**, 53-55.
7. Kuznetsova, S.V., Kuznetsov, S.A., Polyakov, E.G., and Stangrit, P.T. (1989). Cathodic current efficiency during electrodeposition of hafnium coatings, *Rasplavy* **3**, 118-120.
8. Kuznetsov, S.A., Kuznetsova, S.V., and Stangrit, P.T. (1990). Cathodic reduction of hafnium tetrachloride in a molten equimolar mixture of sodium and potassium chlorides, *Elektrokhimiya* **26**, 63-68.
9. Kuznetsova, S.V., Kuznetsov, S.A., and Stangrit, P.T. (1990). Kinetic parameters of the electroreduction of hafnium dichloride in a molten NaCl-KCl mixture, *Ibid* **26**, 102-105.
10. Baraboshkin, A.N., (1976) *Electrocrystallization of Metals from Molten Salts*, Nauka, Moscow.
11. Kuznetsova, S.V., Glagolevskaya A.L., and Kuznetsov, S.A. (1989). Effect of electrolysis parameters and the anionic composition of the electrolyte on the roughness of the hafnium coatings, *Zh.Prikl.Khim.* **63**, 536-539.
12. Glagolevskaya, A.L., Kuznetsova, S.V., and Kuznetsov, S.A. (1989). Corrosion of hafnium in chloride and mixed chloride-fluoride melts, *Ibid* **63**, 2673-2677.
13. Cohen, U. (1981). High rate of electrodeposition of niobium from molten fluoride using periodic reversal steps and the effect on grain size, *J.Electrochem.Soc.* **128**, 731-740.
14. Saltykova, N.A., Kotovskii, S.N., and Portnyagin, O.V. (1990). Effect of the reversible current on the structure of iridium deposits during the electrolysis of a molten chloride, *Elektrokhimiya* **26**, 338-342.
15. Kuznetsova, S.V., Glagolevskaya, A.L., and Kuznetsov, S.A. (1992). Electrodeposition of hafnium coatings from molten NaCl-KCl-$HfCl_4$ mixture using direct and reverse currents, *Rasplavy* **6**, 29-35.
16. Kuznetsova, S.V., Glagolevskaya, A.L., and Kuznetsov, S.A. (1990). Alloy formation in electrodeposition of hafnium from a melt on steel and graphite supports, *Zh.Prikl.Khim.* **63**, 2374-2377.
17. Kuznetsov, S.A., Kuznetsova, S.V., Polyakov, E.G., and Stangrit, P.T. (1990). Alloy formation studies during hafnium electrodeposition on a copper cathode in molten salts, *Elektrokhimiya* **26**, 815-818.

RHENIUM ELECTROCHEMISTRY, CHEMISTRY AND ELECTRODEPOSITION FROM MOLTEN SALTS

S.A.Kuznetsov[1], A.B. Smirnov [2],
A.N. Shchetkovsky[2], A.L. Etenko [2]
1 - Institute of Chemistry Kola Science Centre
14 Fersman Str., Apatity, 184200 Russia
2 - Optimfinish Ltd, 2 Kosmonavtov Str. Rostov-on-Don, 344092 Russia

1. Introduction

There are only a few studies on the electrochemical behaviour of rhenium in molten salts [1-6]. Most investigations dealt with electrodeposition of rhenium and its alloys. The absence of reliable data on the electrochemical reduction of rhenium in molten salts greatly impedes further progress in the field of rhenium galvanoplastics and hinders interpretation of the experimental results.

2. Experimental

For our experiments, we used the linear sweep voltammetry technique supported by a PI-50.1.1 potentiostat LKD-4-003 registrator, and a PO-5122 oscillographic polarograph. The potential scan rate varied between $5x10^{-3}$ and 4.0 V/s.The melt was contained in an ampoule made of SU-2000 glassy carbon that also served as the counter electrode. This container was placed into a hermetically sealed flask of stainless steel (Kh18N10T). The volammetric curves were taken with respect to the silver reference electrode on platinum, glassy carbon, and molybdenum electrodes of 0.8 to 2.0 mm diameter. Alkali chlorides of analytic grade were prerecrystallized and melted in argon gas atmosphere. Sodium fluoride was obtained by the directed crystallization technique. Immediately before starting the experiments, the salt was placed into the glassy carbon container of the electrochemical cell. The background electrolyte in the cell was prepared by stepwise heating in a vacuum to a temperature that was 50-70 degrees lower than the melting point. Then the cell was filled with argon gas and the temperature was raised to the specified level.

Prior to entering the electrochemical cell, the argon gas was passed through a titanium tube heated to 1073K. The quality of prepared background melts where the principal impurities, as a rule, are the O^{2-} and OH^- anions, was monitored by obtaining current-voltage curves on the glassy carbon electrode.

Rhenium metal was introduced into the molten salt either by the direct chlorination of the metal or with the aid of K_2ReCl_6 and K_2ReF_6 salts presynthesized according to [7].

The electrolytic products deposited on the cathode were identified using a DRON-2 diffractometer. Infrared spectra of hardened melts were obtained with the help of a UR-200

D.H. Kerridge and E.G. Polyakov (eds.), Refractory Metals in Molten Salts, 219-226.

and a Specord M-60 instruments.

3. Results and discussion

Cathodic voltammograms, obtained for the purpose of studying the electrochemical behaviour of rhenium in the NaCl melt (the necessary rhenium concentration was secured by the direct chlorination of rhenium metal in sodium chloride), display a well-reproducible distinct peak (Fig.1). The differential volammetric curve also contains only one wave (Fig.2). The cyclic voltammogram include one distinct peak attributable to the reduction of rhenium complexes, the corresponding peak of the metal electrooxidation, and the wave of chlorine oxidation (Fig.3). Potentiostatic electrolysis at the potentials of this wave results in the deposition of metallic rhenium at the cathode. We investigated the dependences of the peak current (I_p) and the peak potential (E_p) on the polarization rate and the rhenium concentration in the melt. The experimental data obtained leads to the following conclusions. 1. The values of $I_p/V^{1/2}$ remain unchanged at polarization rates of up to 1.0 V/s. At polarization rates $V \geq 1.0$ V/s the value $I_p/V^{1/2}$ begins to drop. 2. The peak potential did not change with an increase in the scan rate, and only at $V \geq 1.0$ V/s it shifted slightly in the negative region. 3. The

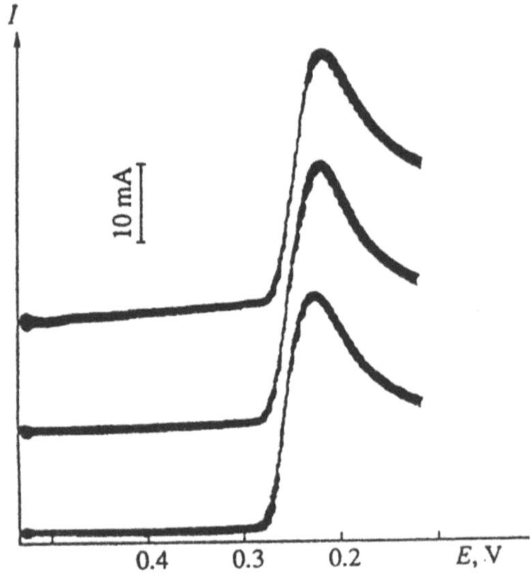

Figure 1. Voltammograms for NaCl-$ReCl_3$ melt (C_{Re}=3.27 x 10^{-5} mol /cm^3; V = 0.5 V/s; T = 1123 K; S = 0.196 cm^2). Curves are shifted along the current axis; they were obtained at 30-s intervals, E is given with respect to silver reference electrode.

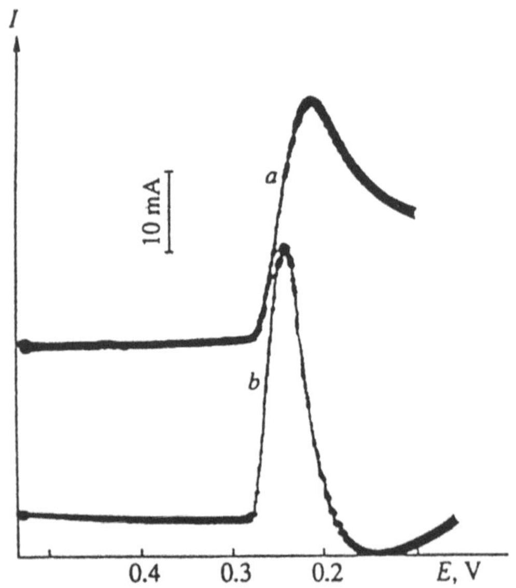

Figure 2. Voltammetric curves for NaCl-$ReCl_3$ melt (C_{rc} = 3.27 x 10^{-5} mol/cm^3; V=0.5V/s; T=1123K; S=0.196 cm^2): (a) integral; (b) differential.

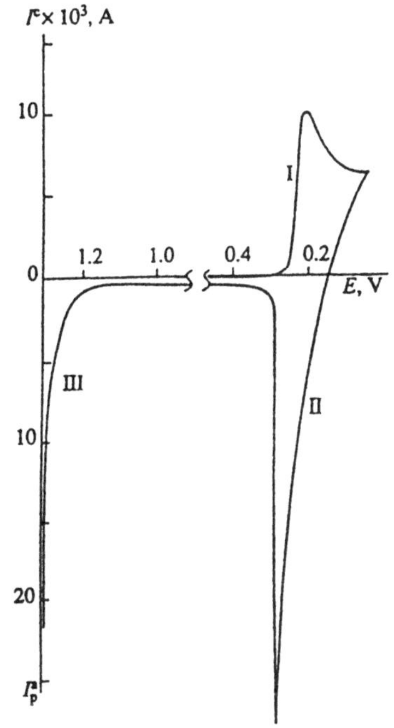

Figure 3. Cyclic voltammogram for NaCl-$ReCl_3$ melt ($C_{Re} = 2.34 \times 10^{-5}$ mol/cm^3; V = 0.5 V/s; T = 1123 K; S = 0.0847 cm^2): (1) Re(III) + 3e → Re; (II) Re - 3e - Re(III); (III) Cl^- - e → 1/2Cl_2.

peak current was proportional to the rhenium concentration in the melt. 4. Upon increasing the rhenium concentration, the peak potential shifted in the positive region.

The above diagnostic criteria derived by Nicholson and Shain [8] make it possible to conclude that, at polarization rates of 1.0 V/s, the reversible reduction of rhenium complexes to the metal phase occurs at the cathode, while at scan rates of 1.0 V/s, the discharge process becomes quasi-reversible.

Thus, an analysis of voltammetric curves (Figs. 1 to 3) points to the fact that only Re(III) complexes are present in the NaCl melt, and that the cathodic reduction process can be written as:

$$ReCl_6^{3-} + 3e \rightarrow Re + 6Cl^- \qquad (1)$$

The results obtained are in agreement with the data of Drobot et al. [9], who showed, using thermal, chemical, and X-ray analysis, that the sodium chloride melt contains only compounds of Re(III)..

The chlorination of rhenium in the KCl and CsCl melts, according to our studies, leads to the formation of Re(IV) complexes in these melts. The cathodic voltammograms shows two reduction peaks (Fig.4). Identical peaks were observed when we introduced the presynthesized K_2ReCl_6 salt into these melts.

Another investigation [5] showed that the reduction of Re(IV) complexes in the KCl and CsCl melts occurs via two successive stages which can be written in the following way:

$$ReCl_6^{2-} + e \rightarrow ReCl_6^{3-} \quad \text{reversible,} \qquad (2)$$

$$ReCl_6^{3-} + 3e \rightarrow Re + 6Cl^- \text{ reversible} \qquad (3)$$

in KCl; irreversible in CsCl.

Interaction of rhenium metal with KCl-K_2ReCl_6 and CsCl-Cs_2ReCl_6 melts at a concentration of (2-8) x 10^{-5} mol/cm^3 causes a change in the shape of voltammograms. Peaks I and II increase in height, and peak I shifts into the anodic region, so that only peak II remains in the cathodic region. This points to the fact that, when the interaction proceeds via the reaction:

$$3Re(IV) + Re \Leftrightarrow 4Re(III), \qquad (4)$$

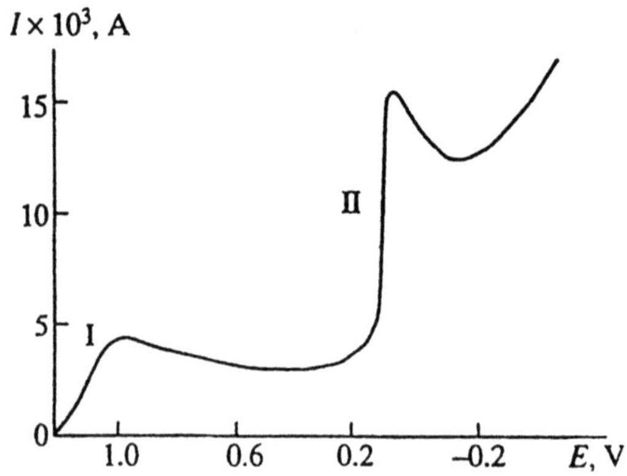

Figure.4. Volammetric curve obtained in CsCl-$ReCl_4$ melt (C_{re} = 3.42x10^{-5} mol/cm^3; v = 0.5 V/s; T = 1023 K; S = 0.236 cm^2).
(1) Re(IV) + e → Re(III) (II) Re(III) + 3e → Re

virtually only Re(III) complexes are present in the melt, which is in equilibrium with the metal. We have determined the diffusion coefficients D for Re(III) complexes.

Coefficients of Re(III) diffusion in chloride alkali metals in the form of the dependence

$$\log D = -A - B/T \pm \Delta \qquad (5)$$

are listed in Table 1.

Our results, as well as numerous experimental data [10], demonstrate that the diffusion coefficients are reduced upon going from sodium chloride to cesium chloride.

TABLE 1.Diffusion coefficients for Re(III) complexes in alkali chlorides

Melt	A	B	Δ	Dx10^5, cm^2/s at 1100 K	U, kJ/mol
NaCl	2.51	1987	0.04	4.83	38.0
NaCl-KCl	2.41	2167	0.02	4.07	41.5
KCl	2.39	2242	0.04	3.73	42.9
CsCl	2.27	2637	0.03	2.15	50.5

At 1100 K, the dependence of the Re(III) diffusion coefficient on the reciprocal cation radius ($\overset{o}{A}$) is given by

$$D = [-1.84 + (6.78 \pm 1.08)/r \pm 0.07] \times 10^{-5}\ cm^2/s \qquad (6)$$

with a correlation coefficient of about unity.

For the investigation of the rhenium electrochemical behaviour in chloride-fluoride melts, we used NaCl-NaF electrolyte of eutectic composition. Rhenium was introduced into the melt in the form of K_2ReCl_6 compound. The presence in such a melt of rhenium as a constituent of fluoride complexes is due to the exchange reaction:

$$ReCl_6^{2-} + 6F^- \Leftrightarrow ReF_6^{2-} + 6Cl^-, \quad (7)$$

because the IR spectrum of the solidified melt contains absorption bands characteristic of ReF_6^{2-} complexes.

The polarization curve for such a melt displays two peaks (Fig.5). The potentiostatic electrolysis at the first-wave potentials yielded no solid phase at the electrode, while the electrolysis at the second-peak potentials led to the deposition of rhenium metal. The mechanism of the Re(IV) electroreduction in a chloride-fluoride melt involves, therefore, two stages. An analysis of the first wave by means of the above criteria shows that the peak potential is independent of the polarization rate while I_p^I is proportional to $V^{1/2}$. The ratio of the heights of the cathodic and anodic peaks is close to unity. This fact, in conjunction with the absence of hysteresis in the cyclic stationary voltammograms testifies to the reversibility of the recharge process. The peak current I_p^{II} of the second stage is proportional to $V^{1/2}$, while the peak potential E_p^{II} becomes more and more negative with increasing potential scan rate. This means that the second stage is controlled by the charge transfer. We calculated the number of electrons for the first stage and determined αn_α using the Matsuda-Ayabe equation [11]:

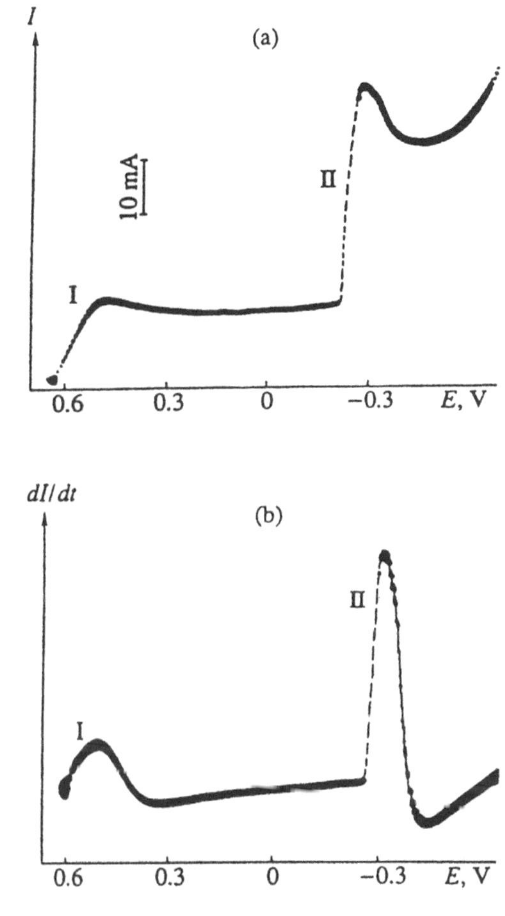

Figure 5. Integral (a) and differential (b) voltammograms obtained in NaCl-NaF melt of eutectic composition ($C_{Re} = 6.12 \times 10^{-5}$ mol/cm^3; V = 0.5 V/s; T = 1023 K; S = 0.293 cm^2).
(I) Re(IV) + e → Re(III); (II) Re(III) + 3e → Re.

$$E_{P/2} - E_p = 1{,}857\ RT/\alpha n_\alpha F \quad (8)$$

Experimental and calculated data on the electroreduction of rhenium complexes are tabulated in Table 2.

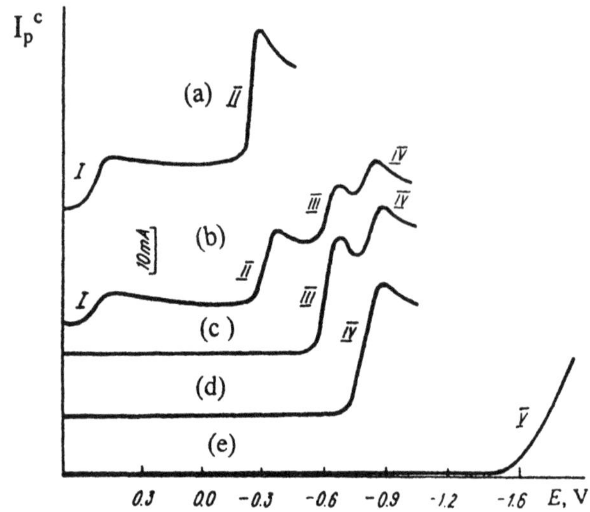

Figure 6. Transformation of volammograms obtained in NaCl-NaF melt of eutectic composition upon introduction of oxygen anions into the melt (C_{Re} = 5.63 x 10^{-5} mol/cm^3; V = 0.25 V/s; T = 1023 K). Mole ratio $[O^{2-}]/[Re]$: (a) 0: (b) 0.32; (c) 0.67; (d) 1.33; and (e) 2.3.

The addition of sodium fluoride in the NaCl and NaCl-KCl melts containing Re(III) chloride complexes causes rhenium to disperse in the bulk melt according to reaction:

$$4ReCl_6^{3-} + 18F^- \Leftrightarrow 3ReF_6^{2-} + Re + 24Cl^- \qquad (9)$$

Concurrently, two peaks appear in the cathodic curve. Thus, the introduction of fluorine anions in the NaCl-$ReCl_3$ melt leads to the emergence of thermodynamically stable complexes ReF_6^{2-} that discharge to the metal phase via a two-stage process. We determined the diffusion coefficients of Re(III) complexes in the NaCl-KCl melt, which is characterized by the mole ratio Re/F = 1:6 under the conditions of an equilibrium with metallic rhenium. These may be described by the equation:

$$\log D = -2.22 - 2462/T \pm 0.03 \qquad (10)$$

TABLE 2. Parameters describing electroreduction of Re(IV) complexes in NaCl-NaF melt at T=1023 K, C = 6.12 x 10^{-5} mol/cm^3, S = 0.293 cm^2

V, V/s	E_p^I, V	$E^I_{p/2}$,V	$E_{p/2}^I$-E_p^I,V	-E_p^{II},V	-$E_{p/2}^{II}$,V	$E_{p/2}^{II}$-E_p^{II},V	n	αn_α
0.125	0.447	0.623	0.176	0.284	0.214	0.070	0.91	2.34
0.250	0.449	0.625	0.176	0.302	0.230	0.072	0.91	2.28
0.500	0.453	0.634	0.181	0.315	0.241	0.074	0.93	2.22
1.000	0.448	0.654	0.206	0.329	0.252	0.077	1.06	2.13
2.000	0.450	0.653	0.203	0.344	0.266	0.078	1.05	2.10

The addition of oxygen anions (in the form of highly soluble calcium oxide or lithium and sodium oxides) in the NaCl-NaF eutectic melt containing rhenium leads to a decrease in the height of waves corresponding to the discharge of rhenium from the fluoride complexes, while at more negative potentials, two extra waves appear (Fig. 6a and 6b).

When the mole ratio of oxygen anions and rhenium is close to the stoichiometry of the reaction:

$$3ReF_6^{2-} + 2O^{2-} \Leftrightarrow 2ReOF_5^{-} + Re + 8F^{-}, \quad (11)$$

waves I and II cannot be observed, and the voltammograms display only waves III and IV with increased heights (Fig. 6 c). Reaction (11) is confirmed by the rhenium dispersion in the bulk melt and by the IR spectra of the hardened melt. The spectra indicate the presence of the $NaReOF_5$ compound in the melt [12] and reveal no traces of fluoride or dioxofluoride rhenium complexes [12,13]. This means that the introduction of oxygen anions into a chloride-fluoride melt, which contains Re(IV) complexes, leads to the stabilization of Re(VI) complexes in the melt.

Further addition of oxygen anions results in the formation of dioxofluoride complexes of rhenium:

$$ReOF_5^{-} + O^{2} \Leftrightarrow ReO_2F_4^{2-} + F^{-}. \quad (12)$$

These can be identified due to the appearance of three groups of doublets in the IR spectra [13,14]. The formation of dioxofluoride complexes causes changes in the voltammograms, namely, wave III disappears while wave IV is enhanced (Fig. 7d). Therefore, waves III and IV can be regarded as corresponding to the discharge of monoxofluoride and dioxofluoride complexes. On the other hand, the presence of wave IV, when only monoxofruoride complexes are present in the melt, is connected with the appearance of oxygen anions in the near-electrode layer due to the discharge of monoxofiuoride complexes and the exchange reaction (12).

Upon further increase in the oxygen anion content in the melt the polarization curve takes on the shape that is characteristic of the background curve (Fig. 6e). Such a shape of the voltammogram may be attributed either to the absence in the melt of soluble rhenium complexes (rhenium happens to accumulate in the bottom portion as a constituent of insoluble compounds) or to the formation of the complexes whose electrochemical activity becomes noticeable at the potentials that are comparable with the decomposition potentials of the supporting electrolyte. An analysis of the samples from the centre portion of a solidified-melt ingot yields rather unexpected results at first glance - it shows that the rhenium concentration in the electrolyte increased. Infrared spectra display only absorption bands that are characteristic of the compound $NaReO_4$ [14]. The results obtained demonstrate that perrhenates are formed in the melt, and that their formation in the presence of a background chloride-fluoride melt is accompanied by the disproportionation reaction:

$$7ReF_4O_2^{2-} + 10\,O^{2-} \Leftrightarrow 6ReO_4^{-} + Re + 28F^{-}. \quad (13)$$

It is the presence of finely divided rhenium particles in the bulk of solidified electrolyte that might lead to higher rhenium concentrations during a chemical analysis.

On the basis of the electrochemical studies and investigation [15] promising compositions $CsCl$-$ReCl_4$ and $NaCl$-NaF-K_2ReCl_6 for rhenium coating electrodeposition were chosen. Rhenium coating up to 2 mm from electrolyte $CsCl$-$ReCl_4$ (C_{Re} = 2.0 w/o) and NaCl-NaF (eutectic) - K_2ReCl_6(C_{Re} = 4.0 w/o) with using rhenium anode at cathodic current densities from 0.01 to 0.1 A/cm^2 on graphite and noble metal substrates were produced.

4. Conclusions

We have studied the electrochemical behaviour of chloride, fluoride, and oxofluoride compounds of rhenium and established the conditions for the stabilization of Re(III), Re(IV), Re(VI), and Re(VII) complexes. For a more rigorous elucidation of interrelations between the redox potentials, the composition of melts, and the oxidation state of rhenium, it is necessary to undertake spectroelectrochemical investigations. Rhenium coatings with a thickness up to 2 mm on graphite and noble metal substrates were produced.

5. Acknowledgments

The author[1] is grateful to the Russian Foundation for Basic Research (grant 96-03-32175) for the financial support of this study.

6. References

1. Bailey, R.A., and Nobile, A.A. (1972) Electrochemical studies on rhenium in fused LiCl-KCl eutectic, *Electrochemical Acta* **17**, 1139-1149.
2. Baraboshkin, A.N., Bychin, V.P., and Vinogradov-Zhabrov, O.N. (1978) Electrochemical behaviour of rhenium in tungstate melt, *Elektrokhimiya* **14**, 155-159.
3. Stepanov, A.D., Shkol'nikov S.N., and Ezrokhina, A.M. (1985) Eguilibrium of rhenium in the eqimole melt potassium chloride and sodium chloride, containing chloride of rhenium, *Izv. Vyssh. Uchebn.Zaved. Tsvetn. Metall.* **5**, 65-67.
4. Affoune, A.M., Bouteillon, J., and Poignet, J.C. (1990) Study of some rhenium electrochemical properties in the fused LiF-NaF-KF eutectic, *J.Electrochem. Soc.* **137**, 172 C.
5. Kuznetsov, S.A. (1994) Electroreduction of rhenium complexes in molten halides and oxohalides, *Russ. J.Electrochemistry* **30**, 1462-1469.
6. Affoune A.M., Bouteillon, J., and Poignet, J.C. (1995) Electrochemical behaviour of perrhenate ions in molten alkal fluorides, *J.Appl. Electrochem.* **25**, 886-889.
7. Braner G., Fehlhammer, W.P., Glemser, O., et.al. (1981) *Handbuch der Praparativen Anorganischem Chemie* **5**, Ferdinand Enke Verlag, Stuttgart.
8. Nicholson, R.S. and Shain, I. (1964) Theory of stationary electrode polarography, *Anal.Chem.* **36**, 706-723.
9. Drobot D.V., Korshunov, B.G., and Sharkadi, B. (1971) Eguilibrium in the systems, containing chlorides Re(III, IV) Izv. *Vyssh. Uchebn.Zaved. Tsveth. Metall.* **3**, 82-84.
10. Komarov V.E., Borodina N.P., and Pakhnutov, I.A. (1986) Eguations for determination of coefficients diffusion of ions in molten chlorides of alkali metals, *Elektrokhimiya* **22**, 478-482.
11. Matsuda, H., and Ayabe, Y. (1955) Zur theorie der Randles-Sevcikschen kathodenstrahl polarographie, *Z. Elektrochem.* **59**, 494-503.
12. Kuhlmann, W., and Sawodny, W. (1977) Vibrational spectra and force constants of some octahedral fluoro- and oxofluorocomplexes of rhenium, *J.Fluorine Chem.* **9**, 341-357.
13. Yagodin, G.A., Dudin, A.S., Rakov, E.G., and Opalovskii, A.A. (1980) Dioxotetra-fluororennates of alkali metals, *Zh. Neorg. Khim.***25**, 170-173.
14. Woodward, L.A., and Roberts, H.L. (1956) The raman and infra-red absorption spectra of osmium tetroxide. Relation to the structure of the perrhenate and tungstate ions in aqueons solutions, *Trans. Faraday Soc.* **52, 615-619.**
15. Kuznetsov S.A. (1997). Revessibility-irrevesibility of processes of electroreduction of complexes refactory metals in molten salts and structure of cathodic deposits, *Russ.J.Appl.Chem.* **70**, 64-72.

OXIDATION OF NIOBIUM IN THE MOLTEN EUTECTIC $NaNO_3$ - KNO_3

YURKINSKY V.P., FIRSOVA E.G., KRASOVITSKAYA O. A.
St Petersburg State Technical University.
Department of Technology and Materials Science.
29, Polytechnicheskaya St., St Petersburg, 195251, Russia.

1. Introduction

Although anodizing in aqueous solutions constitutes an important and long-established industrial metal finishing process, the application of molten salts in this field represents a recent innovation. The nitrate melts of alkali metals are the most promise for anodizing of metals. Improved dielectric properties, enhanced corrosion resistance and better adhesic bonding have been claimed to result from anodizing in these nonaqueous media as compared with aqueous electrolytes. First and foremost it is concerned with processes of electrochemical oxidation of valve-metals, such as tantalum, aluminium, titanium, niobium [1-3].

Previously we reported results of the study of mechanism and kinetics of anodic oxidation of titanium, tantalum, tungsten and molybdenum in nitrate melts [4].

In the present investigation, the kinetics of growth of anodic oxide films on niobium were studied in molten eutectic $NaNO_3$ - KNO_3 (50 m/o) over the temperature interval 530 650 K.

2. Experimental

Previously the eutectic mixture of sodium and potassium nitrates was carefully dehydrated under vacuum at 670 K during 8-12 hours.

Experiments were carried out in quartz tube under nitrogen atmosphere.

Specimens under study were cut from sheet of niobium. The surface of the specimens was treated by electrochemical polishing in the mixture of H_2SO_4 (86-92 wt %) - HF (1.5-4 wt %) - H_2O (5.5-10 wt %) at temperatures of 288 to 323 K and at voltages of 9 to 15 V.

The oxidation were carried out at the constant current stage (C.D. range 10-70 $A\,m^{-2}$) as well as at the constant voltage stage (voltage was variable from 10 to 30 V). The electrochemical cell was comprised of three electrodes. The working electrode was a specimen of niobium (surface $2\cdot10^{-4}$ m^2), the counter electrode was a graphite rod (surface $6\cdot10^{-4}$ m^2) and the reference electrode was an Ag/Ag (I) (0.07 M).

D.H. Kerridge and E.G. Polyakov (eds.), Refractory Metals in Molten Salts, 227-230.

The thickness of oxide films was determined by using the capacitance measurements and ellipsometry.

The phase constitution of oxide layer was established by X-ray diffraction.

3. Results and Discussions

It is found experimentally that at the anodic oxidation of niobium under constant current stage the shape of the potential versus time plots at various current densities was linear. Therewith the potential increased with time as the films grew.

It is shown that the rate of the growth of anodic potential ($d\varphi/dt$) is given by the empirical equation adequately

$$\frac{d\varphi}{dt} = n \cdot i^{m} \qquad (1)$$

where j is the current density and n, m are constants. The values of n and m are given in Table.

The study of the current efficiency (η) at constant current was shown that η is less than 100% and decreases monotonously with a rise in temperature. These results suggest that passing through the oxide film current presents the sum of ionic ($j\eta$) and electronic currents. The share of electronic current increases with a rise in temperature. The values of η are given in Table.

The ionic current density ($j \cdot \eta$) depends on the electric field strength (E) through the ionic transport equation [5,6]

$$j \cdot \eta = j_o \cdot \exp(-W/k \cdot T) \cdot \exp(q \cdot a \cdot E/k \cdot T) \qquad (2)$$

W is the activation energy for the ionic transport, q - the ionic charge, a - the activation distance and j_o - the constant, depending the nature of metal. The rest of the symbols have the usual meaning.

Experimental results on the oxidation of niobium showed that the relation between $\ln(j\eta)$ and E was linear as predicted by Eq.2. By using Eq.2 the main kinetic parameters (a, W, j_o) were calculated. Their values are given in Table.

TABLE . Kinetic parameters of anodic oxidation of Nb in molten eutectic $NaNO_3$ - KNO_3.

T,K	n	m	η, %	a, nm	j_o, $A \cdot m^{-2}$	W, eV
538	-5.63	1.02	86±7	0.36±0.03	$8.56 \cdot 10^9$	1.18±0.16
568	-6.11	1.14	75±6	0.36±0.03	$8.56 \cdot 10^9$	1.18±0.16
603	-6.35	1.25	67±6	0.36±0.03	$8.56 \cdot 10^9$	1.18±0.16
621	-6.92	1.41	54±5	0.36±0.03	$8.56 \cdot 10^9$	1.18±0.16
653	-7.86	1.79	37±4	0.36±0.03	$8.56 \cdot 10^9$	1.18±0.16

The kinetics of growth of anodic films on niobium was also studied at a constant voltage stage.

It was established that the thickness of oxide films (δ) varied in direct proportion to the voltage of oxidation (V_{ox}) according to the equation

$$\delta = \alpha \cdot V_{ox} \qquad (3)$$

where α is coefficient of proportionality.

The coefficient α increases with the rise in temperature. At the anodization of niobium during 15 minutes the temperature dependence of α is linear and is given by the empirical relation

$$\alpha = 0.1313 \cdot T - 64.001 \qquad (4)$$

The coefficient of correlation r was equal 0.993.

Study of current density versus time curves, obtained at the anodization of niobium in the constant voltage stage, show the availability of the dependence

$$j = A \cdot t^{B} \qquad (5)$$

It is known that if passing through the oxide current is caused by a migration of ions in the electric field the index of degree B is equal -1.0. Essentially in the case of niobium coefficient B was variable from (-0.76±0.05) at 538 K to (-0.43±0.04) at 653 K. Thus we can conclude that the main reason of $\eta<100\%$ at the oxidation in molten nitrates is considerable contribution of electronic current. Based on the obtained values of B it may be deduced that the share of electronic current increases with the rise in temperature. These results are in excellent agreement with the evidence from galvanostatic measurements.

The availability the electronic current during the process of oxidation of niobium is responsible for side reaction on the surface of anode [7,8]

$$NO_3^- = NO_2 + \frac{1}{2} O_2 + \bar{e} \qquad (6)$$

which is associated with electro-decomposition of nitrate melt. The formation of NO_2 on the niobium surface was observed visually.

The formation of oxide Nb_2O_5 was established by X-ray diffraction at the anodic oxidation of niobium just as at the constant current stage so at the constant voltage stage.

4. References

1. Inman, D., and Lovering, D.G. (1983) Electrochemistry in molten salts, *Comprehensive Treatise Of Electrochemistry* 7, New York, London, pp..593-640

2. Nigam, R.K., and Chaudhary, R.S. (1970) Steady state kinetics of formation of anodic oxide films on tantalum in molten electrolytes of $LiNO_3$ - $NaNO_3$ - KNO_3, *Ind. J. Chemistry* **8**, 343-346
3. Yurkinsky, V.P., Vorobyeva, N.B., Morachevsky, A.G., Ivanov, A.E., and Riskina, E.Y. (1979) Oxidation of titanium in molten salts, *Prikl. Khim (Rus.)*. **52**, 241-245
4. Yurkinsky, V.P. (1988) Anodic oxidation of metals in nitrate melts, *Rasplavy* **2**, 74-82
5. Gьntherschulze, A., and Betz ,H. (1952) *Electrolyt-kondensatoren*, Cram, Berlin
6. Young, L. (1961) *Anodic oxide films*, Academic Press Inc., New York
7. Swofford, H.S., and Laitinen, H.K. (1963) An electrochemical investigation of the reduction of nitrate in $NaNO_3$-KNO_3 eutectic at 250°C, *J. Electrochem. Soc.* **110**, 814-820
8. Triaca, W., and Arvia, A. (1965) The kinetics of anodic reaction occuring at electrolysis of molten nitrates on platinum electrode at different temperatures, *Electrochim. Acta* **10**, 409-427

EVALUATION OF HIGH-FREQUENCY SUPERCONDUCTIVITY OF NIOBIUM COATINGS PREPARED BY ELECTRODEPOSITION IN MOLTEN SALTS

V.N. KOLOSOV and E.S. MATYCHENKO
Institute of Chemistry, Kola Science Centre RAS,
184200, Apatity, Murmansk region, Russia

1. Introduction

To prepare high-frequency (hf) structures from niobium, massive turned bars are usually used [1]. The drawback to this process are great losses of the superconducting material and, consequently, high cost of the article. Moreover, it is difficult to produce intricately shaped structures by this method.

A more economical method of making hf structures involves electrochemical deposition from molten salts of a thick niobium layer of a sufficiently high mechanical strength onto a matrix reproducing the inner surface of the article followed by removing the matrix [2,3]. The aim of this work is investigation of physico-mechanical characteristics of niobium coatings obtained by electrodeposition to evaluate the hf superconductivity of the articles which can be made on their basis.

2. Theoretical Aspects and Requirements for the Superconducting Material

The principal requirements for high-frequency cavities are the following: the capacity to withstand accelerating hf fields up to several MV/m, low power losses, the capability for long operating without deterioration of performance.

Theoretically, the maximum attainable accelerating field in the superconducting cavity is determined by a magnetic breakdown which takes place when the magnetic component of the hf field is equal to the critical field of the superconductor. However, in practice the breakdown is caused by other effects such as electronic load or, more often, a thermal runaway related to the formation of weak zones of superconductivity caused by defects and limited thermal conductivity of the cavity walls.

To achieve low power losses one needs a high quality factor Q of the cavity. Besides, the high Q value helps to solve the problem of limiting the thermal runaway, since the amount of heat removed from the cavity system in one cycle becomes smaller. Since

$$Q = \Gamma / R_s(\nu, T) \quad (1)$$

D.H. Kerridge and E.G. Polyakov (eds.), Refractory Metals in Molten Salts, 231-238.

(Γ is a factor depending on the specific cavity geometry and the operating mode, ν is frequency, T is temperature) then according to (1) the low surface resistance gives a high Q. Thus, the superconducting material used in a hf cavity should have low surface resistance and high thermal conductivity.

To determine the hf characteristics of the cavity material it is convenient to express the superconductor properties (surface resistance, critical fields, etc.) using three basic and easily measured values: the critical temperature T_c, the residual resistivity ρ_o and the Sommerfeld constant γ.

For niobium and its compounds at $T < 0.5\ T_c$, where for the mean free path of an electron l_o and the coherence length ξ_o the ratio $l_o >> \xi_o$ is true, these bonds may be written in a simple form [4]. Thus for the thermodynamic critical field H_c and for the field of magnetic breakdown H_{mb} they have the form:

$$H_c = 2.43\gamma^{1/2}\,T_c \qquad (2)$$

$$H_{mb} = 0.75\,H_c = 1.82\,\gamma^{1/2}\,T_c \qquad (3)$$

The magnetic field penetration depth λ in the same approximation at which equation (2) and (3) apply can be represented in the form

$$\lambda = c\,(\rho_o / T_c)^{1/2}, \qquad (4)$$

where c is a constant depending on the superconducting material. Surface resistance R_s of the superconductor can be split into two components:

$$R_s = R_{BCS} + R_o \qquad (5)$$

where R_{BCS} is the contribution to the resistance determined by the Bardeen-Cooper-Shrieffer theory of superconductivity, R_o-residual resistance. The relationship between resistance R_{BCS} and temperature can also be expressed by ρ_o and T_c:

$$R_{BCS} = A\,\nu^{\alpha}\rho_o^{1/2}\exp(-1.76\,T_c / T) \qquad (6)$$

where A and α are constants of the superconducting material. The second member R_o in equation (5) is resistance independent of temperature, but its origin has not been fully explained theoretically yet. The experimental data gives evidence that it is associated with the trapped magnetic flux and impurities. Another source of increasing R_o is imperfection of the superconducting surface. The roughness, however microscopic, may also result in a noticeable increase of the R_o value.

So a good superconducting hf cavity is characterised by low surface resistance R_s and high critical fields.To sum it up, in order to attain a high quality factor Q in superconducting hf structures it is necessary to use materials with high T_c and low ρ_o. On the other hand, the selection of the materials should be determined by the stability of the cavity structure in the process of operation. In this case the mechanical strength of the hf structure and its thermal conductivity play an important part.

3. Experimental

The coatings in question were obtained by electrochemical deposition in (Li-Na-K)F-K_2NbF_7 and (K-Na)Cl-K_2NbF_7 melts with a soluble anode. Their thicknesses were 5-200 μm. The current cathode density i_c was varied over the interval 50-1000 A/m^2. The deposition was conducted at 970-1070 K onto flat pieces (30x50 mm) and cylinders (∅ 6x60 mm) of copper and molybdenum substrates, which were separated from the coatings to be studied.The equipment, the reactants and preparation of the latter were basically the same as in earlier papers [5,6].

The structure of the sample cross sections was studied by the method of metallographic analysis. It included the investigation of the total structure of the coatings and determining their thickness.

The tensile strength of coatings σ_B at thicknesses δ = 50-200 μm were determined by the conventionial method of stretching till rupture at temperature 290 K. For stretching, a testing machine was used with a sensitivity to the change in length and load of ±5 μm and ±1 N, respectively. The rate of stretching was 0.3 mm/min. The σ_B value of coatings 5-40 μm thick was measured at temperatures 290 K and 77 K by the pneumatic method. The latter is described in detail in [6,7].

Specific resistivity ρ and critical temperature T_c were measured by the four-probe method. For the T_c value the point in the middle of the curve of transition to the superconducting state was taken. As a temperature sensor in the cryogenic zone a germanium thermometer of the VG-type was used. The values of the lower H_{c1} and the upper H_{c2} of the critical magnetic field at the temperature of 4.2 K were determined from differential curves of magnetization similarly to [8].

4. Results and Discussion

In studies of the coating mechanical characteristics the strongest niobium coatings were those obtained at high cathodic current densities i_c.Thus, on increasing i_c from 50-100A/m^2 to 1000 A/m^2 the tensile strength σ_B increases 2-fold (Fig. 1a). The increased tensile strength of the coatings is due to diminishing of the mean grain size and enhancement of the dislocation density N [9]. For ordinary metals increasing strength with growth of the dislocation density is observed up to the values 10^{13}-10^{14} cm^{-2}. Further increasing N results in intensive formation of microcracks reducing the metal strength. Since microcracking is mostly caused by point defects, in the case of electrolytical coatings the formation of microcracks initiated at N ≈ 10^{12} cm^{-2}. So, it can be suggested that over the entire i_c interval studied the density of dislocations in the resulting niobium coatings did not exceed 10^{13}-10^{14} cm^{-2}.

The fall in the strength of the coatings obtained at a constant cathodic current density with increasing thickness (Fig. 1b) is probably caused by two factors. Firstly, by increased mean grain size d_g, because the previous studies of electrolytic niobium coatings have demonstrated that the magnitude d_g is proportional to the layer thickness [10]. Secondly, the thinner the coating, the more defective it is and subjected to greater plastic deformation after electrolysis while cooling down from the temperature of the deposition to room temperature.

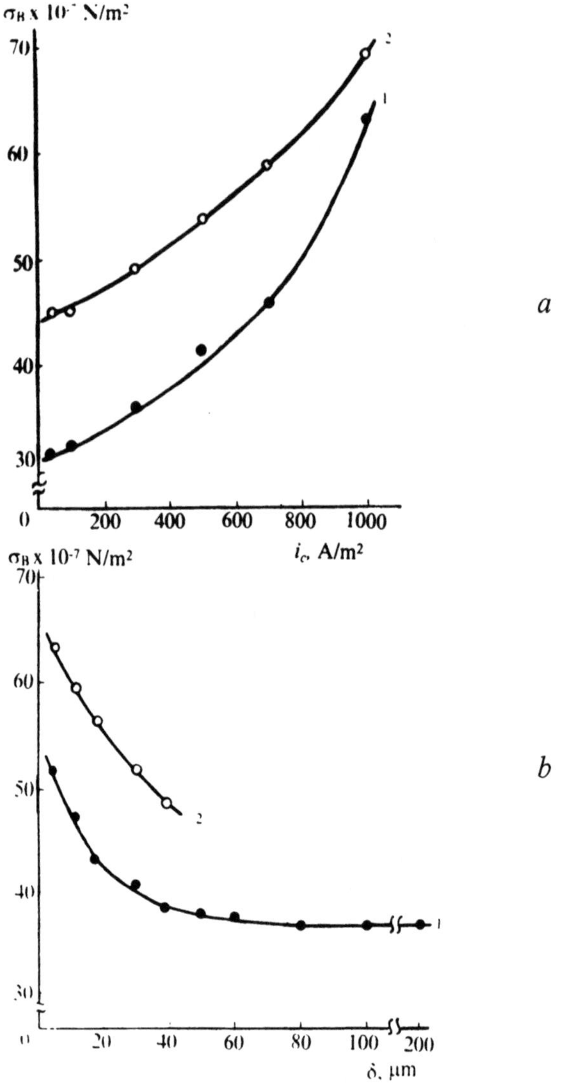

Figure. 1. The plot of the niobium coating tensile strength σ_B vs cathodic current density i_c*(a)* and thickness δ *(b)*. 1 - T = 290 K, 2 - T = 77 K.

One of the most sensitive indicators of impurities and other defects of the metal's crystalline lattice is the residual resistivity ratio:

$$RRR = \rho_r / \rho_o \tag{7}$$

where ρ_r is the sample resistivity at room temperature, ρ_o is residual resistivity at 4.2 K. The plot of the RRR value vs. current density for niobium coatings obtained in the fluoride melt is shown in Fig. 2. These data correlate well with the aforementioned strength characteristics of the samples. Because of numerous defects and smaller mean grain size the RRR value decreases as i_c increases.

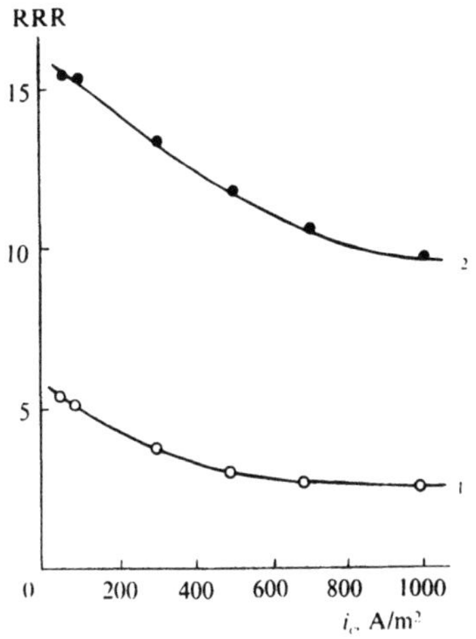

Figure.2. The plot of the RRR residual resistivity ratio of niobium coatings vs cathodic current density i_c: (1) 10 and (2) 100 μm.

In operation the hf structure becomes heated by the high-frequency field. Therefore for a stable operation of the superconducting cavity this heating and heat withdrawal must be balanced. In this connection thermal conductivity of the cavity walls is an important feature. The heat balance equation can be written in the following form [11]:

$$0.5\ \mu^2_0\ R_s(T,H)\ H^2 = k_t / d\ \Delta T \tag{8}$$

where μ_0 is magnetic permeability of vacuum, k_t is thermal conductivity coefficient of the superconducting material from which the cavity is made, d is the cavity thickness, ΔT is the difference in temperature of the inside and outside walls. Equation (8) shows that the higher the R_s surface resistance the higher coefficient of thermal conductivity k_t should be in order to withdraw the heat generated by the hf field. The thermal conductivity of the samples was evaluated by using the Wiedemann-Franz law. In pure metals at low temperatures almost the whole of the heat flow is transferred by electrons and therefore equation (9) applies [12]:

$$k_t = L\ T / \rho \tag{9}$$

where T is temperature, ρ is specific resistivity, $L = 2.45 \times 10^{-8}$ W Ω / K is the Lorenz constant. Fig. 3 represents the k_t dependences obtained from equation (9) for the best samples studied by us in the working interval of temperatures (curves 1,2). For comparison, curve 3 from [13] for pure niobium with RRR =113 is shown. It is evident that the thermal conductivity of electrolytical niobium obtained both in fluoride and fluoride-chloride melts is insufficiently high and these coatings cannot be used in hf structures immediately after electrolysis. To increase their thermal conductivity an extra purification from impurities is needed.

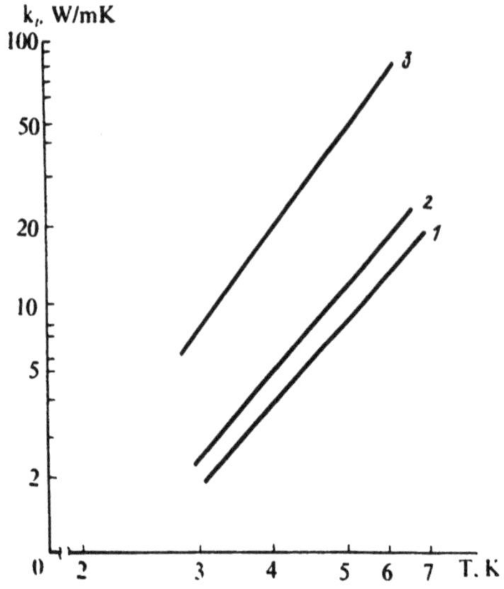

Figure.3. The plot of niobium coefficient of thermal conductivity k_t vs temperature;(1) niobium coatings produced in a fluoride melt; (2) niobium coatings produced in a fluoride-chlorde melt; (3) niobium used for hf cavities in [13].

Earlier [14] a ratio connecting the Ginsburg-Landau parameter χ of niobium with the value of specific electroresistivity ρ was obtained:

$$\chi = \chi_o + 2.37 x 10^6 \gamma^{1/2} \rho \tag{10}$$

where $\chi_o = 1.22$. Using this equation the χ value for electrolytical coatings at 4.2 K was found to change from 1.5 to 1.8 .With the help of ρ the mean free path of an electron in niobium can be evaluated [15]:

$$l_e = 3.7 x 10^{-12} \Omega\ cm^2 / \rho \tag{11}$$

The l_e value for electrolytical coatings at 4.2 K determined from equation (11) is 300-500 A.

Our measurements have shown that the critical temperature T_c of coatings obtained both in fluoride and fluoride-chloride melts depends neither on the electrolysis conditions nor the coating thickness and is 9.2±0.1 K . At the same time for the coatings obtained in both of the electrolytes the critical magnetic fields are observed to depend on the niobium coating thickness. These dependences are presented in Fig. 4. Using the above parameters of a superconducting material let us evaluate its hf characteristics. Firstly we calculate the fields of magnetic breakdown H_{mb}. For these coatings the H_{mb} values are in the range 1400-1600 Oe and equal to 70 % of the best values for niobium presently attained. The depth of penetration for the magnetic field λ for the coatings studied is 450-480 A.

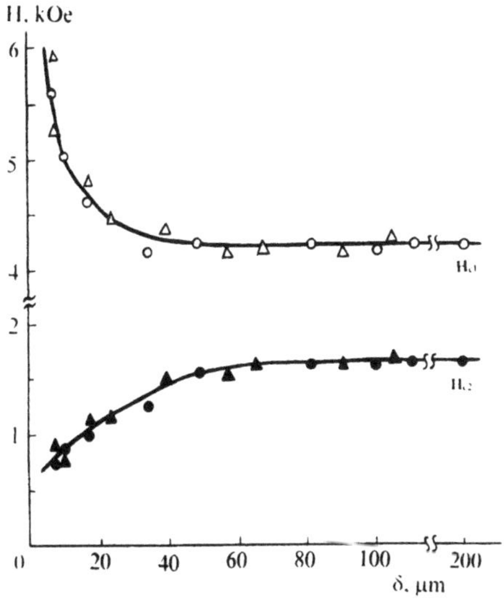

Figure. 4. Dependence of critical magnetic fields H_{c1} and H_{c2} of niobium coatings produced in a fluoride melt (▲,Δ) and a fluoride-chloride melt (●, o) on the layer thickness.

Fig.5 represents the calculated dependence $R_{BCS} = f(i_k)$ for niobium coatings produced from the fluoride melt. It was of interest to compare our results with the data obtained by other researchers. Earlier [16] the coatings for hf cavities were

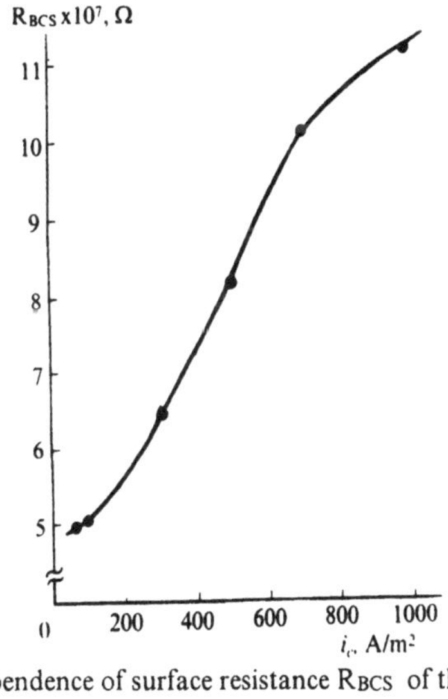

Figure.5. Calculated dependence of surface resistance R_{BCS} of the electrolytical coatings on current cathodic resistance i_c: T = 4.2 K, ν = 350 MHz.

also deposited and studied under conditions close to ours. The measurements showed that the quality factor Q of cavities obtained in [15] by the mode of E_{110} type was ~ 10^5, whereas it was ~ 10^6 with the mode of H_{011} type (ν = 9 GHz). Assuming $R_o = 0$ in equation (5) we evaluated the maximum possible Q for a cavity produced from the coating in question. It was 10^6-10^7. Actually with a real

R_o value the real system high-quality factor will be less by an order of magnitude and close to the values obtained in [16]. At the same time , the best Q values attained for niobium cavities are at present much higher than these values [17].

Thus, the determination of hf characteristics of electrolytical coatings has shown their inapplicability for real hf structures due to their poor performance immediately after the electrolysis. The low hf characteristics are caused by insufficient purity of the latter. Therefore, to be applied in the hf engineering, these coatings need further purification.

References

1. Kumar, P. (1988) High purity niobium for superconducting applications, *J. Less-Common Metals* **139**, 49-58.
2. Meyerhoff, P.W. (1969) Fabrication of niobium rf cavities, *J. Applied Physics* **40**, 2011-2014.
3. Kolosov, V.N. and Shevyryov, A.A. (1996) *Electrolytical superconducting materials (Russ.),* KSC RAS, Apatity.
4. Di Leo, Nigro, A., Nobile, G. and Vagilo, R. (1990) Niobium-titanium nitride thin films for superconducting rf accelerator cavities, *J. Low Temperature Physics* **78**, 41-50.
5. Kolosov, V.N. (1994) Physical properties of superconducting niobium films prepared by electrodepositing process in molten salts, *Programme and Abstracts of Papers Euchem Conference on Molten Salts,* Bad Herrenalb, Germany, p. A24.
6. Kolosov, V.N., Matychenko, E.S. and Novichkov, V.Yu. (1996) Properties of niobium coatings electrodeposited from fluoride melts, *Russ. J. of Electrochemistry* **32**, 1354-1359.
7. Kolosov, V.N. and Matychenko, E.S. (1996) Investigation of electrolytical copper coatings to be applied in a Nb_3Sn -Nb-Cu superconducting compositions, *Fizika i khimija obrabotki materialov(Russ.)* **3,** 23-29 .
8. Kolosov, V.N., Gel, R.P., Drobotenko, G.A., Karpenko, O.A. and Shevyryov, A.A. (1993) Magnetic shielding by means of superconducting electrolytical niobium coatings, in *The problems of Effective Utilisation of Mineral Raw Materials of the Kola Peninsula* (Russ.), KSC RAS, Apatity, pp 21-33.
9. Povetkin, V.V. and Kovarsky, A.V. (1989) *The Structure of Electrolytical Coatings (Russ.)*, Metallurgia, Moscow.
10. Gel, R.P., Shevyryov, A.A. and Belyaevskii, A.T. (1987) Relief and structure of electrolytical niobium coatings, in *Chemical and Technological Investigations of Raw Mineral from the Kola Peninsula (Russ.),* Leningrad, Nauka, pp103-107.
11. Didenko A.N., Sevryukova L.M. and Yatis A.A. (1981) *Superconducting Accelerating RF structures(*Russ.), Energoisdat, Moscow.
12. Rubalko, A.S. and Sterin, M.B. (1996) Electrical and thermal conductivity of the conical thermal contacts used at ultra low temperature, *Fizika Nizkikh Temperatur(Russ.)* **22,** 1095-1099.
13. Lengeler, H., Weingarten, W., Muller, G. and Piel, H. (1985) Superconducting niobium cavities of improved thermal conductivity, *IEEE Transaction on Magnetics* ***MAG*-21,** 1014-1017.
14. Goodman, B.B. (1962) The magnetic behaviour of superconducting of negative surface energy, *IBM J. Res. Development* **6**, 63-67
15. Goodman, B.B. and Khuhn, G. (1968) Influence des defaunts etendus sur les properties superconducories du niobium, *J. Physique* **29**, 240-252.
16. Litvinov, V.N. (1973) Superconducting niobium cavities with a galvanic coatings, *Trudy Radiotekhnicheskogo Instituta (Russ.)* **15**, 164-172.
17. Benveniti, C., Calatroni, S., Chiaveri, E., Orlandi, G. and Weingarten, W. (1994) Superconducting cavities for particle accelerations: achievements and problems, *Cryogenics* **34**, 57-63.

EVALUATION OF HIGH-FREQUENCY SUPERCONDUCTIVITY OF Nb_3Sn COATINGS OBTAINED BY JOINT ELECTROCHEMICAL DEPOSITION

V.N.KOLOSOV
Institute of Chemistry, Kola Science Centre RAS,
184200, Apatity, Murmansk region, Russia

1. Introduction

For making high-frequency (hf) structures from Nb_3Sn compounds the method of diffusion from the gaseous phase is conventionally used [1]. However, this method is unproductive and does not always yield high-quality working layers of superconductors. Therefore searching for alternative methods of producing Nb_3Sn coatings for operation at high frequencies is highly topical.

In earlier works [2,3] it was shown that electrodeposition combined with galvanoplastics can be used to make monophase high-purity Nb_3Sn coatings on intricately shaped articles with low surface roughness, which meet on the whole the requirements placed upon the hf structures. The main demands for the superconducting material for a hf cavity have been discussed in detail [4].

Moreover, for hf cavities with thin-film coatings the low magnetic field penetration depth λ is also a considerable advantage since the coating thickness should be $d \geq 10\ \lambda$ to avoid a significant hf field penetration into the substrate, and the low d values should improve the coating adhesion and the article heat conductivity. The aim of this work is studying the physico-mechanical characteristics of Nb_3Sn coatings to evaluate hf superconductivity of the articles prepared by this method.

2. Experimental Methods

The coatings in question were obtained by joint electrochemical deposition of the components onto molybdenum and niobium substrates. This method makes it possible to produce monophase coatings of the Nb_3Sn compound throughout the entire field of its existence in the phase diagram (18-25 at.%) [5]. The phase composition was determined using a DRON-2 diffractometer (Cu K_α radiation). The niobium to tin ratio in the alloy was determined using a X-ray quantometer. The coating continuity and homogeneity were studied with a microprobe analyzer Cameca. The structure of Nb_3Sn coatings cross section was studied by metallographic analysis, which determined the total structure of the coatings and measuring their thickness. Coatings of 5-20 μm thickness were investigated.

The measured superconducting characteristics included critical temperature T_c, critical magnetic fields H_{c1} and H_{c2}. Moreover, residual

D.H. Kerridge and E.G. Polyakov (eds.), Refractory Metals in Molten Salts, 239-244.

resistivity ρ_o was measured. The methods of measuring these values were described in detail [6,7].

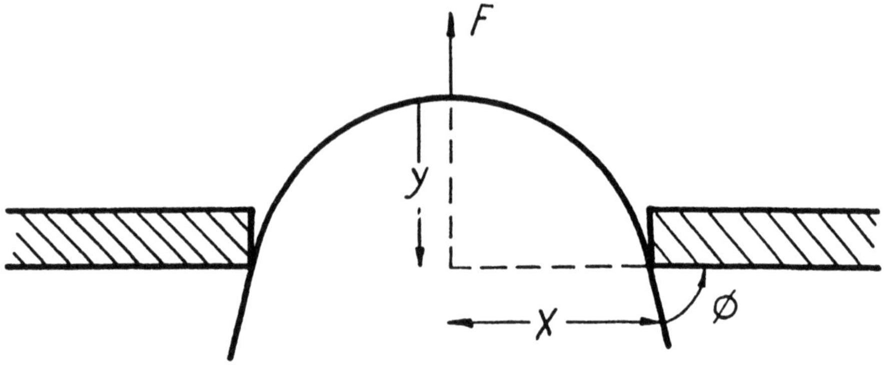

Figure 1. Schematic of experimental arrangement for measuring the Young's modulus of the coatings by the bending method.

The specific features of electrodeposited niobium stannide coatings consist in that they are brittle foils, therefore the methods of measuring the strength characteristics adopted for plastic materials are inapplicable. So to measure the Young's modulus E_{Nb3Sn} at room temperature the method of three- point bending was used [8], when the sample in the form of a symmetrical rectangular strip made of Nb_3Sn-Nb-Nb_3Sn composite was bent under the action of a force (Fig.1). In this case for an effective modulus of elasticity E_{eff} the following ratio is calculated:

$$E_{eff} = \frac{3F(x^2 + 1.88y^2)}{wt^3 \sin\phi} \quad (1)$$

where w and t are, respectively, width and thickness of the Nb_3Sn-Nb-Nb_3Sn composite. If the E_{Nb} value is known then the E_{Nb3Sn} value can be derived from the equation:

$$E_{Nb_3Sn} = \frac{E_{eff} - E_{Nb} d^3 / t^3}{1 - d^3 / t^3} \quad (2)$$

where d is thickness of the niobium layer in the sample.

3. Results and Discussion

For the investigations, samples of the Nb_3Sn coating with a tin concentration of 21-25 at.% were selected. This concentration range was selected because at C_{Sn} <21 at.% the critical temperature of the Nb_3Sn compound diminishes drastically [5]. The studies, carried out using the microprobe microanalyzer Cameca, have shown the coatings' continuity and homogeneity. Fig. 2 shows the experimental dependances of the critical temperature T_c and the upper critical field H_{c2} of niobium stannide on composition. It is evident that as the tin concentration is reduced in comparison with the stoichiometric composition, the H_{c2} value grows considerably. However, this growth is initially accompanied by reduction in the critical temperature. Therefore the selection of optimum composition of niobium stannide layers intended for work in hf structures should also be based on other characteristics on which the high-frequency properties of the superconducting material depend. As was shown in [4], one of such characteristics is residual resistivity ρ_o. Thc experimental plots of ρ_o vs tin concentration C_{Sn}. are given in Fig. 3.

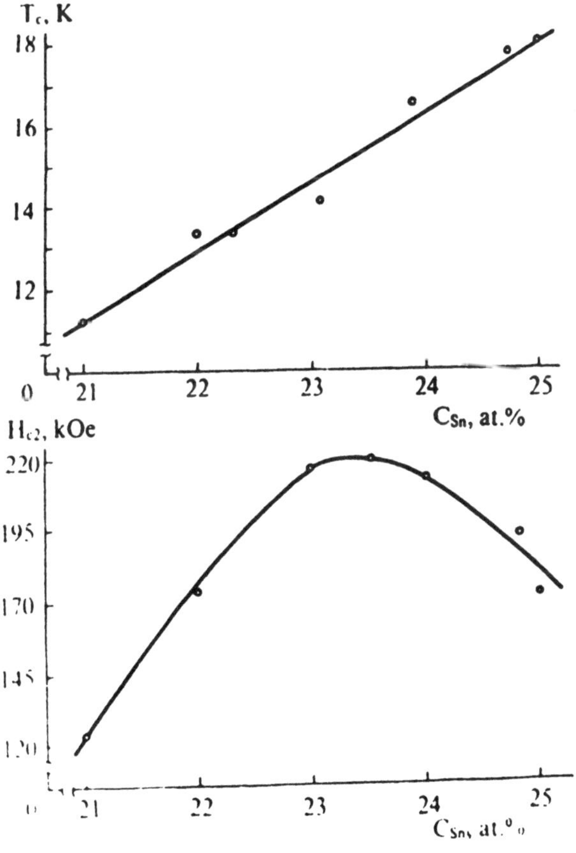

Figure 2. Experimental dependences of the critical temperature T_c and upper critical field H_{c2} of electrolytical niobium stannide on composition.

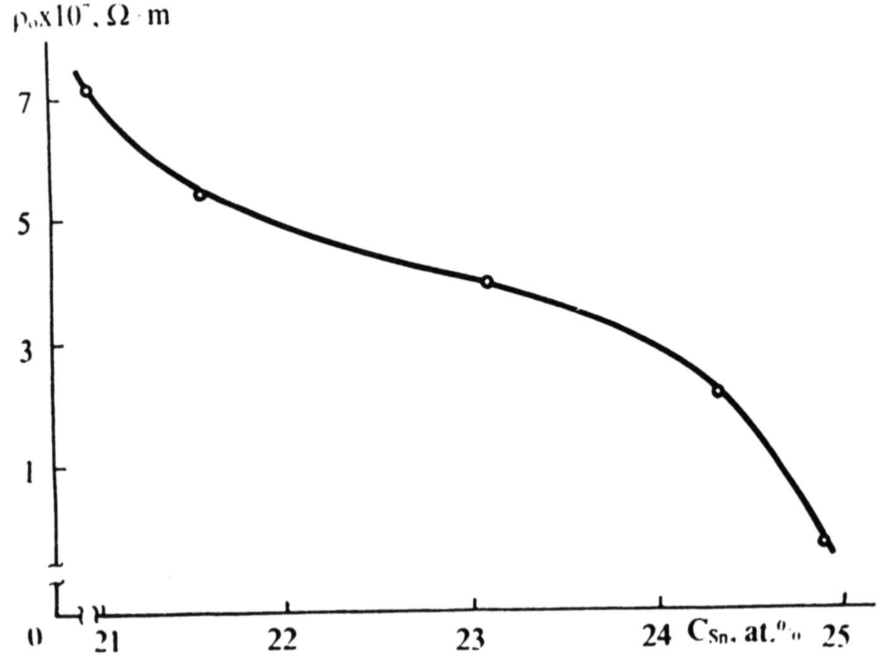

Figure 3. Experimental dependences of residual resistivity ρ_o on niobium stannide composition.

The above experimental findings were used to calculate the dependences of the magnetic field penetration depth λ and the field of magnetic breakdown H_{mb} for the coatings studied. It follows from the calculated data (Fig. 4) that the minimum λ value ~1200 A is true at a composition close to the stoichiometric one. At the same time the maximum H_{mb} value is observed at the tin concentration in the coating C_{Sn} ~ 24 at.%. To evaluate the coating optimum composition we calculated the dependence of the surface resistance $R_{BCS} = f(C_{Sn})$ which is given in Fig. 5. Thus, by comparing the dependences shown in Fig.3-5 it can be inferred that the best combination of the operating hf characteristics is observed at a tin concentration in the coating of 23.5-24.5 at.%.

For the optimum composition samples the Young's modulus was measured by the three-point bending method. For this purpose the Nb_3Sn coatings with the thickness d=10-15 μm deposited onto a niobium foil 50 μm thick were used. The selection of such a thickness was determined by the fact that at lower d the method sensitivity deteriorates, whereas at d>15 μm niobium stannide may crack on testing. Within the measurement error the E_{Nb3Sn} value was $(130\pm10)\times10^9$ H/m^2. This value is close to that of the Young's modulus for Nb_3Sn obtained by the method of chemical deposition [9].

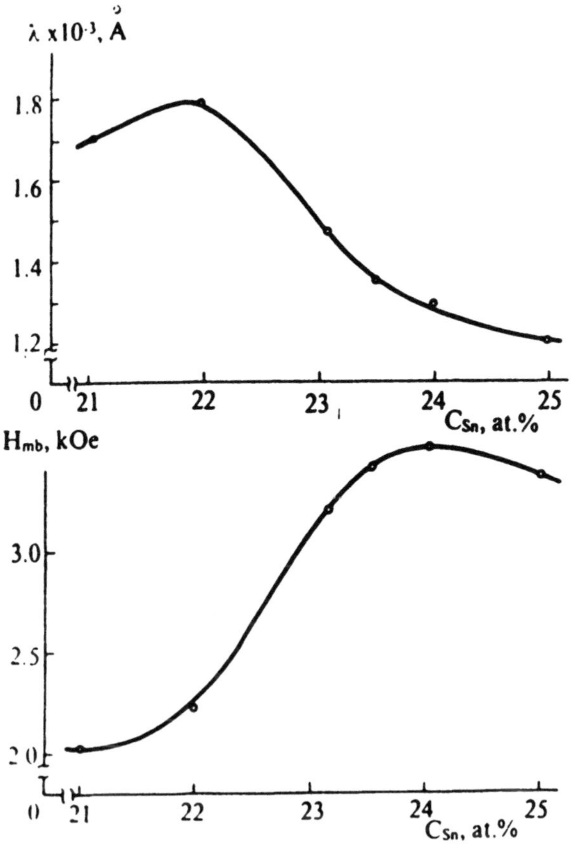

Figure 4. Calculated dependences of the magnetic field penetration depth λ and critical field H_{mh} of niobium stannide electrolytical coatings on composition.

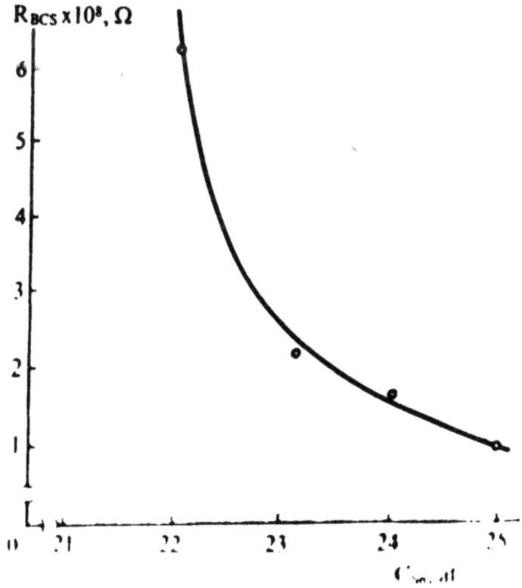

Figure 5. Calculated dependence of resistance R_{BCS} of niobium stannide coatings on composition

The best results for superconducting hf structures from Nb_3Sn were attained when this compound was made by diffusion from the vapour phase including the nucleus formation using $SnCl_2$ [10]. The maximum quality factor Q at the frequency of 3 GHz at temperature 4.2 K equals $7x10^9$. Assuming in equation (5) [4] that Ro=0, we evaluated the maximum quality factor Q for a hf cavity made of electrolytic niobium stannide. The resulting value was Q (4.2 K, 3 GHz) ~ 10^{10}, with different values for R_0 it is possible to reduce Q 10-15-fold. As a result, the quality factor of a real Nb_3Sn structure produced by joint electrochemical deposition should be not less than 10^8 - 10^9.

Thus, the findings have shown the applicability of joint electrochemical deposition in production of Nb_3Sn coatings with sufficiently high hf characteristics suitable for use in real structures.

References

1. Dasbach, D., Muller, G., Peininger, M., Piel, H., and Roth, R.W. (1989) Nb_3Sn coating of high purity Nb cavities, *IEEE Transaction on Magnetics* ***MAG-25***, 1862-1864.
2. Drobotenko, G.A., Kolosov, V.N., Shevyryov, A.A. and Gel, R.P. (1992) Creation of superconducting composites by galvanoplasty, in *Physical Chemistry and Electrochemistry of Rare and Non-ferrous Metals (Russ.),* KSC, RAS, Apatity, p.35.
3. Kolosov, V.N. (1995) The evaluation of the composition effect on RF characteristics of the Nb_3Sn compound produced by joint electrochemical deposition, *Abstracts of International Conference Rare Earth Metals: Raw Material, Processing, Technology of Compound and Related Products,* Krasnoyarsk, Russia, p.215.
4. Kolosov, V.N. and Matychenko, E.S. (1998) Evaluation of high-frequency superconductivity of niobium coatings prepared by electrodeposition in molten salts, *This conference, proceeding paper.*
5. Kolosov, V.N., Gel, R.P., Drobotenko, G.A. and Shevyryov, A.A. (1994) On possibility of electrolytical niobium A-15 crystall structure stabilization by tin, *Fizika metallov i metallovedenie (Russ.)* **77**, 79-82.
6. Kolosov, V.N., Gel, R.P., and Drobotenko, G.A. (1993) Magnetic shielding properties of superconducting electrolytical Nb_3Sn coatings , in *Low Temperature Electronics and High Temperature Superconductivity, The Elecrtochem. Soc., Softbound Ser.* **93-22**, 397-410.
7. Kolosov, V.N. and Matychenko, E.S. (1996) Investigation of electrolytical copper coatings to be applied in a Nb_3Sn -Nb-Cu superconducting compositions, *Fizika i khimija* obrabotki *materialov (Russ.)* **3,** 23-29 .
8. Bussiere, J.F., Welch, D.O. and Suenaga, M. (1980) Young's modulus of polycrystalline Nb_3Sn between 4.2 and 300 K, *J. Applied Physics* **51**, 1024-1030.
9. Bussiere, J.F., Le Huy, H. and Faucher, B. (1984) Elastic behaviour of polycrystalline Nb_3Sn and Nb_3Ge, *Advanced in Cryogenic Engineering Materials*, **30**, 859-866.
10. Peininger M. and Piel H. (1985) A superconducting Nb_3Sn coated multicell accelerating cavity, *IEEE Transaction on Nuclear Science*, **32**, 3610-3612.

THE OXIDATION OF ALLOYS Fe-Si-(Ta, Nb, W, Sn) BY SODIUM AND CALCIUM SULPHATE MELTS

S.A. KRASIKOV, V.V. SHULGIN, V.M. CHUMAREV, V.P. MARYEVICH
Institute of Metallurgy, Ural Division of
Russian Academy of Sciences
101, Amundsen St., Ekaterinburg, 620016, Russia

Abstract

The oxidation of polymetal iron-based alloys containing tantalum, niobium, tungsten and tin by sodium and calcium sulphates was investigated between 1473 and 1623 K. Control of the sulphate charge allowed the removal up to 99% of rare metals into the slag keeping up to 70% of iron and up to 90% of tin in the alloy. The results of this work indicate that Na and Ca sulphates can be successfully applied to processing of poor Ta-Nb-W raw materials by oxidising alloys containing refractory rare metals.

1. Introduction

An effective processing of poor raw materials containing tantalum, niobium, tungsten, usually accompanied by tin, can be carried out by pyrohydrochemical technology including the reduction fusion of ore concentrates [1,2]. With high-temperature reduction processing, up to 99% of rare metals and tin can be transfered into an iron-based alloy, as well as part of the silicon and other impurities. The subsequent extraction and division of target components by hydrochemical methods is possible after selective oxidation of the alloy by oxidizer salts, e.g. using sulphates.

Known reaction thermodynamics and kinetics show the oxidizer distribution between alloy components at the contact of liquid phases and the external diffusion process are conditioned by the sulphate concentrations in the surface layer of metal melt and the equilibrium constants for interactions of metals with sulphate. In such cases the oxidation of liquid alloys enables the selective concentration of a target component in one of the phases - slag or alloy . In particular, it has been found experimentally that in processing of iron-tin-arsenic melts by sodium and calcium sulphates the rate of oxidation of iron allowed the concentration up to 90% Sn and up to 90% As in alloy with the loss of not less than 60% Fe [3]. For short short

D.H. Kerridge and E.G. Polyakov (eds.), Refractory Metals in Molten Salts, 245-250.

process times (up to 5 minutes) the limiting role of mass transfer in the oxidizer phase was marked.

The purpose of the present work was to study the possibility of selective oxidation of metals by sodium and calcium sulphates in liquid iron-based alloys containing tantalum, niobium, tungsten and tin.

2. Experimental

The sodium and calcium sulphates were placed in alumina crucibles positioned in the hot zone of a vertical resistance furnace, with alloys of various compositions (Table 1). The graphite heater of the furnace was isolated from the reaction chamber by a corundum tube to avoid any reduction effect. The weight of the alloy samples (size fraction, 0,1-10 mm) was 0,1-0,3 kg. The temperature of the tests was varied within the range 1473-1623 K. Contact between alloys and oxidizers was varied as follows: melting of mixed solid reagents, introduction of alloy crystal particles into sulphate melt, mixing of a liquid alloy into the sulphate melt, and adding the charge of crystal sulphates to the alloy melt. The duration of phase contact did not exceed 20 minutes.

TABLE 1. Composition (wt.%) of alloys

Alloy	Sn	Ta	Nb	W	Mn	P	Si	Fe	C
W-Sn-Si	2.0	-	-	6.8	-	-	22.0	68.2	-
W-Sn-P	10.8	-	-	8.8	-	2.5	0.1	76.3	1.6
Nb pig-iron	-	-	1.2	-	-	1.1	3.0	-	4.3
Ta-Nb	-	5.8	4.3	-	24.3	-	12.1	-	-

3. Results and discussion

The method used to contact the alloys and the sulphates did not greatly influence the final distribution of metals between oxidation products. The most important characteristics determining the processing parameters were the sulphate composition and the initial composition of the alloys. The average total oxidation rates of components were close to previous values [3] and were satisfactorily described by a equation characterising the processes with a predominant role for mass transfer in the liquid phase:

$$V = (1\text{-}3) \times 10^{-4}\, C_s, \qquad g\ cm^{-2}\ s^{-1},$$

where C_s(g cm^{-3}) is the sulphate concentration in oxide melt.

The oxidizing reactions of silicon and tungsten are primary for the treatment of W-Sn-Si alloy by sodium sulphate (Figure 1). Here the parallel removal of silicon and tungsten took place after initial oxidation of silicon only. The derivatives $d(C_W)/d(C_{Fe})$ and $d(C_W)/d(C_{Si})$ equalled 1.5 and about 10 for the stages of parallel oxidation of Si and W and the primary oxidation of silicon. Slags with tungsten contents of 3.0 to 4.5% were obtained within the range 1573-1623 K during phase contact of up to 15 minutes, without consecutive processing of the alloy by portions of sodium sulphate. Under similar experimental conditions the interaction between W-Sn-P alloy and sodium/calcium sulphate mixture (1:1) with alloy:oxidizer ratios of more than 5:1, caused predominantly by of phosphorus and tungsten into the slag (Figure 2). When W extraction into the slag equalled about 80%, and up to 70% Fe and up to 90% Sn of the initial metal remained in alloy, the tungsten concentration in the slag and the tin content in the alloy were about 22% and about 18 %, respectively. With decrease of the sulphate ratio up to 5:3, tungsten and phosphorus were oxidized almost completely and the maximum amount of iron in the slag was 60% and of tin 48%.

The oxidation of elements depended only on the oxidizer charge in the processing of Nb pig-iron by sodium sulphate between 1473 and 1573 K with liquid phase contact of up to 20 minutes. As in the case with W-Sn-Si alloy, primary silicon oxidation occurred first, followed by joint transfer of Si and Nb into the slag (Figure 3). Despite the high slag basicity, phosphorus and carbon transfer into slag during this period was insignificant. The oxidation of phosphorus and carbon began only after practically complete transfer of silicon and niobium into the slag. The parallel oxidation of Si and Nb took place while their concentrations in the metal phase were equal. However, joint transfer of silicon and niobium began at Si/Nb=0.5 after consecutive processing of pig-iron by sodium sulphate portions (up to 4 % of initial weight of alloy in each cycle), and , therefore, here slag containing up to 13 % Nb_2O_5 and only 9-12 % SiO_2 was formed.

This laboratory research was confirmed by large-scale experiments conducted in an induction furnace with 20 kg charges of Nb-Fe-C alloys. The results of Table 2 indicate that the treatment of liquid alloys by sodium sulphate at 1473 K for 10 minutes provided about 90% niobium extraction into the slag.

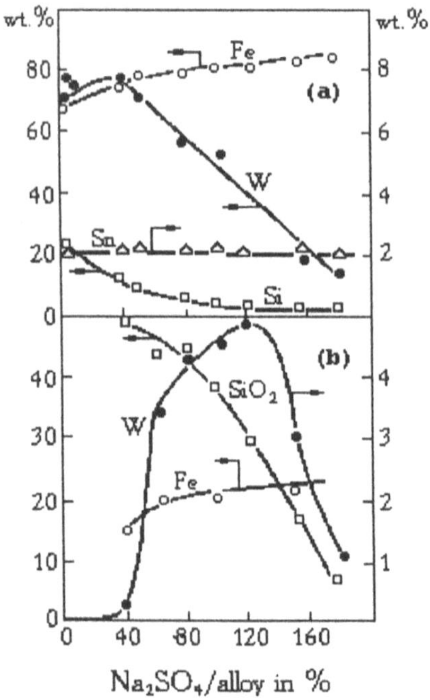

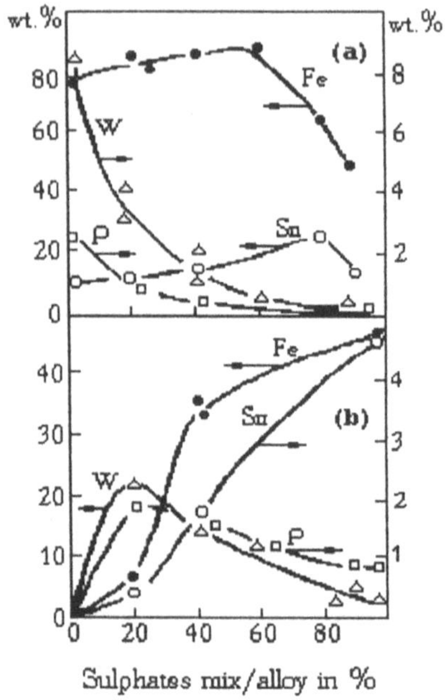

Figure 1. Effect of the sodium sulphate charge on the component concentration in alloy (a) and slag (b) for the oxidation of W-Sn-Si alloy.
Figure 2. Effect of the charge of sodium and calcium sulphates mix (1:1) on the component concentration in alloy (a) and slag (b) for the oxidation of W-Sn-P alloy.

The behaviour of Ta-Nb alloy was studied in an analogous way. To achieve a greater degree of selectivity the processing of this alloy was made by consecutive sodium sulphate additions of 10 % by weight of initial metal. The test results (Figure 4) show that oxidation of tantalum and niobium proceeded in parallel, and significant quantities of manganese in the alloy do not effect the slagging of rare metals, but reduced the selective division of the alloy components. However, a slag containing more than 15 % total of tantalum and niobium pentoxides was formed for all that and the extraction of Ta and Nb was up to 99 %.

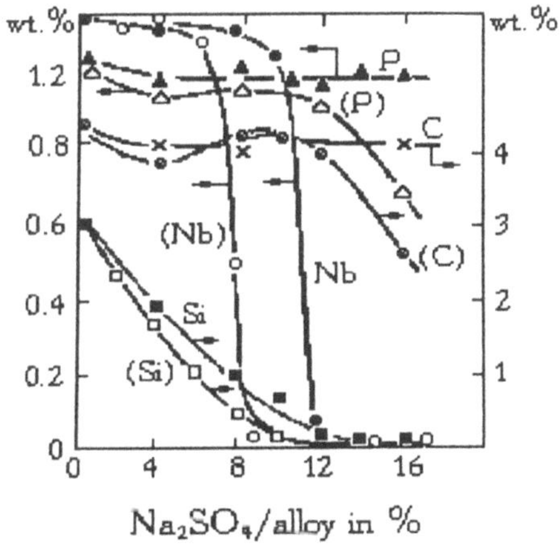

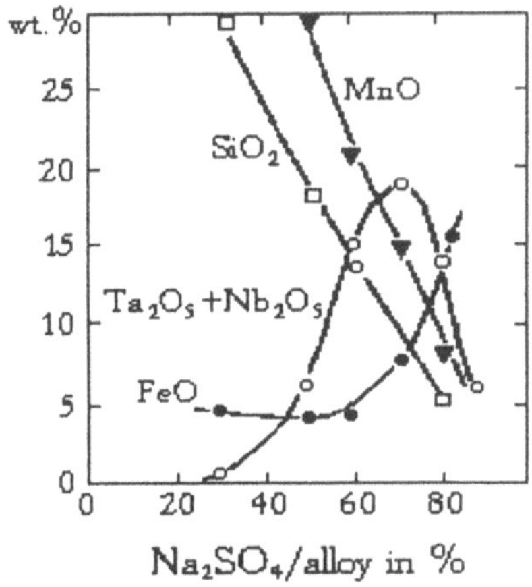

Figure 3. Effect of the sodium sulphate charge on the element concentration in alloy for the oxidation of Nb pig-iron. The symbols ( ) refer to consecutive processing of the alloy by sulphate portions.
Figure 4. Effect of the sodium sulphate charge on the metal oxide concentrations in the slag for the oxidation of Ta-Nb alloy.

TABLE 2. Results of Nb-Fe-C alloy refining

Test No.	Na_2SO_4 to initial alloy (%)	Product	Product to initial alloy (%)	Composition (wt.%) Slag Nb_2O_5	Slag SiO_2	Slag Fe	Slag Na_2O	Slag S	Alloy Nb	Alloy Si	Alloy C	Alloy S
1	15	Initial alloy	100.0	-	-	-	-	-	1.25	3.20	4.4	-
		Slag 1	11.0	5.7	22.0	22.9	22.7	6.0	-	-	-	-
		Slag 2	11.0	2.7	11.2	27.9	22.0	9.3	-	-	-	-
		New alloy	80.0	-	-	-	-	-	0.14	0.16	4.0	0.4
2	20	Initial alloy	100.0	-	-	-	-	-	1.88	1.97	3.9	0.1
		Slag 1	10.0	11.5	19.5	19.6	21.6	8.7	-	-	-	-
		Slag 2	10.0	8.5	19.4	19.6	20.2	6.1	-	-	-	-
		New alloy	82.0	-	-	-	-	-	0.24	0.19	4.1	0.4
3	4	Initial alloy	100.0	-	-	-	-	-	0.60	0.70	4.5	0.2
		Slag 1	6.0	6.2	18.1	25.5	18.9	-	-	-	-	-
		New alloy	90.0	-	-	-	-	-	0.15	0.13	4.5	0.3

After fusion the cooled slags were investigated by X-ray analysis. The X-ray results have shown that Ta, Nb and W were in the oxide products, predominately as tantalate, niobate and tungstate of sodium and calcium. Such compounds of rare metals in the slags permitted their extraction and division in subsequent processing stages by known hydrochemical methods.

4. CONCLUSION

(1) The processing of iron-based alloys containing tantalum, niobium, tungsten and tin by sodium and calcium sulphates at 1473-1623 K transferred into the slag up to 99 % of rare metals and left in the alloy up to 90 % of tin and up to 70 % of iron. The enrichment of the target components can be adjusted by the sulphate ratio and the method of treatment.
(2) An increasing rate of transfer of silicon into the slag was observed during the first stages of the oxidation of Si-bearing alloys. Then parallel slagging of silicon and rare metals occurred, which allowed slags richer in a target component to be obtained by cyclic sulphate processing of the initial alloys.
(3) The phase structure of the products, the high rates and the selective character of these interactions pointed to the use of sodium and calcium sulphates as oxidizers of multicomponent iron-based alloys in the processing of raw materials containing refractory rare metals.

References

1. Rockenbauer, W. and Starck, H.C. (1984) Die verarbeitung von zinnshlacken mit verschiedenen tantal- und niobgehalten auf synthetische konzentrate und tantal-niob-produkte, *Metal* < *W.-Berlin* > **38**, No. **2**, 156-159.
2. Chumarev, V.M., Okunev, A.I., Krasikov, S.A., Fedorov, V.D. and Safonov, A.V. (1995) The technology for complex processing of tin-rare earth raw materials, *Colour Metals*, No.**2**, 22-24.
3. Chumarev, V.M. and Shashmurin, V.A. (1980) The oxidation kine-tics of alloys Fe-Sn-As by melts Na_2SO_4-$CaSO_4$, *The Researches of Redox Processes in Oxide-Sulphide and Metal Systems*, Ural Scientific Center OF USSR Academy of Sciences, Sverdlovsk, pp. 76-86.

THE ELECTROCHEMICAL AND THERMODYNAMIC CHARACTERISTICS OF CHROMIUM COMPLEXES AGAINST THE BACKGROUND OF LiCl-KCl AND NaCl-KCl MELTS ACCORDING TO THE LINEAR SWEEP VOLTAMMETRY DATA

S.A.Kuznetsov and P.T.Stangrit
Institute of Chemistry Kola Science Centre,
14 Fersman Str., Apatity, 184200 Russia

1. Introduction

We have shown earlier that it is possible to calculate the equilibrium constant of a metal-salt reaction from linear sweep voltammetry data [1]. The results of [1] also enable us to determine the thermodynamic properties of compounds during their formation from elements.

Flengas and coworkers [2,3] have measured the equilibrium potentials of chromium in a molten equimolecular mixture of sodium and potassium chlorides at various concentrations of its ions, based on the ratio to a silver reference electrode. They determined [2,3] the standard electrode potentials $E^o_{Cr(II)/Cr}$, $E^o_{Cr(III)/Cr}$ and the redox potential $E^o_{Cr(III)/Cr(II)}$.

The merits of the linear sweep voltammetry method are:

1) rapidity since it does not require a long holding of the metallic electrode in the melt of its salts;

2) the melt under investigation contains ions of a higher valence, the measurements are not complicated by disproportionation reactions, and they do not require any knowledge of the concentrations of the lower- and higher-valence cations.

2. Experimental

The study was performed employing the linear voltammetry method. The apparatus used included a PI-50-1.1 potentiostat, an LKD-4-003 recording device, and a PO-5122 oscilloscopic polarograph. The potential scan rate was varied from 0.005 to 4.0 V s^{-1}. The products that formed at the cathode during electrolysis were identified with the aid of a DRON-2 diffractometer. The melt was placed in an ampoule fabricated of glassy carbon of the SU-2000 type and placed into a hermetically sealed retort made of Kh18N10T stainless steel. The ampoule also served as the counter electrode. The voltammograms were recorded at the glassy carbon (SU-2000) electrode 0.8-1.0 mm in diameter and platinum electrode 0,4mm in diameter with respect to the silver reference electrode Ag/NaCl-KCl-AgCl (2 w/o). The potentials were recalculated for a chlorine electrode. Salts were obtained by the directed crystallization technique. The argon gas,

D.H. Kerridge and E.G. Polyakov (eds.), Refractory Metals in Molten Salts, 251-258.

prior to entering the electrochemical cell, was passed through a titanium sponge heated to a temperature of 1073K.

3. Results and discussion

If the reduction of complexes of the higner oxidation state Me^{n+} to metal takes place in two sequential stages with the formation of complexes of the Me^{m+} lower oxidation state by the reaction

$$Me^{n+} + (n\text{-}m)e \rightarrow Me^{m+}, \qquad (1)$$

$$Me^{m+} + me \rightarrow Me, \qquad (2)$$

then there will be two reduction peaks E_p^{I} and E_p^{II} on voltammograms at high polarization rates corresponding to the nonstationary processes, via reactions (1) and (2).

It the case of chromium trichloride, in the molten equimolecular mixture of NaCl-KCl and eutectic LiCl-KCl the most frequent case is when in practice where both stages are reversible. The first stage corresponds to the formation of the reduced form in the melt and the second stage to the formation of the metal at the cathode [4].

The voltammetric curve for molten NaCl-KCl-$CrCl_3$ and LiCl-KCl-$CrCl_3$ melts are presented in Fig.1 and Fig.2. They have two clear, well reproducible peaks, which correspond to the one- and two-electron processes [4]:

$$Cr(III) + e \rightarrow Cr(II) \qquad (3)$$

$$Cr(II) + 2e \rightarrow Cr^{o}, \qquad (4)$$

i.e., n=3 and m=2 in our case.

According to the theory of linear sweep voltammetry and polarography, the following relation holds for stationary electrodes for the first stage of electrochemical reduction between the peak potential and half-wave potential [5]:

$$E_p^{I} = E^{I}1_{/2} - 1{,}11\,(RT/F), \qquad (5)$$

where

$$E^{I}_{1/2} = E^{o}_{Cr(III)/Cr(II)} + \frac{RT}{F}\ln\sqrt{\frac{D_{ox}}{D_{red}}} + \frac{RT}{F}\ln\frac{\gamma_{ox}}{\gamma_{red}} \qquad (6)$$

D_{ox} and D_{red} are diffusion coefficients of the oxidized and reduced forms; γ_{ox} and γ_{red} are the activity coefficients of the oxidized and reduced forms.

In the concentration range of metals with mole fraction N less than $(3\text{-}5)\cdot10^{-2}$, the activity of coefficients in the molten salts remains constant and they can lead to the values of the conventional formal redox potentials [6]:

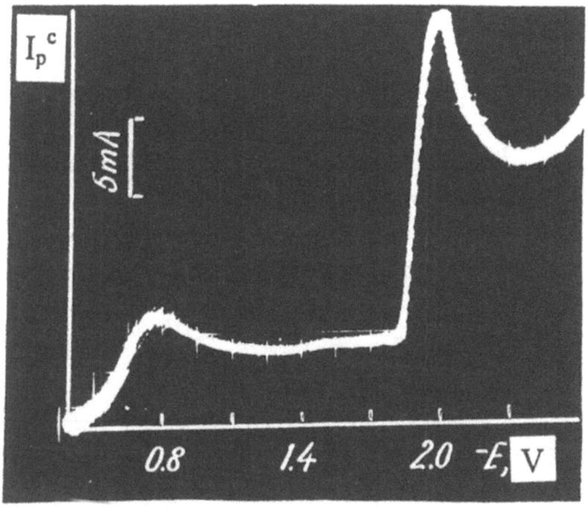

Figure 1. Voltammetric curve for NaCl-KCl-$CrCl_3$ melt at a glassy carbon electrode. V=0,66 V/s; T=1023K; N_{CrCl_3} =1.57 10^{-3} mole fract.; E is measured relative to a chlorine reference electrode

$$E^{*}_{Cr(III)/Cr(II)} = E^{o}_{Cr(III)/Cr(II)} + \frac{RT}{F}\ln\frac{\gamma_{ox}}{\gamma_{red}} \qquad (7)$$

One can regard $\sqrt{D_{ox}/D_{red}} \approx 1$,where $E^{*}_{Cr(III)/Cr(II)} \approx E^{I}_{1/2}$. Thus, the conventional formal redox potential can be calculated from the equation:

$$E^{*}_{Cr(III)/Cr(II)} = E^{I}_{p} + 1{,}11\frac{RT}{F} \qquad (8)$$

In our previous investigation [1] we used equation (8) for determination of $E^{*}_{Cr(III)/Cr(II)}$. For more precise calculation of values $E^{*}_{Cr(III)/Cr(II)}$,coefficients of diffusion Cr(III), Cr(II) by methods of linear sweep voltammetry (LVA) and reversive chrono-potentiometry (RCP) were determined [7]. Our experemental data and the results of investigation [8], are presented in Tables 1, and 2.

It can be seen from Tables 1 and 2 that the diffusion coefficients complexes decrease when the chromium oxidation state increases and that the values found by different methods are in a good agreement.

The refined values of Cr^{*}(III)/Cr(II) were calculated by the following equation:

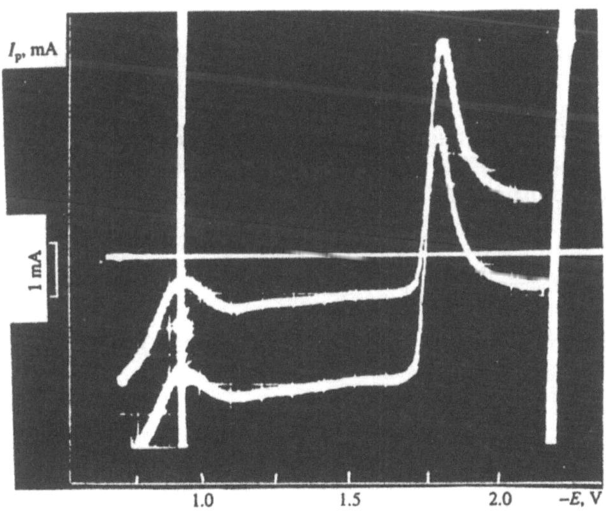

Figure 2. Voltammograms for a LiCl-KCl-$CrCl_3$ melt at a platinum electrode. N_{CrCl_3}=4.1x10^{-3} mole fract.; V=0.5 V/s; T=723K. The curves were taken at 5-min intervals and shown to be shifted along the current axis. E is given relative to a chlorine reference electrode

Table 1. Diffusion coefficients of Cr(III) in NaCl-KCl melt

Source	Method	$D \cdot 10^5$, cm^2/s				Activation energy of diffusion, kJ/mole
		973K	1023K	1073K	1100K	
[7]	LVA	1,63	2,08	2,64	2,98	43,9
[7]	RCP	1,46	1,96	2,61	2,92	48,5
[8]	CP	1,52	2,08	2,64	3,00	48,1

Table 2. Diffusion coefficients of Cr(II) in NaCl-KCl-melt

Source	Method	$D \cdot 10^5$, cm^2/s				Activation energy of diffusion, kJ/mole
		973K	1023K	1073K	1100K	
[7]	LVA	2,47	3,16	3,94	4,41	40,5
[7]	RCP	2,26	2,98	3,82	4,32	45,6
[8]	CP	2,24	2,99	3,85	4,37	46,5

$$E^{*}_{Cr(III)/Cr(II)} = E^{I}_{p} + 1{,}11\frac{RT}{F} + \frac{RT}{F}\ln\left(\frac{D_{Cr(III)}}{D_{Cr(II)}}\right)^{1/2} \quad (9)$$

The Berzins and Delahay's equation [9] for the second stage of the electroreduction can be written as:

$$E^{II}_{p} = E^{o}_{Cr(II)/Cr} + \frac{RT}{2F}\ln N - 0{,}854\frac{RT}{2F} \quad (10)$$

Since $E^{*}_{Cr(II)/Cr} = E^{o}_{Cr(II)/Cr} + \frac{RT}{2F}\ln\gamma$, the magnitude of the conventional formal electrode potential is related to the magnitude of the peak potential by the following equation:

$$E^{*}_{Cr(II)/Cr} = E^{II}_{p} - 2{,}303\frac{RT}{2F}\lg N + 0{,}854\frac{RT}{2F} \quad (11)$$

The conventional formal electrode potential $E^{*}_{Cr(III)/Cr}$ was calculated from Luter's equation

$$3E^{*}_{Cr(III)/Cr} = E^{*}_{Cr(III)/Cr(II)} + 2E^{*}_{Cr(III)/Cr} \quad (12)$$

For determination of conventional formal potentials it is possible to use the results of stationary voltammograms. As has been shown above, $E^{I}_{1/2} \approx E^{*}_{Cr(III)/Cr(II)}$ and it is easy to show by using Kohlthoff-Lienhein equation [9] that:

$$E^{*}_{Cr(II)/Cr} = E^{II}_{1/2} - 2{,}303\frac{RT}{2F}\lg\frac{N}{2} \quad (13)$$

Compared to the results obtained at high polarization rates, the stationary voltammograms have poor reproducibility in determining the half-wave potentials, especially in the second stage. This is true for the deposition of chromium on the cathode, which changes the electrode surface area, which, in turn, makes it difficult to determine the limiting diffusion current in the second stage.

The dependences of the conventional formal electrode potentials and conventional formal oxidation-reduction potential are described by the following equations:

$$E^{*}_{Cr(III)/Cr(II)} = -1,302 + 6,4 \cdot 10^{-4} T \pm 0,005 \quad (14)$$

$$E^{*}_{Cr(II)/Cr} = -2,118 + 5,04 \cdot 10^{-4} T \pm 0,003 \quad (15)$$

$$E^{*}_{Cr(III)/Cr} = -1,849 + 5,52 \cdot 10^{-4} T \pm 0,003 \quad (16)$$

The $E^{*}_{Cr(II)/Cr}$ values found by us are in good agreement with the data [2,3], but the values of $E^{*}_{Cr(III)/Cr(II)}$ markedly differ from the results of the investigations [2,3].

In our opinion the reason of these disagreements is using in works [2,3] of platinum electrode for determination of redox potentials.

The study of electroreduction of chromium trichloride in the NaCl-KCl melt conducted in [10] showed that the equilibrium potential of platinum in this melt overlaps partially the region of the recharge potentials intrinsic to Cr(III) complexes. As a result, platinum can dissolve via the reaction

$$2Cr(III) + Pt \Leftrightarrow 2Cr(II) + Pt(II). \quad (17)$$

Thus, the platinum electrode in the NaCl-KCl melt containing Cr(III) is not indifferent. The voltammogram recorded at the platinum electrode in the NaCl- KCl-$CrCl_3$ melt (Fig. 3) exhibits a single reversible cathodic wave II according to reaction (4) and wave I, corresponding to platinum dissolution:

$$Pt - 2e \rightarrow Pt(II) \quad (18)$$

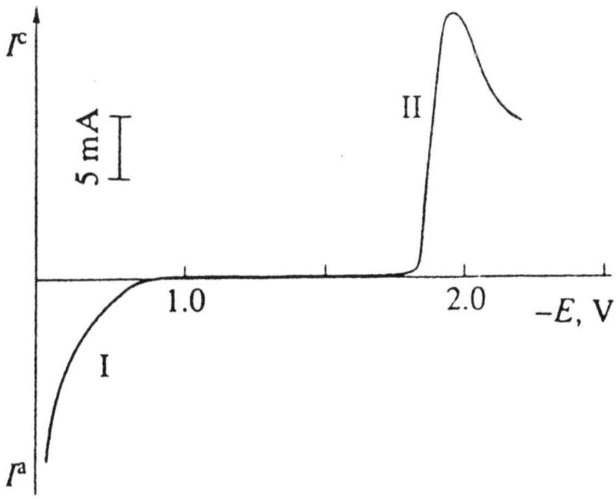

Figure 3. Voltammogram for the NaCl-KCl-$CrCl_3$ molten salt at platinum electrode C_{CrCl_3} = 4.21x10^{-5} mol cm^{-3}; V=0.5V s^{-1}; T=1023K; S=0.244 cm^2; (I) Pt-2e → Pt(II) and (II) Cr(II)+2e → Cr

The reversible process of recharge [10] of the Cr(III) complexes to Cr(II) ones is not observed at the platinum electrode, and is clearly observed at the cathode prepared from glassy carbon (Fig. 1).

The values of redox potentials $E^{*}_{Cr(III)Cr(II)}$ obtained by us in [1] and in this investigation are in close agreement with the data obtained by electronic absorption spectra measurements [11].

On the basis of experimental data Ep^{I}, Ep^{II} and by employing equation (9,11,12) we established the following dependences in LiCl-KCl eutectic melt:

$$E^*_{Cr(III)/Cr(II)} = -1,473 + 8,65 \cdot 10^{-4}\,T \pm 0,009 \qquad (19)$$

$$E^*_{Cr(II)/Cr} = -1,870 + 3,08 \cdot 10^{-4}\,T \pm 0,001 \qquad (20)$$

$$E^*_{Cr(III)/Cr} = -1,736 + 4,90 \cdot 10^{-4}T \pm 0,003 \qquad (21)$$

The values, calculated according to equations (19-21) for the temperature of 723 K are in good agreement with data of investigation [12].

Knowing the values (15,16) and (20,21) we can determine the change in the Gibbs energy for the formation of chromium chlorides from the elements in NaCl-KCl and LiCl-KCl melts.

For the formation of chromium chlorides in molten NaCl-KCl, obtained the following changes in the Gibbs energy:

$$\Delta G_{CrCl_2} = -97690 + (23,25 + 4,576\ \lg[Cr(II)])T,\ \text{cal/mole} \qquad (22)$$

$$\Delta G_{CrCl_3} = -127786 + (38,40 + 4,576\ \lg[Cr(III)])T,\ \text{cal/mole} \qquad (23)$$

in the melt LiCl-KCl:

$$\Delta G_{CrCl_2} = -86251 + (19,21 + 4,576\ \lg[Cr(II)])T,\ \text{cal/mole} \qquad (24)$$

$$\Delta G_{CrCl_3} = -120106 + (33,90 + 4,576\ \lg[Cr(II)])T,\ \text{cal/mole} \qquad (25)$$

The literature data [13] on the thermodynamics of the reaction of liquid chromium dichloride formation in the temperature range 1088-1123K allow us to represent the change $\Delta G_{CrCl_2(l)}$ with temperature in the form of the equation:

$$\Delta G_{CrCl_2(l)} = 84235 + 18,86\,T \qquad (26)$$

The change in the Gibbs energy during the mixing of chromium dichloride and molten equimolecular mixture of NaCl-KCl, when dilute solutions are formed, is equal to

$$\Delta G_{mix} = \Delta G_{CrCl_2} - \Delta G^{o}_{CrCl_2(l)} \qquad (27)$$

$$\Delta G_{mix} = -13455 + (4,39 + 4,5761\ \lg[CrII)] \cdot T,\ \text{cal/mole} \qquad (28)$$

in the melt LiCl-KCl:

$$\Delta G_{mix} = -2016 + (-4,65 + 4,576\ \lg\,[Cr(II)]) \cdot T,\ \text{cal/mole} \qquad (29)$$

The values of ΔG_{mix} found show that the mixing of chromium dichloride with the NaCl-KCl and LiCl-KCl melts are exothermic processes as a result of the complex-formation reaction:

$$Cr^{2+} + 4Cl^{-} = CrCl_4^{2-} \quad (30)$$

at the same time comparative stability of complexes $CrCl_4^{2-}$ is higher in the NaCl-KCl melt.

The dependences of the conventional formal potentials determined here enable us to calculate the conventional equilibrium constants for the reaction:

$$2Cr(III) + Cr \leftrightarrow 3Cr(II) \quad (31)$$

from the equation [1]:

$$\log K^{*} = \frac{n(n-m)}{1{,}984x10^{-4}xT}(E_p^{I} - E_p^{II} + 2{,}303\frac{RT}{mF}\lg N - 0{,}854\frac{RT}{mF} + 1{,}11\frac{RT}{(n-m)F} \quad (32)$$

The calculated values of the equilibrium constants logarithm of the linearly depend upon the reciprocal of temperature in the NaCl-KCl melt:

$$\log K^{*} = 1{,}41 + 8158/T \pm 0{,}72 \quad (33)$$

on the melt LiCl-KCl:

$$\log K^{*} = 5{,}55 + 4073/T \pm 0{,}17 \quad (34)$$

Since $\Delta G^{*} = -2{,}303RT \log K$ for the metal-salt reaction we got in the melt NaCl-KCl:

$$\Delta G^{*} = -37332 - 6{,}45T, \text{ cal/mole} \quad (35)$$

in the melt LiCl-KCl:

$$\Delta G^{*} = -18638 - 25{,}40\ T, \text{ cal/mole} \quad (36)$$

As can be seen from equations (33,34) that on contact with metallic chromium of chloride melts, containing complexes Cr(III) the only complexes existing in the melt are $CrCl_4^{2-}$

4. Conclusions

The conventional formal electrode potentials $E^{*}_{Cr(II)/Cr}$, $E^{*}_{Cr(III)/Cr}$ and conventional formal redox potentials $E^{*}_{Cr(III)/Cr(II)}$ are determined from linear sweep voltammetry data. From the experimental data the thermodynamics of formation of a dilute solution of chromium di- and trichloride in an equimolecular mixture of NaCl-KCl and LiCl-KCl eutectic melt were calculated. The effect of temperature on the equilibrium constants of the metal-salt reaction were determined.

This investigation hase shown that linear sweep voltammetry is not only an effective method for study the electrode process kinetics, but an express and simple method for determining a number of thermodynamic parameters as well.

5. Acknowledgements

The authors wishes to acknowledge their thanks to the Russian Foundation for Basic Research for the financial support of this work (grant 96-03-32175).

6. References

1. Kuznetsov, S.A., and Stangrit P.T. (1990). Thermodynamics of the formation of chorimium di - and trichloride from the elements in a molten equimolecular mixture of sodium and potassium chlorides from linear-voltammetry data, *Rasplavy* **4**, 44-48.
2. Flengas, S.N., and Ingraham, T.R. (1959). Electromotive force series of metals in fused salts and activities of metal chlorides in 1:1 molar KCl-NaCl solutions, *J.Electrochem.Soc.* **106**, 84-91.
3. Tumidajski, P.J., and Flengas S.N. (1991). Potential measurements of reactive metal chloride in alkali halide solution. II.Chromium, *J.Electrochem.Soc.* **138**, 1659-1665.
4. Baimakov, A.N., Kuznetsov, S.A., and Stangrit P.T. (1983). Mechanism of cathodic reduction and coordination of chromium in molten chloride-fluoride mixtures, *Izv.Vyssh.Uchebn.Zaved., Tsvetn.Metall.* **1**, 79-82.
5. Matsuda H., and Ayabe Y. (1955) Zur Theorie der Randles-Sevcikschen kathodenstrahl polarographie, *Z.Elektrochem.* **59**, 494-503.
6. Smirnov, M.V. (1973). Electrode Potentials in Molten Salts, Nauka, Moscow.
7. Kuznetsov, S.A. (1993). Electroreduction of refractory metal complexes in molten salts: specific features and relationships, *Russ.J.Electrochemistry* **29**, 1154-1160.
8. Volkov, A.M.(1970). The study of chromium and some impurities in process of electrolytic refinement of chromium in molten chlorides, *Thesis*, Leningrad Polytechnic Inst., Leningrad.
9. Delahay, P. (1954) *New Instrumental Methods in Electrochemistry*, Interscience Publishers, New York.
10. Kuznetsov, S.A. (1996). The use of the glassy-carbon electrode for studing electroreduction of refractory metal complexes,*Russ.J.Electrochemistry*. **32**, 1215-1219.
11. Smirnov, M.V., and Potapov, A.M. (1994). Redox potentials and electronic absorption spectra of dilute solutions of nickel and chromium chlorides in molten alkali chlorides, *Electrochim.Acta*, **39**, 143-149.
12. Laitinen, H.A., Liu, C.H. (1958). An electromotive force series in molten lithium chloride-potassium chloride eutectic, *J.Amer.Chem.Soc*, **8**, 1015-1020.
13. Wicks, C.E., and Block F.E. (1965). *Thermodynamic Properties of 65 Elements, Their Oxides, Halides, Carbides, and Nitrides* (Russian translation), Metallurgia, Moscow.

VOLTAMMETRIC OXYGEN DETERMINATION IN MOLTEN FLUORIDES

L.P. POLYAKOVA [1], E.G. POLYAKOV [1], G.A. BUKATOVA [1], N.J. BJERRUM [2]
[1] *Institute of Chemistry, Kola Science Center, 14 Fersman St., Apatity, 184200, Russia*
[2] *TechnicalUniversity of Denmark, Lyngby, Denmark*

1. Introduction

Since the seventies it has been known, that the mechanism of electrode reactions in the rare metal halide containing melts is considerably affected by oxide-ion contents [1]. Voltammetry, contrary to other analytical methods , enables to realize oxygen determination directly in the melt despite its aggressiveness.

As was stated in our earlier work [2], the oxide-ion oxidation current peak height was directly proportional to the amount of oxide added to FLINAK-K_2TaF_7-Na_2O melt in a rather wide concentration range. An attempt was also made to determine the residual content of oxygen in FLINAK background mixture. For this purpose a plot of peak current height of oxide-ion oxidation vs. oxygen concentration was constructed for FLINAK-K_2TaF_7 melt. This plot was used as a calibration curve to determine initial oxide contents in the melt by extrapolation on abscissa. The distance between the point of intersection and zero in terms of concentration units was taken for oxide-ion residual content in FLINAK melt. Such a procedure is not completely correct, being based on the assumption of oxide-ion diffusion coefficients equality in different melts: FLINAK-Na_2O and FLINAK-K_2TaF_7-Na_2O. That remains to be either confirmed or invalidated.

The aim of the present work is to estimate the linear voltammetry possibilities in oxide-ion analytical determination in FLINAK melt with and without addition of tantalum and boron complex fluorides. Sodium oxide was used as a source of oxygen in the melts.

2. Experimental

Initial salts underwent careful purification and then were kept under conditions, excluding oxy- phase formation. The experiments were carried out in a three-electrode cell, where glassy-carbon served both as a container for the electrolyte and as a counter electrode. A glassy-carbon rod, attached to the molybdenum lead, was used as working electrode; the same rod of a larger diameter served as a quasi-reference

D.H. Kerridge and E.G. Polyakov (eds.), Refractory Metals in Molten Salts, 259-262.

electrode. The Ag/AgCl reference electrode was rejected to avoid inevitable pollution of the melt with oxygen caused by the silver electrode ceramic shell.

The experiments were carried out in the atmosphere of thoroughly purified argon. Potentiostat PI-50-1.1 with PR-8 programmer and PDA-1 recorder were used in linear sweep voltammetry measurements.

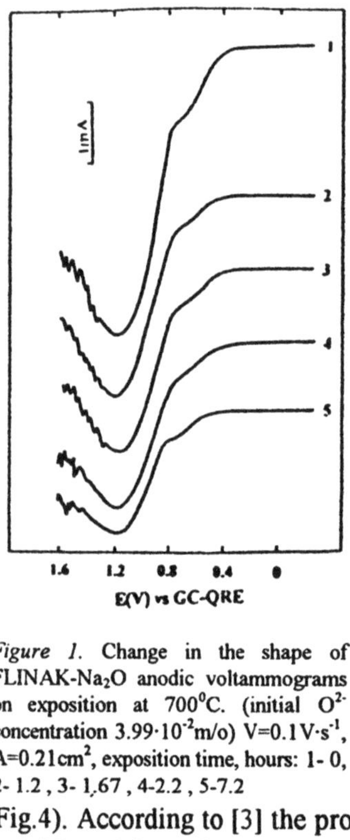

Figure 1. Change in the shape of FLINAK-Na_2O anodic voltammograms on exposition at 700^0C. (initial O^{2-} concentration $3.99 \cdot 10^{-2}$m/o) $V=0.1 V \cdot s^{-1}$, $A=0.21 cm^2$, exposition time, hours: 1- 0, 2- 1.2 , 3- 1.67 , 4-2.2 , 5-7.2

3. Results and discussion

The absence of noticeable volatility of a substance is an indispensable condition for the correct determination of its concentration at high temperatures. Thus the thermal stability of FLINAK - Na_2O melt has been examined by periodic measuring the heights of the anode peaks after exposure of the melt at a certain temperature. Further on the time dependence of oxide-ion oxidation current peak was plotted. Figures 1,2 demonstrates significant diminution of sodium oxide in the melt due to its volatility. The oxide-ion concentration attained $1.7 \cdot 10^{-2}$ m/o in six hours and further remained stable. At temperatures below 600^0C the melt is more stable and the concentration range of correct oxygen determination extends up to 0.5 m/o (Fig.3). Voltammograms at this temperature, since O^{2-} concentration is about 10^{-1} m/o, are characterized by two peaks with specific oscillations (Fig.4). According to [3] the products corresponding to the more negative peak are CO and CO_2, while at the more positive potential CF_4 was found alongside to them.

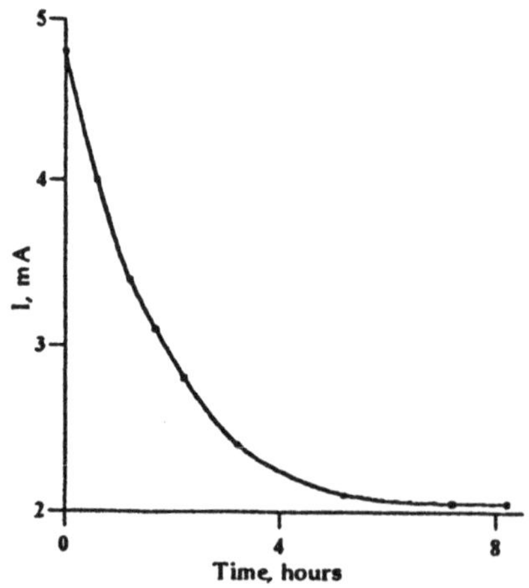

Figure 2. Decrease in oxide-ion concentration on exposition of FLINAK-Na_2O melt at 700^0C. Initial O^{2-} concentration $3.99 \cdot 10^{-2}$m/o; final $1.7 \cdot 10^{-2}$ m/o

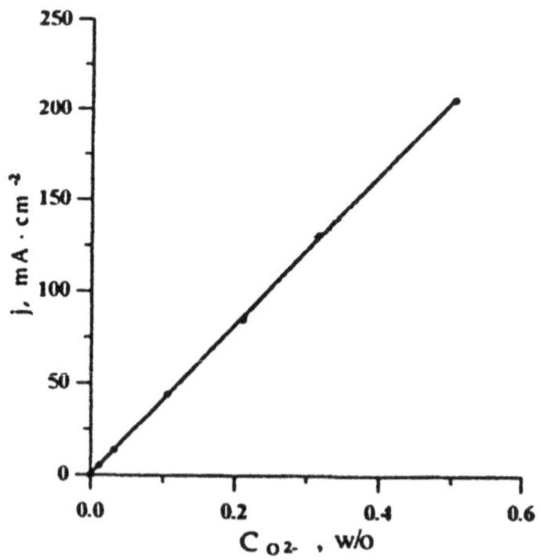

Figure 3. Concentrational dependence of O^{2-} oxidation peak current. $T=600^0$C; $V=0.1 V \cdot s^{-1}$

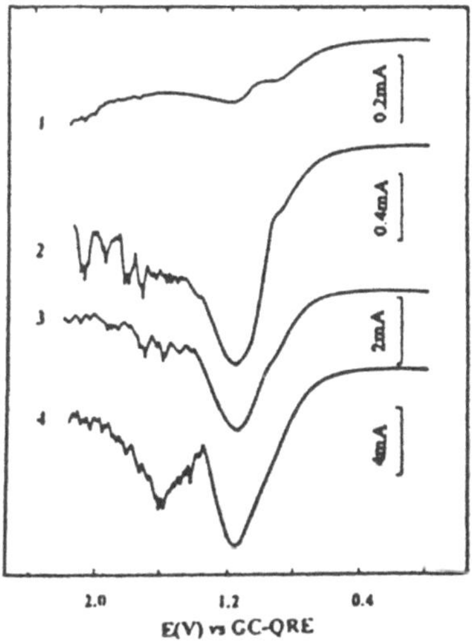

Figure 4. Voltammograms of FLINAK-Na_2O melt at 600°C. V=0.1V·s^{-1}, A = 0.165cm^2, Na_2O concentration (·10^{-2}) m/o : 1 - 0.27, 2- 1.92, 3- 6.2 , 4- 15.7

The authors assumed a joint discharge of oxygen and fluorine and COF_2 as a possible gaseous product. The experimental identification of reaction products was hindered by their instability at high temperatures [4].

The low melting point of FLINAK (454°C) allows to carry out the experiments at 550°C which makes the results more reliable. Then the plot of oxide-ion oxidation peak current, which consists of both the first peak height and that of the forewave, against concentration is characterized by the expression:

$$J_{550}=309.1 \cdot C_{O^{2-}} \qquad (1)$$

The necessity often arises to compare the data on oxide-ion residual contents calculated by various authors at different temperatures. That was the reason to carry out the experiments to get a temperature dependence of the oxide-ion oxidation peak current (Fig.5). To avoid the losses of sodium oxide due to its volatility at high temperatures its concentration was maintained at no more than 1.1 10^{-2}m/o. The equation obtained was:

$$J = -7.845 + 0.205 \cdot 10^{-1}\, t \qquad (2)$$

To bring the peak current value to the desirable temperature, for example to 550°C, the expression is proposed:

$$J_{550} = J_t / (5.98\ 10^{-3} \cdot t - 2.287) \qquad (3)$$

where j_{550} and j_t are current densities of oxide-ion oxidation at 550°C and at any other temperature t.

The same investigations were carried out in FLINAK-Na_2O-K_2TaF_7 and FLINAK - Na_2O - KBF_4 melts with tantalum and boron fluoride concentrations within 0.5 m/o and the molar ratio Ta(B)/O > 1.

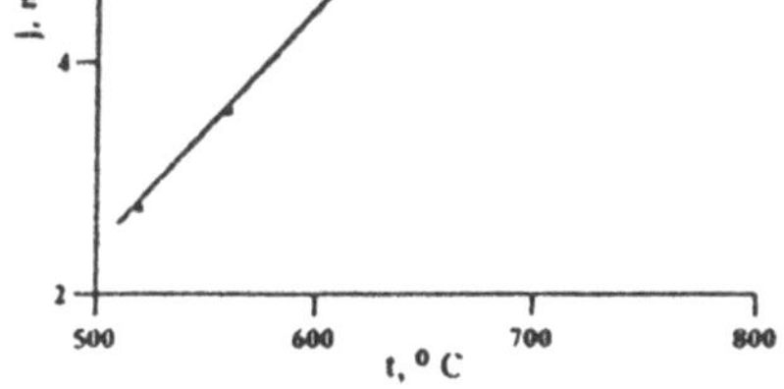

Figure 5. O^{2-} oxidation peak current thermal dependence in FLINAK-Na_2O (1.1·10^{-2} m/o) melt. V=0.1V·s^{-1}

The next equations for FLINAK-K_2TaF_7-Na_2O were obtained:

$$J_{550} = 239.93 \cdot C_{O^{2-}} \qquad (4)$$

$$J = -115.05 + 0.233 \cdot t \qquad (5)$$

$$J_{550} = J_t / (1.78\ 10^{-2} \cdot t - 8.782) \qquad (6)$$

For FLINAK-KBF_4-Na_2O:

$$J_{550} = 262.40 \cdot C_{O^{2-}} \qquad (7)$$

$$J = -349.72 + 0.709 \cdot t \qquad (8)$$

$$J_{550} = J_t / (1.76\ 10^{-2} \cdot t - 8.693) \qquad (9)$$

The essential difference in current thermal coefficients in FLINAK-Na_2O on the one hand and in FLINAK-Na_2O-K_2TaF_7, FLINAK - Na_2O - KBF_4 - on the another, if electrode process mechanisms are just the same, suggests the difference in diffusion activation energies of oxygen-bearing species in the mentioned media. From the complexation theory point of view it means a qualitative modification of the melt structure when adding complex fluorides of tantalum or boron to the FLINAK-Na_2O mixture.

Ta(V) and B(III) possess a stronger complexation potential, than the alkali cation; thus the oxofluoride complexes of tantalum and boron predominate in oxide-ion transport to the anode.

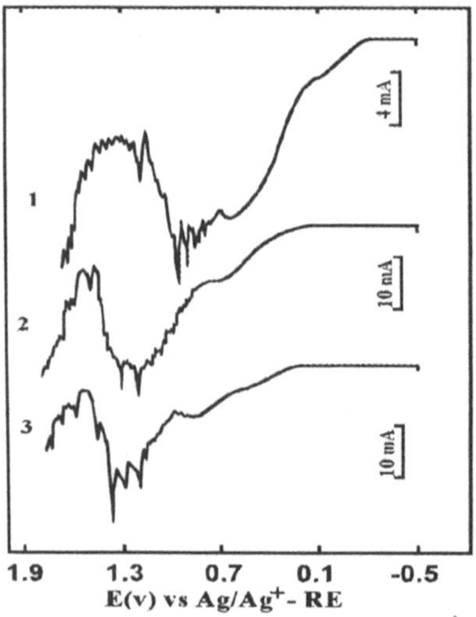

Figure 6. Anodic voltammograms. T=665°C; V=0.1V·s^{-1}, A=0.29 cm^2. The melts composition (m/o): (1) FLINAK-Na_2O(0.11); (2) FLINAK - Na_2O(0.11) - K_2TaF_7 (0.18); (3) FLINAK - Na_2O(0.11) - K_2TaF_7 (0.18) - KBF_4(0.18)

Transition from FLINAK-Na_2O to FLINAK - Na_2O - K_2TaF_7 and FLINAK - Na_2O - KBF_4-K_2TaF_7 melts (Fig.6) caused a positive shift of anodic potentials and a change in a shape of voltammogram. That points out the discrete discharge of "free" and "complexed" oxide-ions. Earlier the discrete discharge was known as applied to the cathodic reduction of cations but unknown for the anodic oxidation of anions in molten salts. It is interesting to note the O^{2-} oxidation current decreases on the addition of fluoborate to the FLINAK - Na_2O - K_2TaF_7 melt. Negligible amount of the addition can not change significantly the melt viscosity. It seems plausible to suggest the formation of massive low-mobile oxygen-bearing particles in this medium, such as heteronuclear complexes, with boron and tantalum central cations being bonded through oxygen bridges.

4. Acknowledgements

Authors[1] wish to acknowledge their thanks to the Russian Foundation for Basic Research for the financial support of this work (grant 96-03-32175)

5. References

1. Konstantinov, V.I., Polyakov, E.G., and Stangrit, P.T. (1978) Cathodic electrolysis of chloride-fluoride and oxyfluoride melts-chloride melts of tantalum, *Electrochim.Acta* **23**, 713-716.
2. Polyakova, L.P., Polyakov, E.G., Matthiessen, F., Christensen, E., and Bjerrum, N.J. (1994) Electrochemical study of tantalum in fluoride and oxofluoride melts, *J. Electrochem. Soc.* **141**, 2982-2988.
3. Nekrasov, V.N., Cherepanov, V.B., and Ivanovsky, L.E. (1986) *High Temperature Electrochemistry: electrolytes, kinetics,* USC RAS, Ekaterinbourg
4. Amphlett, J.C., Dacey, J.R., and Pritchard, G.O. (1971) An investigation of the reaction $2COF_2 = CO_2 + CF_4$ and the heat of formation of carbon fluoride, *J.Phys. Chem.***75**, 3024-3029.
5. Nekrasov, V.N., Cherepanov, V.B., and Ivanovsky, L.E. (1986) *Investigation of anodic process on a glassy carbon electrode in LiF-NaF-KF melt,* Inst. of Electrochemistry USC RAS, Ekaterinbourg. Dep. VINITI № 3420-84.
6. Nekrasov, V.N., Barbin, N.M., and Ivanovsky, L.E. (1989) Voltammetric investigation of the anodic process on a glassy carbon electrode in NaCl-KCl-Li_2O melt, *Rasplavy* (Rus.) **6**, 51-58.

TANTALUM ELECTROCHEMISTRY IN CHLORIDE-FLUORIDE AND OXOCHLORIDE-FLUORIDE MELTS.

L.P. POLYAKOVA, Z.A. KONONOVA, V.G. KREMENETSKY, E.G. POLYAKOV
Institute of Chemistry KSC RAS
14 Fersman St., 184200, Apatity, Russia

1. Introduction

Two papers dedicated to studying tantalum electrodeposition from NaCl-KCl-K_2TaF_7 melt are known in the literature [1,2]. These authors showed that tantalum reduction occurred through a single reduction step: Ta(V) + 5e $\leftrightarrow$ Ta(0). In the first of them Lantelme et al. established a spontaneous formation of tantalum and oxygen compounds on a tantalum wire emmersed in a molten electrolyte. The Na_2O and BaO additions to the NaCl-KCl-K_2TaF_7 melt [1] at a molar ratio O/Ta=1 resulted in formation of an oxohalide $TaOCl_xF_y^{z-}$ complex, where z=x+y-3, to which a single peak of tantalum reduction corresponded on voltammograms. The nature of this complex has not been established yet. Increase of the 2<O/Ta<1 molar ratio did not cause any change in voltammograms.

In work [2], by the method of linear voltammetry the feasibility of existence of $TaOF_6^{3-}$, $TaO_2F_4^{3-}$, TaO_3F^{2-} - complexes in NaCl-KCl-K_3TaOF_6 melt depending on the O/Ta molar ratio was shown. It follows from the above, rather contradictory data, oxide-ions are a critical impurity for halide melts and the ignorance of their presence in even small quantities may result in erroneous interpretation of the data obtained.

The aim of the present work was study electrochemical behaviour of tantalum in chloride-fluoride and oxochloride-fluoride melt in the presence of increasing amounts of sodium oxide. We investigated KCl-NaCl-K_2TaF_7, KCl-NaCl-K_2TaF_7-Na_2O ($0<O/Ta\leq2.4$), KCl-NaCl-K_3TaOF_6 and KCl-NaCl-$K_3TaO_2F_4$ melts. Cyclic voltammetry, potentiostatic electrolysis , IR spectroscopy, optical crystallography and X-ray diffraction were used as methods of research and identification of initial salts, solidified electrolytes and products of electrolysis.

2. Experimental

The equipment and the electrochemical cell were basically similar to those described

D.H. Kerridge and E.G. Polyakov (eds.), Refractory Metals in Molten Salts, 263-268.

earlier [1]. The experiments were carried out in a three-electrode cell in a temperature range 670-800^{0}C in inert atmosphere. A glassy carbon can SU-2000 (O 46 mm, height 270 mm) was used simultaneously as a container for electrolyte and a counter electrode. The voltammetric curves were recorded using glassy carbon, platinum and tantalum electrodes. A glassy carbon quasi reference electrode was mostly used in experiments. A silver reference electrode Ag/KCl-NaCl-AgCl (2 w/o) whose construction was described in [3] was employed only when an exact measurement of potentials was needed. Sweep rate of potential were varied from 0.005-1.0 V-s^{-1}.

As the using melts were very sensitive to oxygen containing impurities, oxide ceramics were eliminated completely from the cell structure for avoidance of pollution of the electrolyte. The purity of background electrolyte was controlled by cyclic voltammetry at a glassy carbon electrode. The oxide ion content in it determined by the technique described in [4] was (0.5-1.0) 10^{-3} m/o.

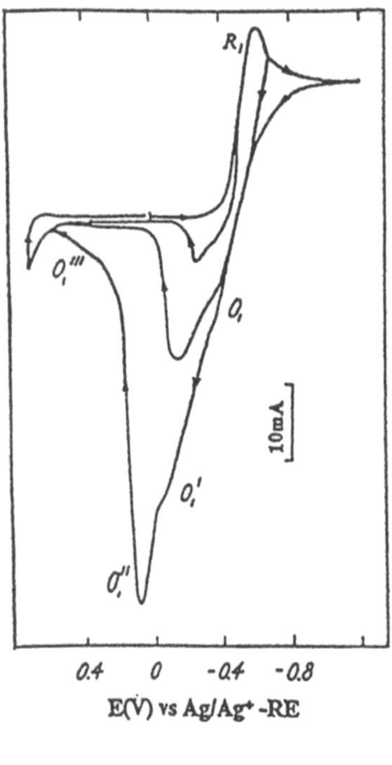

Figure 1. Cyclic voltammograms of KCl-NaCl-K_2TaF_7 melt at a platinum electrode. V=0.1 B·c^{-1}, A_{Pt}=0.32 cm^2, $C_{K_2TaF_7}$ =0.127 m/o, T =750 ^{0}C.

3. Results and discussion

KCl-NaCl-K_2TaF_7 melts On addition of potassium heptafluotantalate to the NaCl-KCl melt a peak of tantalum reduction to metal (R1) from fluorochloro complex (as will be shown below) was observed on voltammograms (Fig. 1). On the anodic part of the curve several peaks of the deposited tantalum oxidation correspond to it. The anodic peaks O_1'' and O_1''' should be referred to tantalum dissolution respectively in the form of Ta(IV) and Ta(V) chloro complexes, formed in the absence of excess fluoride ions. Recording of the cyclic voltammetric curve with a potential reverse at more positive potential -0.48 or -0.6 V (Fig.1), when the amount of tantalum deposited at the electrode is small and there is not enough time for diffusion of a considerable quantity of the liberated fluorine into the melt bulk, shows no such peaks, which corroborates our assumption. Peaks O_1 and O_1' were caused by tantalum dissolution with the formation of fluoride and mixed fluoro-chloride complexes of tantalum (V). As was established by analysis of the samples of solidified electrolytes, they contain tantalum in the form of NaK_2TaClF_7 [5], from which it may be surmised that the tantalum complex composition in the melt corresponds to formula $TaClF_7^{3-}$.

A typical quasi - steady state voltammogram is reported in Fig.2. Only one reduction wave Ta(V)→Ta(0) was also observed. According to X-ray analysis the

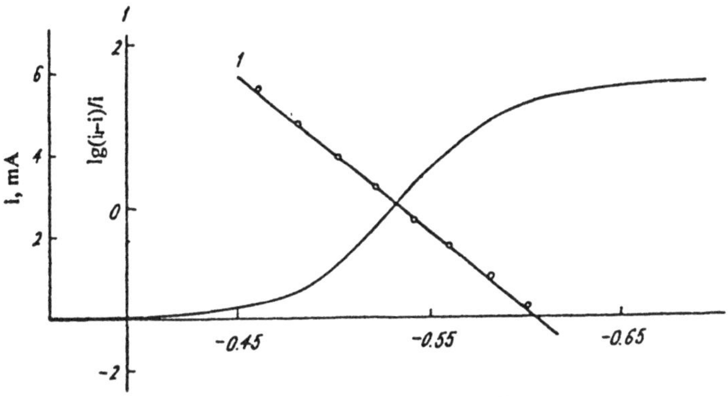

Figure 2. Quasi steady state voltammetric curve of NaCl-KCl-K_2TaF_7 melt. A_{Pt}=0.31 cm^2, V=0.005 B·c^{-1}; T=670 ^{0}C, $C_{K_2TaF_7}$ =0.069 m/o.

deposit obtained during potentiostatic electrolysis at the half-wave potential of this wave consisted of tantalum metal. An increase in the sweep rate caused a shift of the peak potential of tantalum reduction to the cathodic region. Linear dependence of the peak current on a square root of the scan rate and on K_2TaF_7 concentration was established. In compliance with diagnostic criteria of the linear sweep voltammetry (LVA) method the process of tantalum reduction to metal from the $TaClF_7^{3-}$ complex was controlled by the rate of charge transfer. All the kinetic parameters of reduction Ta(V)→Ta(0) are summarized in Table 1.

TABLE 1. Kinetic parameters of $TaClF_7^{3-}$ +5 $\bar{e}$ →Ta +7 F^- + Cl^- process in KCl-NaCl-K_2TaF_7 (A_{Pt}=0.307 cm^2, V=0.1 V·s^{-1}).

T, K	E_p, V	$I_p \cdot 10^3$, A	$C_{K_2TaF_7} \cdot 10^5$, mol·cm^3	αn	$D \cdot 10^5$, cm^2·s^{-1}	$K^o_{fh} \cdot 10^{10}$, cm·s^{-1}
943	-0.636	12.8	3.38	2.47	0.87	0.18
973	-0.624	13.6	3.34	2.50	1.03	0.40
1003	-0.612	14.5	3.31	2.52	1.22	0.94
1023	-0.606	15.0	3.28	2.53	1.35	1.56
1053	-0.595	15.9	3.25	2.55	1.58	3.32
1073	-0.587	16.2	3.22	2.56	1.69	5.50

The diffusion coefficients of $TaClF_7^{3-}$ complexes and αn values were determined by the equation for the case of irreversible electrochemical reduction with the formation of insoluble compounds at the electrode surface[6]. The activation energy of diffusion was equal 43.9 kJ·mol^{-1}.

KCl-NaCl-K_3TaOF_6 u KCl-NaCl-K_2TaF_7-Na_2O (0<O/Ta≤1) melts. On addition of Na_2O to NaCl-KCl-K_2TaF_7 melt an extra peak R_2 of tantalum reduction at potentials of approximately 0.2-0.25V more negative than peak R_1 was observed (Fig. 3a, curves 2-5). At the same time the height of peak R_1 was observed to decrease proportionally to

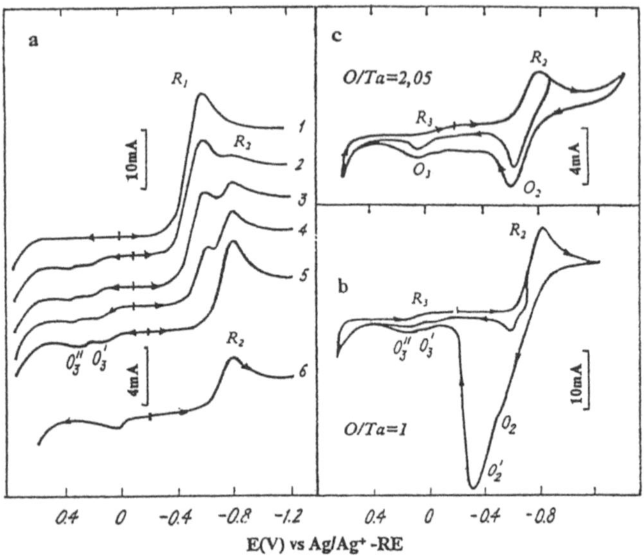

Figure 3. Transformation of voltammograms curves while titration of NaCl-KCl-K_2TaF_7 (0.127 m/o) melt with sodium oxide. V=0.1 $V \cdot s^{-1}$, T=750 °C, A_{Pt}=0.32 cm^2. Molar ratio O/Ta: (a) 1-0; 2-0.39; 3-0.61; 4-0.78; 5-1, 6- 2.05; (b) - 1; (c) - 2.05.

Na_2O addition. At a molar ratio of O/Ta=1 only one peak R_2 of tantalum reduction from the tantalum oxohalide complex remained (Fig. 3a, curve 5). Small peaks O_3' and O_3'' on the anodic part of curves 2-5 in Fig. 3a were due to oxide ion discharge. When the oxide ion concentration was increased, peaks O_3' and O_3'' merged in one O_3 peak and their growth practically ceased .

After the dissolution of K_3TaOF_6 compounds in KCl-NaCl melt identical voltammograms were obtained. As the solubility of the K_3TaOF_6 compound in KCl-NaCl melt is unknown, we undertook a series of experiments to determine it. K_3TaOF_6 was found to dissolve in KCl-NaCl melt at least up to 3.01 m/o at 750°C. Nevertheless, neither X-ray diffraction analysis nor IR spectroscopy corroborated the presence of K_3TaOF_6 in a solidified NaCl-KCl-K_3TaOF_6 sample but indicated the existence of an unknown phase. As a result of synthesis of a number of compounds such as $K_3TaOF_6 \cdot nKCl$ (n=1;2;3), $K_3TaOClF_5$ and $K_2NaTaOClF_5$ and comparing their IR spectra (Fig 4) and X-ray patterns with the melts under study it was found that in solid samples $K_2NaTaOClF_5$ is present. Thus we may suggest that interaction of K_3TaOF_6 and Na_2O with KCl-NaCl and KCl-NaCl-K_2TaF_7 melts occurs by the reactions:

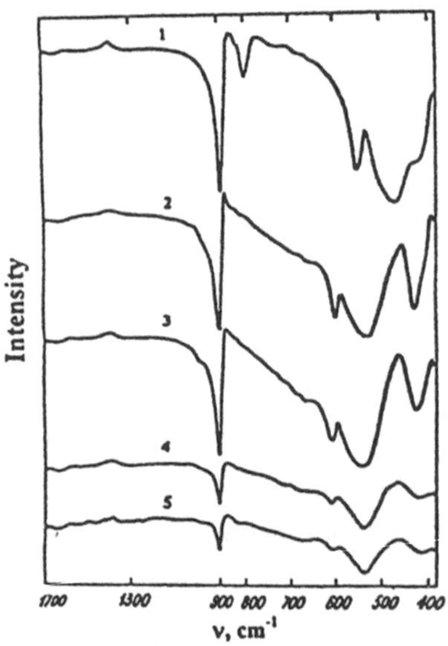

Figure 4. Infrared spectra of solid samples: 1 - K_3TaOF_6; 2 - K_3TaOF_5Cl; 3 - $K_2NaTaOF_5Cl$; 4 - NaCl-KCl-K_3TaOF_6 (3.01 m/o); 5 - NaCl-KCl-K_2TaF_7 (2.83 m/o)-Na_2O (2.83 m/o).

$$K_3TaOF_6 + 2NaCl = \\ = K_2NaTaOClF_5 + NaF + KCl \qquad (1)$$

$$K_2TaF_7 + Na_2O + NaCl = K_2NaTaOClF_5 + 2NaF \quad (2)$$

The emergence of bands corresponding to NaF is recorded on X-ray patterns of KCl-NaCl -K_3TaOF_6 and KCl-NaCl -K_2TaF_7-Na_2O (Na_2O/K_2TaF_7=1) solidified melt samples.

In KCl-NaCl-K_3TaOF_6 and KCl-NaCl-K_2TaF_7-Na_2O (O/Ta=1) melts tantalum reduction occurred in one stage from the only available $TaOClF_5^{3-}$ complex. X-ray analysis of the deposit obtained at a potential close to the foot of the cathodic wave indicated that pure tantalum was generated. At potential scan rate $V \geq 1 V \cdot s^{-1}$ $TaOClF_5^{3-}$ complexes discharge irreversibly. At 750^0C the mean value diffusion coefficient of $TaOClF_5^{3-}$ complex is found to be about $1.1 \cdot 10^{-5}$ $cm^2 \cdot s^{-1}$. At $V \geq 1$ $B \cdot c^{-1}$ the reduction process was complicated by formation of tantalum oxocarbides on a glassy carbon electrode and an intermetallic Pt-Ta compounds on platinum electrode.

KCl-NaCl-$K_3TaO_2F_4$ u KCl-NaCl-K_2TaF_7-Na_2O (1<O/Ta≤2) melts. Further titration of KCl-K_3TaOF_6 and NaCl-KCl-K_2TaF_7 melts with sodium oxide (Fig. 1) at a molar ratio O/Ta>1 no additional peaks were observed on voltammograms. At this the peak of tantalum reduction from a monoxohalide complex (R_2). reduced. When O/Ta=2 the peak height comprised about third from initial. X-ray diffraction of solidified melts gave presence of $K_{0.4}TaO_{2.4}F_{0.6}$ compound. Tantalum concentration in the middle part of cooling melt was reduced about 3.5 times. In at attempt to synthesise $K_3TaO_2F_4$ "in situ" by the addition of K_2TaF_7 and Na_2O to KCl-NaCl melt, compound, identified as $K_{0.4}TaO_{2.4}F_{0.6}$ by XRD, was observed to precipitate onto the crucible bottom, while in the middle of the solidified electrolyte X-ray diffraction spectra showed the existence of $K_2NaTaOF_5Cl$, NaCl, KCl and NaF compounds. The similar results were obtained when previously synthesized $K_3TaO_2F_4$ was added to KCl-NaCl melt. Based on the data obtained the reaction mechanism of dissolved substance in KCl-NaCl melt at O/Ta=2 was offered:

$$7K_3TaO_2F_4 + 17NaCl \rightarrow 2K_2NaTaOF_5Cl + 5K_{0.4}TaO_{2.4}F_{0.6}\downarrow + 15NaF + 15KCl \quad (3)$$
$$7K_3TaOF_6+7Na_2O+17NaCl \rightarrow 2K_2NaTaOF_5Cl+5K_{0.4}TaO_{2.4}F_{0.6}\downarrow+29NaF+15KCl \quad (4)$$
$$7K_2TaF_7+14Na_2O+10NaCl \rightarrow 2K_2NaTaOF_5Cl+5K_{0.4}TaO_{2.4}F_{0.6}\downarrow+36NaF+8KCl \quad (5)$$

The results of thermodynamic calculations for reactions (3-5) are given in Table 2.

TABLE 2. The calculating thermodynamic data for reactions (3-5)

Reactions	T, K	ΔH, $kJ \cdot mol^{-1}$	ΔS, $J \cdot mol^{-1} \cdot K^{-1}$	ΔG, $kJ \cdot mol^{-1}$
(3)	298	- 590	- 121	- 554
	1023	- 626	- 277	- 343
(4)	298	- 3235	128	- 3273
	1023	- 3057	168	- 3229
(5)	298	- 4895	- 94	- 4866
	1023	- 4514	323	- 4845

The large negative ΔG^0 values testified to the possibility of these reactions occurring in the melt. The proximity of ΔG^0 values for temperature 298 K and 1023 K is evidence that composition of solidified electrolytes may well correlate with the melted one.

KCl-$NaCl$-K_2TaF_7-Na_2O ($2<O/Ta\leq 2,4$) melt. Further titration of this melt with sodium oxide to $O/Ta=2,4$ resulted in reduction of peak R_2. At $O/Ta\geq 2.4$ peak R_2 was entirely suppressed, tantalum concentration practically reduced to zero and the cathodic branch of the voltammogram corresponded to a curve characteristic for the background electrolyte. Onto the crucible bottom $KTaO_3$ was observed to precipitate on the crucible bottom, together with $K_{0.4}TaO_{2.4}F_{0.6}$. It is attributable to occur the reaction:

$$5M_3TaOF_5Cl + 7M_2O = 5K0_{.4}TaO_{2.4}F_{0.6}\downarrow + 22MF + 5MCl \qquad (6)$$

and consecutive interaction of $K_{0.4}TaO_{2.4}F_{0.6}$ with excess Na_2O by reaction:

$$5K_{0.4}TaO_{2.4}F_{0.6\,(solid)} + 7M_2O = 5KTaO_{3\,(solid)} + 3MF \qquad (7)$$

4. Conclusions

Cyclic voltammetry indicates that the reduction of Ta(V) in $NaCl$-KCl-K_2TaF_7 melt involved a single irreversible five electron step $Ta(V)+5e\leftrightarrow Ta(0)$. $TaOClF_5^{3-}$ complexes are present in the melt at the molar ratio $O/Ta=1$. The concentration of $TaOClF_5^{3-}$ complexes drops and insoluble $K_{0.4}TaO_{2.4}F_{0.6}$ precipitates on the bottom of crucible when the molar ratio $O/Ta>1$. Neither linear voltammetry nor XRD nor IR data show the formation of $TaOF_6^{3-}$, $TaO_2F_4^{3-}$, TaO_3F^{2-} complexes in the melts studied.

5. Acknowledgements

The authors wish to acknowledge their thanks to the Russian Foundation for Basic Research for the financial support of this work (grant 96-03-32175).

6. References

1. Lantelme, F. and Barhoun, A. and Li, G. (1992) Electrodeposition of tantalum in $NaCL$-KCl-K_2TaF_7 melts, *J.Electrochem. Soc.* **139**, 1249-1255.
2. Kuznetsov, S.A. and Glagolevskaya, A.L. (1995) The electrochemical behaviour of tantalum oxofluoride complexes in an equimolar mix of potassium and sodium chlorides, *Tr. VIII Kola Seminar on Electrochemistry. KSC RAS. Apatity*, 44
3. Polyakova, L.P., Polyakov, E.G., Sorokin, A.I., and Stangrit, P.T. (1992) Secondary processes during tantalum electrodeposition in molten salts, *J. Appl. Electrochem* .**22**, 628-637.
4. Polyakova, L.P., Elizarova, I.R., Kononova, Z.A., and Polyakov, E.G. (1994) Voltammetric determination of oxide-ion in halide melts, *Zh.Analyt.Khim.* (Rus), **49**, 1228-1232.
5. Polyakova, L.P., Kononova, Z.A., Kremenetsky, V.G., and Polyakov, E.G. (1996) Effect of caesium chloride on the diffusion coefficient value of tantalum complexes in $NaCl$-KCl-K_2TaF_7 melts, *Zhurn. Prirkl. Khimii* (Rus.), **69**, 1307-1313.
6. Delahay, P. (1957) *New equipment and methods in electrochemistry*, Publ. House of Foreign Litr., Moscow.

SUBJECT INDEX

MIX
Papier aus verantwortungsvollen Quellen
Paper from responsible sources
FSC® C105338

If you have any concerns about our products,
you can contact us on
ProductSafety@springernature.com

In case Publisher is established outside the EU,
the EU authorized representative is:
Springer Nature Customer Service Center GmbH
Europaplatz 3, 69115 Heidelberg, Germany

Printed by Libri Plureos GmbH
in Hamburg, Germany